AF619338

VARIABLE DENSITY FLUID TURBULENCE

# FLUID MECHANICS AND ITS APPLICATIONS
Volume 69

*Series Editor:* R. MOREAU
*MADYLAM*
*Ecole Nationale Supérieure d'Hydraulique de Grenoble*
*Boîte Postale 95*
*38402 Saint Martin d'Hères Cedex, France*

*Aims and Scope of the Series*

The purpose of this series is to focus on subjects in which fluid mechanics plays a fundamental role.

As well as the more traditional applications of aeronautics, hydraulics, heat and mass transfer etc., books will be published dealing with topics which are currently in a state of rapid development, such as turbulence, suspensions and multiphase fluids, super and hypersonic flows and numerical modelling techniques.

It is a widely held view that it is the interdisciplinary subjects that will receive intense scientific attention, bringing them to the forefront of technological advancement. Fluids have the ability to transport matter and its properties as well as transmit force, therefore fluid mechanics is a subject that is particulary open to cross fertilisation with other sciences and disciplines of engineering. The subject of fluid mechanics will be highly relevant in domains such as chemical, metallurgical, biological and ecological engineering. This series is particularly open to such new multidisciplinary domains.

The median level of presentation is the first year graduate student. Some texts are monographs defining the current state of a field; others are accessible to final year undergraduates; but essentially the emphasis is on readability and clarity.

*For a list of related mechanics titles, see final pages.*

# Variable Density Fluid Turbulence

by

P. CHASSAING
*Institut de mécanique des fluides de Toulouse,*
*Toulouse, France and*
*École nationale supérieure d'ingénieurs de constructions aéronautiques,*
*Toulouse, France*

R.A. ANTONIA
*University of Newcastle,*
*Newcastle, N.S.W., Australia*

F. ANSELMET
*Institut de recherche sur les phénomènes hors équilibre,*
*Marseille, France*

L. JOLY
*École nationale supérieure d'ingénieurs de constructions aéronautiques,*
*Toulouse, France*

and

S. SARKAR
*University of California at San Diego,*
*Department of Mechanical & Aerospace Engineering,*
*La Jolla, U.S.A.*

Springer-Science+Business Media, B.V.

A C.I.P. Catalogue record for this book is available from the Library of Congress.

DOI 10.1007/978-94-017-0075-7

---

*Printed on acid-free paper*

## Table of Contents

*CHAPTER 1*

# VARIABLE DENSITY FLUID TURBULENCE: PREAMBLE

## 1.1. Introduction

Turbulent flows of variable density fluids are widely present in various domains of human activity and natural environment. In actual fact, density changes arise in low-speed flows and high-speed motions as well. Aeronautics is probably one of the most commonly quoted situations, since compressible subsonic, transonic, and supersonic motions are present in many applications, such as high-speed aircraft flight or supersonic combustion ramjet engines, for instance. However, as far as industrial applications involving fluids motions are concerned, variable density fluid flows are part of chemical engineering, thermal engineering, energetic, etc.
Density variations also occur in many examples of natural fluids motions. In geo-fluids, for instance, such density changes can be due to temperature and salinity effects (ocean) or moisture effects (atmosphere), not to mention astrophysics.

Faced with such a variety of natural and practical situations of variable density fluid motion, it is not easy to define nor analyze the scope of all density effects on turbulence. In fact, one is faced with a set of strategic questions, that are :

*Why* and *How* to deal with such a variety of situations as a whole?

So far, and to a wide majority of the scientific community, the first question was rather taken as irrelevant, and, as a direct consequence, the answer to the second one became obvious. Such a point of view can be easily understood, from theoretical, historical, and practical grounds. At first glance, the Navier-Stokes model for a fluid with variable physical properties looks rather complex in the turbulent regime, due to the multiplicity of extra sources of non-linearities (as compared with the incompressible case). Similarly, owing to practical needs, research on variable density turbulence was historically basically developed for aeronautics applications, and thus mainly devoted to compressible, high speed gas flows.
According to the latter statement, other variable density flows, in which density variation originates from a different source, are addressed separately, thus inducing some specific analysis.

On the other hand, as it is often the case in physics, one can wonder whether a better understanding of general features (or mechanisms) of

turbulence in variable density fluids motions could not emerge from a comparative analysis of the various situations where density variations may occur.
Finally, as far as industrial applications are concerned, it is now clear that in many engineering problems, turbulence can be affected by density variations resulting from different origins acting simultaneously: inhomogeneous species mixing and heat release in addition to compressible effects in supersonic combustion for instance).

Thus, in addition to detailed analysis of specific situations where variable density turbulent flows are involved, synthetic overview could contribute to improving our understanding of variable density effects in turbulence.

This is the guess made in the present monograph. Now, if one agrees, the next question is how to make the presentation tractable from an educational point of view?

This monograph is intended as a first attempt to addressing the previous questions.
The underlying idea is that an understanding picture should emerge from a rational approach, developed on a gradual analysis of the complexity resulting from density fluctuations in turbulence. It is clearly apparent from the outline of the monograph detailed in section 1.3.

In order to illustrate the position adopted here, let us first comment on three examples of expected effects of density variations, whatever their origin. They concern transition, compressibility and mixing.

## 1.2. Some expected incidences of density variation in turbulence

### 1.2.1. STABILITY AND TRANSITION

The fully developed turbulent regime is the ultimate consequence of instability mechanisms which originate from imbalance between stabilizing and destabilizing forces in the fluid motion. In constant density, homogeneous, single phase flows, such forces mainly[1] refer to viscous and inertial forces respectively. This doesn't mean that transition is a universal process. Various routes to turbulence are actually observed, depending on the boundary and initial conditions applied to the flow field. From a phenomenological point of view, instability mechanisms correspond to the emergence of various types of structures, such as Taylor-Couette vortices, Von Kármán billows, Tollmien-Schlichting waves...
In thin shear layer flows for instance, the basic instability process which takes place in incompressible flows is the so-called Kelvin-Helmholtz instability, associated with an inflexional velocity profile.

[1] when discarding gravity, capillarity effects...

In variable density fluid motions, the situation is radically changed due to external body forces in the gravitational field, and/or new mechanisms of vorticity generation by density gradients in presence of pressure gradients (baroclinic torque). The general consequences of the new mechanisms due to density variations can be to (i) enhance or (ii) reduce the incompressible transition process when they both compete, and (iii) introduce new routes to turbulence.

### 1.2.2. COMPRESSIBILITY

In the incompressible regime, without concentration and temperature variations, the kinetic energy of the fluctuating motion can only be "produced" by the turbulent stresses acting through the mean velocity gradient. In low speed buoyant flows, temperature velocity correlations in the presence of temperature gradients, are adding a new production mechanism. However in both cases, pressure fluctuations do not take a dominant part in the energy transfer between mean and fluctuating motions. It can be easily inferred that this will no longer be the case in compressible, high speed flows where noise radiation and shock waves for instance are well known specific consequences of energy transfer.

In aerospace applications, shock interaction with a turbulent boundary layer is an ubiquitous phenomenon of high speed flows. In this case, turbulence is enhanced by shock interaction with temperature fluctuations.
Such interactions also occur in supersonic wakes and jets. As shown by Hussaini *et al.* [228] the interaction of a localized temperature disturbance with a shock can be an important mechanism of vorticity production, particularly relevant to supersonic wake mixing, where large density fluctuations (40% of the mean value) can be observed. Similarly, the mixing of hot and cold jets of oxydant and fuel in the combustion chamber of a scramjet engine is enhanced through shock interaction [263]. Another example is the hot rocket exhausts with oblique shock waves, where a predominant source of noise generation results from entropy-shock interaction.

### 1.2.3. MIXING

One of the most important features of turbulence is likely mixing. Physically speaking, mixing in fluid mechanics is the process of diffusion of substances across intermaterial surfaces. By extension, this *mass* mixing definition also applies to all flow properties, such as momentum and heat.
Mixing can be driven by a wide variety of motions, the characteristic length scales of which are ranging from molecular to continuous values. In the first case, according to simple gradient schemes, mixing is driven by gradients in the flow field, and depends on physical properties of the fluid

material. In high turbulence Reynolds number flows, it is mainly dominated by the characteristics of the motion itself, and significantly competes with molecular mixing only in near wall regions.
Turbulent mass transfer is a very important dilution process in various natural and engineering situations, involving non reactive free jets and diffusion flames for example. It is also of considerable importance in confined flows, as those occurring in internal combustion engines for instance, where it is one key to improving the engine efficiency and reducing the generation of pollutants. In this case, it can be drastically modified by the compression of turbulence.

Large density variations can induce significant changes in the physical properties of the fluids, with direct consequences on the transfer coefficients at a solid boundary, even in low speed flows. In compressible flows, compressibility effects and inhomogeneous composition effects can be coupled in order to enhance mixing, as it is the case in supersonic combustion applications with the shock-induced mixing in non-uniform density jets [212], [348].

## 1.3. Monograph roadmap

As sketched in table 1.1, the different chapters of the present monograph can be grouped into three main parts:

- Theoretical elements,
- Physical analysis,
- Modeling for industrial applications.

The orientation of each part can be depicted as follows:

***Part 1: Theoretical elements*** The goal of the first part is to provide a general approach to the analysis of density variations in fully turbulent flows. It is devoted to the presentation of the "theoretical material" needed to deal with density effects in turbulence. All possible origins of density variations are addressed, restricting to non-reactive flows.
Its scope is deliberately educational and mainly concerned with (i) the formulation of the general instantaneous balance equations, (ii) the derivation of various "weak compressible" or linearized asymptotic limits to the general model, (iii) the statistical averaging of the transport equations of single point, first and second order moments and (iv) the presentation of basic mechanisms of the physics of variable density fluid turbulence.

Especially for the reader who is not familiar with variable density turbulent flows, some striking features of density and/or compressibility effects are presented in the second chapter, as an illustration of some aspects to be discussed in the theoretical analysis. Chapter 2 is also intended as introducing the second part of the monograph.

TABLE 1.1. Overall Table of contents.

| | | |
|---|---|---|
| Preamble | Chapter 1 | Introduction |
| Part I | Chapter 3 | Approximate models for turbulent, variable density fluid motions |
| | Chapter 4 | General equations and classification of turbulent, variable density fluid motions |
| | Chapter 5 | Statistical averaging and transport equations in variable density fluid turbulence |
| | Chapter 6 | Basic variable density mechanisms in turbulent flows |
| Part II | Chapter 2 | Examples of variable density effects in turbulent flows |
| | Chapter 7 | Relative behaviour of velocity and scalar structure functions in turbulent flows |
| | Chapter 8 | The structure of some variable-density, low-speed shear flows |
| | Chapter 9 | The high-speed turbulent shear layer |
| Part III | Chapter 10 | First-order modeling |
| | Chapter 11 | Second-order modeling |

***Part 2: Physical analysis*** In this second part, basic data on dynamical and scalar properties of variable density turbulent flows are presented. The discussion is based on experimental measurements and/or results from numerical simulations.
It aims to bring out the elements that are presently known on distinct characteristics of variable density turbulent flows. Such results are required to improving our intrinsic knowledge of density effects and providing a comparison basis with model predictions. This part is rather concerned with a research audience. Three basic situations are addressed: (i) scalar flow field properties, in the limit of passive contaminant, (ii) low-speed free turbulent variable density shear flows and (iii) high-speed shear layer mixing with uniform and non-uniform composition.

***Part 3: Modeling for industrial applications*** The last part is more directly devoted to an engineering audience. Thus the prediction methods which are presented, are restricted to single point modeling of the statistical equations derived in the physical space. Both first and second order closure levels are detailed, with special emphasis on the capability of including some specific variable density/compressibility effects.

## 1.4. Outline of the Chapters

***Chapter 1: Preamble.*** The goal of the present chapter is simply to introduce the subject and present the general outline of the monograph.

***Chapter 2: Examples of variable density effects in turbulent flows.*** To give a flavor of the subject, the two-fold goal of this chapter is : (i) to provide some illustrative examples of the rather wide range of situations concerned with the matter of the monograph, (ii) to bring out some quantitative information about some salient features of variable density effects in turbulence.

***Chapter 3: Approximate models for turbulent, variable density fluid motions.*** This chapter aims at giving a comparative overview of some of the various models which deriving from the general Navier-Stokes equations, account for density variations according to several types of approximations.
The role of the pressure is first examined. The Helmholtz decomposition is introduced and the linear analysis of Kovasznay compressible modes is presented. Then several models are discussed, referring to Boussinesq's approximation, and other approximations which are concerned with (i) filtering acoustic effects, (ii) incorporating density variations in pseudo-incompressible formulations and (iii) deriving weakly compressible limits to the general compressible equations.

***Chapter 4: General equations and classification of turbulent, variable density fluid motions.*** The first goal of this chapter is to recall the "general, instantaneous, local" equations governing variable density fluid motions, according to the classical approach of continuum thermo-mechanics and using the local equilibrium hypothesis. On this basis, the classical numbers associated with density changes in a fluid motion are introduced and the departure from the solenoidal condition is discussed. The final part of the chapter is devoted to the presentation of a general time-scale analysis which is relevant to distinguishing and classifying several turbulent flows of variable density fluid.

***Chapter 5: Statistical averaging and transport equations in variable density turbulence.*** This chapter is devoted to deriving statistical transport equations by averaging local, instantaneous, single point equations. It is only concerned with first (mean) and second order moments. Two different formulations will be mainly considered, referring to a mean mass-conservative and non conservative evolution. It is not intended for extensive and detailed presentations of the different formalisms — using either the classical formulation of the equations ('standard', 'mass-weighted' averaging,...) or the specific volume formulation —, but rather focuses on physical interpretation of density fluctuation correlations, according to mass conservative and non conservative mean flow analysis.

***Chapter 6: Some basic variable density mechanisms in turbulent flows.*** The aim of this chapter is to point out the existence and emphasize the understanding of some of those properties which make variable density turbulent flows different from the incompressible ones.
The governing equation of the instantaneous vorticity is first derived in the general case: fluid with variable density *and* non constant physical properties. Then, new vorticity generation mechanisms, as compared with the constant density situation, are discussed.
The second section deals with correlations with density fluctuations (d.f.c.) which are necessarily introduced in any statistical treatment of the instantaneous Navier-Stokes equations. The "diffusive" role of such d.f.c. is analyzed and discussed in low speed flows.
The last part of the chapter is devoted to specific mechanisms associated with dilatation fluctuations in pressure-correlation and dissipation terms. They are analyzed in relation with their contributions in various energy balance equations.

***Chapter 7: Relative behaviour of velocity and scalar structure functions in turbulent flows.*** This chapter reviews in a critical manner the existing analytical framework for describing the behaviour of velocity and scalar structure functions in turbulent flows. The assumptions which underpin this framework are only likely to be validated at very large Reynolds numbers and for relatively homogeneous and isotropic flows. These conditions are unlikely to apply in the laboratory. The major emphasis is on the likely dependence of second-order structure functions (or equivalently spectra) on both the Taylor micro-scale Reynolds number $R_\lambda$ and other parameters, such as the large scale anisotropy or the dissipation time scale ratio or, more generally, the initial conditions of the flow. Measurements strongly indicate that the influence of $R_\lambda$ and of the other parameters cannot be ignored. The retention of the non-homogeneity of the flow in the Navier-Stokes and heat transport equations provides a better idea of how large the magnitude of $R_\lambda$ should be before the "asymptotic" results of Kolmogorov and Yaglom may be attained. Special attention is given to a suitable framework which allows velocity and scalar fluctuations to be compared meaningfully. The analogy between scalar and energy structure functions (or spectra) appears to work well for flows with a continuous injection of turbulent energy and scalar variance.

***Chapter 8: The structure of some variable-density, low-speed shear flows.*** This chapter focuses on the influence of density contrasts on the development of some basic low-speed shear flows. The specific features of these variable-density flows are best accounted for as seen from their vorticity dynamics. The baroclinic torque, connecting misaligned pressure and density gradients, reorganizes the vorticity field according to the fluid inertia. It is introduced after a short literature survey. Then the particular cases of the mixing layer and the jet are examined. The two-dimensional and some three-dimensional aspects are documented based on temporally and spatially developing numerical simulations.

***Chapter 9: The high-speed turbulent shear layer.*** In high speed flows, significant density fluctuations can be generated by the flow itself. In this Chapter, compressibility effects in free shear flows with uniform and non uniform density are discussed in detail. The free shear layer is chosen as a useful benchmark for evaluating such effects, since, unlike the jet, neither the Mach number nor the density ratio are decreasing with distance. Recent DNS results are used to illustrate important aspects of the compressible free shear layer, including Mach number effects and variable density effects.

***Chapter 10: First order modeling.*** This chapter opens the last part of the monograph, which is more specifically dedicated to an engineering audience. It begins with a general presentation and discussion of prediction methods for turbulent flows, based on statistical — or Reynolds — averaged Navier-Stokes equations (RANS). Then, the incidence of density changes and the incorporation of variable density and compressibility effects in first order closure models are analyzed with respect to (i) "modifications" to incompressible schemes and (ii) introduction of additional "specific contributions" to non-constant density flows. At last, some zero-, one-, two- and three-equation models are reviewed

***Chapter 11: Second order modeling.*** In addition to first order closure models reviewed in the previous chapter, second order closure schemes are discussed in the present one to complete the review on single point modeling. Accordingly, this chapter is intended as providing some insights on where second order turbulence models have reached in accounting for several distinct effects due to density variation in low speed motions and compressibility in high Mach number flows.

## 1.5. The authors

The book has been co-authored by different contributors, under the editorship of P. Chassaing. The authors of the different contributions are quoted in table 1.2. The remaining chapters have been written by P. Chassaing.

TABLE 1.2. Authors' contributions.

| Author | Contribution | |
|---|---|---|
| R.A Antonia & F. Anselmet | Chapter 7 | Relative behaviour of velocity and scalar structure functions in turbulent flows |
| L. Joly | Chapter 8 | The structure of some variable-density, low-speed shear flows |
| S. Sarkar | Chapter 9 | The high-speed turbulent shear layer |

***Acknowledgements*** R. A. Antonia acknowledges the support of the Australian Research Council, both through a large grant and an IREX grant which has facilitated the collaboration between the University of Newcastle and the I.R.P.H.E.
F. Anselmet acknowledges the support of the C.N.R.S. for the IREX program. We are also grateful to Professor M. Coantic, Drs T. Zhou, L. Danaila, B. Pearson, J-J. Lasserre and to Mr. R. Smalley for their collaboration, discussions and contributions to various aspects of the work described in this chapter. Finally, we are most happy to acknowledge the inspiration that Dr L. Fulachier has provided us, partly through his own seminal research on the spectral analogy and his continuing interest in our work.

Chapter 8 by L. Joly has directly benefited from the results obtained by J. Reinaud during his PhD thesis and from discussions with V. Chapin, P. Chassaing and J. Micallef, whom he warmly thanks.

It is a pleasure for S. Sarkar to acknowledge discussions with G. Erlebacher, T.B. Gatski, M. Y. Hussaini and C. G. Speziale in the area of compressible turbulence when he commenced work in the subject at ICASE, NASA Langley Research Center, as well as later interactions at UCSD with C. Pantano, his former Ph.D. student. The sponsorship of his research in this area by NASA Langley and AFOSR is gratefully acknowledged.

P. Chassaing is much indebted to professors L. Fulachier and J. Lumley for having suggested the idea of such a monograph to Kluwer Academic and having proposed his name to the editing committee. He would like to gratefully acknowledge Professor R. Moreau for his kind and well-advised pieces of advice when planning the monograph. He deeply acknowledges professors R.A. Antonia and S. Sarkar and his French colleagues F. Anselmet and L. Joly for their contributions to this book. This work has benefited from the results and scientific activity of his research groups, both at IMFT and ENSICA, over the past ten years. It is a great pleasure for him to acknowledge all his colleagues and students interacting on the topic, with a special mention to J. Borée and J.-B. Cazalbou.

*CHAPTER 2*

# EXAMPLES OF VARIABLE DENSITY EFFECTS IN TURBULENT FLOWS

*To give a flavor of the subject, the two-fold goal of this chapter is: (i) to provide some illustrative examples of the rather wide range of situations concerned with the matter of the monograph, (ii) to bring out some quantitative information about some salient features of variable density effects in turbulence.*

## 2.1. General definitions

### 2.1.1. DIFFERENT TYPES OF VARIABLE DENSITY FLUID MOTIONS

To distinguish among the various sources of density variations, let us start simply from the definition of the density $\rho$ as the ratio between the amount of mass $M$ of a given body of fluid to its volume $V$:

$$\rho = M/V . \tag{2.1}$$

From eq.(2.1), it is clear that two specific situations can be considered separately, according to whether (i) the *volume* of a constant mass fluid element varies or (ii) the *mass* of a given volume of fluid is changed.
Adopting a lagrangian analysis, i.e., following a moving body of fluid with a given constant amount of mass, the first type of variable density fluid motions that can be depicted, includes three distinct configurations, where the volume variation can be associated with:

- **Geometrical effects**, when the volume of a confined mass of fluid is changed in response to boundary changes. It occurs typically in combustion chambers of reciprocating engines;
- **Mach effects**, as occur, for instance, in high speed flows past solid obstacles. The compressibility of fluid particles gives rise to various velocity induced pressure effects;

- **Dilatation effects**, as is observed when the volume of a fluid element changes due to the thermal expansion of the fluid. This volume variation, which results in a given amount of heat release to the fluid particle (by thermal convection, radiation, chemical reaction...) leads to buoyancy effects in non zero gravity situations.

The second class of density variation associated with eq.(2.1) can be more easily conceived within an eulerian approach. It is obtained when the mass of a given control volume changes due to variable mass fluxes depending on spatial inhomogeneity in composition or temperature. This origin of density variation is called:

- **Mixing effects**. They can be easily observed when two (ore more) non reactive, different mass species mix, with constant temperature and pressure. Analogous to this case is the mixing of different temperature bodies of the same fluid, under constant pressure.

Excepting Mach effects, it is important to emphasize that all other variable density effects can be encountered *in low speed* flows.

According to Lele [291], some of the previous variable density situations can be grouped, from a slightly different way, into two classes, where compressibility effects are associated with (i) the volume changes of fluid elements, and (ii) *inertia* effects as a result of either variable composition or volume changes due to heat transfer.

### 2.1.2. VARIABLE DENSITY FLUID TURBULENCE

As it is usual in fluid turbulence study, a statistical description is generally introduced to deal with the random character of this flow regime. This point will be examined in detail in Chapter 5. For the present discussion, we just recall that some kind of "mean motion" can be introduced, from which any actual flow realization departs by turbulent fluctuations.
Thus, when the density changes, the following questions, at least, arise:
- what are the effects of density variations on the mean flow field?
- what are the effects of density fluctuations on the statistical characteristics of the fluctuating motion?
- what are the specific turbulence mechanisms due to fluid density variations?

Considering the different situations previously introduced, the last question can be addressed by referring to:
- compressed turbulence,
- compressible turbulence,

– dilatational turbulence,

– inhomogeneous composition fluid turbulence,

where the last class is restricted here to single phase flow.

Dealing with *variable density turbulence* means that one is mainly concerned with (turbulent) density fluctuations. In this respect, a distinction between *high-speed* and *low-speed* flows is generally introduced, referring to a given Mach number. The distinguishing feature comes from the role of pressure fluctuations.
In low speed flows, pressure fluctuations cannot be taken as responsible for significant density fluctuations. Consequently, turbulent density fluctuations basically occur from temperature or concentration fluctuations, obviously associated with velocity and vorticity fluctuations.
On the other hand, the high speed flow condition allows the compressible nature of the fluid to produce significant density fluctuations from the pressure ones. Obviously, in the same type of motion, density fluctuations can also be generated by temperature fluctuations, as a consequence, for instance, of the friction heating in a compressible boundary layer with a no slip condition at the wall. The viscous dissipation is then responsible for entropy (temperature) fluctuations, which, in turn, generate density fluctuations.
According to Kovasznay's terminology [259], the terms "*acoustic*" mode[1] and "*entropy*" mode will be adopted to distinguish between pressure-induced and temperature-induced density fluctuations.

To illustrate various effects of density variation on turbulent flows in different situations, some examples will be presented now, starting with the transition to turbulence in variable density fluid motions.

## 2.2. Specific density effects in the transition to turbulence

Turbulence develops whenever incipient instabilities, which are not dissipated quickly enough by the stabilizing forces acting in the fluid motion, became able to extract continuously energy from the flow. In incompressible, isothermal motion, the driving forces are the inertia ones (advective terms), and the stabilization action only results from the molecular viscosity. Then, the basic mechanism from which turbulence can originate (Kelvin-Helmholtz instability in free shear layers, Tollmien-Schlichting waves in boundary layers, for instance) is dominated by inertia effects, with or without competition with molecular diffusion.

[1]The acoustic denomination does not necessarily involve here linearizing assumptions, as it is the case in classical acoustic and Kovasznay modes decomposition.

2.2.1. RAYLEIGH-TAYLOR AND RICHTMYER-MESHKOV INSTABILITIES

When the density is no longer constant, new additional driving and/or stabilizing forces can be present, introducing new instability mechanisms. Front instability developing between fluid layers of different density submitted to a normal gravity acceleration was first studied by Taylor [456] in 1950, in the context of the Rayleigh-Taylor instability (RT) in the linear regime.
As sketched in Fig.2.1, the RT instability occurs when a fluid accelerates another one of higher density in a direction perpendicular to the initial surface front.

The Richtmyer-Meshkov [2] (RM) instability develops when a shock wave passes through a perturbed "interface" between two fluids of different densities. As mentioned in Brouillette *et al.* [59] for example, the shock-induced RM instability plays important roles in technological applications (inertial confinement fusion, laser-matter interactions, pressure waves and flame fronts interactions in super and hypersonic combustion, for instance) and in astrophysics (supernova explosions). The RM model can also be used to describe bubble motion in an incompressible, inviscid fluid with a free surface, as proposed by Hecht *et al.* [211], for instance.

In the limit $\delta \to 0$ (sharp density interface, see Fig.2.1), the equation governing the amplitude $\eta(t)$ of periodic perturbations at a given wave number $k = 2\pi/\lambda$ in the transverse plane can be obtained from the classical RT and RM theories, as recalled, for instance, by Brouillette *et al.*[59] and Mikaelian [328]. The corresponding equations read:

$$RT \ : \ \frac{d^2\eta(t)}{dt^2} = kgA\eta(t) \qquad\qquad RM \ : \ \frac{d\eta(t)}{dt} = k\Delta uA\eta_0 \, .$$

In these equations, $A = (\rho_2 - \rho_1)/(\rho_2 + \rho_1)$ is the Atwood number, $g$ the modulus of the gravitational acceleration vector, $\Delta u$ the change of interface velocity induced by the shock and $\eta_0 = \eta(t = 0)$.
By analytical integration, it is concluded that the growth of the amplitude perturbation is *exponential* in time for the RT instability, and *linear* for the RM instability, since:

$$RT \ : \ \eta(t) = \eta_0 e^{\sqrt{Akg}.t} \qquad\qquad RM \ : \ \eta(t) = \eta_0\left(1 + Ak\Delta u.t\right) .$$

Thus, as long as $\eta(t) \ll \lambda$, the growth rates $\eta^{-1}d\eta/dt$ and $\eta_0^{-1}d\eta/dt$ of the RT and RM instabilities are respectively given by $\sqrt{Akg}$ and $Ak\Delta u$,

[2] As quoted by Mikaelian [329], for instance, this denomination refers to Richtmyer's theoretical work in 1960 and Meshkov's experiments in 1969. The references can be found in [329]. As noticed by Brouillette *et al.* [59], the case of an interface under shock acceleration was previously considered by Markstein in 1957.

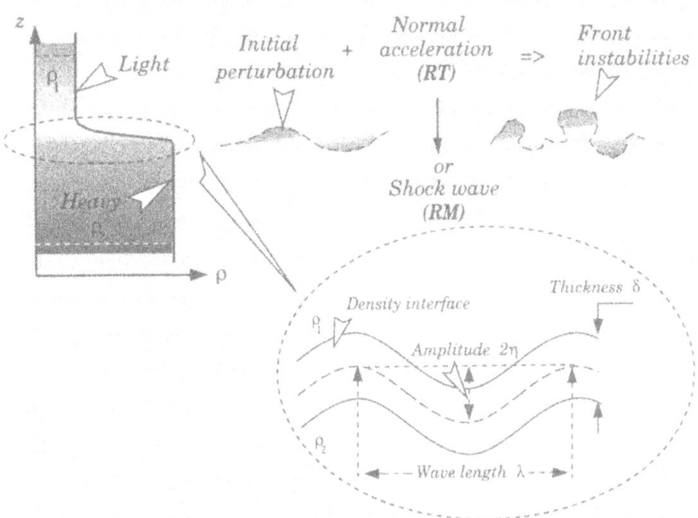


*Figure 2.1.* Sketch of Rayleigh-Taylor ($RT$) and Richtmyer-Meshkov ($RM$) instabilities.

where the Atwood number accounts for the relative amplitude of the density variation.

The previous classical RM analysis can be considered as an "incompressible" simplified analogy of an inherently compressible and complex process. In fact, the basic mechanism for the amplification of perturbations at the density interface is the baroclinic generation of vorticity resulting from the misalignment of the density gradient and the pressure gradient driven by the interfering shock wave. At the same time as vorticity is generated, the interface becomes more and more distorted, so that shear (Kelvin-Helmholtz) instabilities can develop, leading to the intensification of turbulent motions and mixing. The analysis of such complex mechanisms is now possible from direct numerical simulations (Mikaelian [330], 1996).

### 2.2.2. DENSITY EFFECTS ON THE TRANSITION IN LOW SPEED FREE SHEAR FLOWS

As recalled by Huerre *et al.* [227] in 1996, it has been recognized that transition to turbulence, in free shear flows at low speed, such as jets and mixing layers, can be strongly affected by density variation.

***Mixing layer.*** The nonlinear evolution of weakly amplified waves in two-dimensional free shear layers, when the Reynolds number is large and the critical layer viscous, can be analyzed within the frame of weakly non linear perturbations to a parallel basic flow state with a hyperbolic tangent velocity profile [222], [226] and [223], for instance. In temporally evolving two-dimensional mixing layers, the kinetic energy of the fluctuations over the flow field $<e>$ is governed by the Landau nonlinear equation model:

$$\frac{1}{2}\frac{d<e>}{dt} = \sigma <e> + L <e>^2 ,$$

where $\sigma$ is the linear temporal growth rate and $L$ the so-called Landau coefficient.
As reported in [227], results obtained in the constant density situation can be extended to a variable temperature mixing layer by using numerical simulations. The following conclusions can be drawn:

- For *isothermal* mixing layers, the long-time dynamics obtained from numerical simulations [227] give rise to a *supercritical* bifurcation as the wave number crosses its neutral value ($L < 0$);
- In *non-isothermal* mixing layers, a different behavior can be observed, corresponding to a *subcritical* bifurcation ($L > 0$), provided a sufficient cooling of the initial shear layer is applied: $T_c - T_\infty = -0.8$ for the example given in [227], where $T_c$ and $T_\infty$ are the values of the temperature along the center and outside the mixing layer respectively.

***Jet.*** Free jet is likely the low speed flow configuration which exhibits the most striking density effects to the transition. It is well known and extensively documented in the literature, that constant density jets are very sensitive to low level external forcing. As pioneered out by Huerre, Monkewitz and their colleagues [224], [336], [493], when the density of the jet is lowered with respect to that of the ambient fluid, finite regions of *absolute instability* can appear in the flow field near the exit.
In this case, density variation is directly responsible for the onset of global mode oscillations or self-sustained periodic fluctuations, beating at a well defined frequency, a feature which makes the situation definitively different from the constant density jet which is *convectively* unstable. Such peculiarities can be observed either in round or plane jets as shown by Binder's group in Grenoble [208], [378], [395].

From the experimental data of Kyle and Sreenivasan [266], the distinction between the two kinds of transition in a round, variable density jet is found to depend solely upon (i) $D/\theta$, where $D$ is the diameter of the jet and $\theta$ the momentum thickness of the boundary layer at the nozzle exit, and (ii) the density ratio at the exit $s_0 = \rho_e/\rho_\infty$, where $\rho_e$ and $\rho_\infty$ are the densities of the jet at the exit and of the ambient fluid respectively. Experimentally, the "incompressible" behavior is observed as long as $s_0 > 0.6$.
Now, as the inlet density of the jet is sufficiently decreased, such global

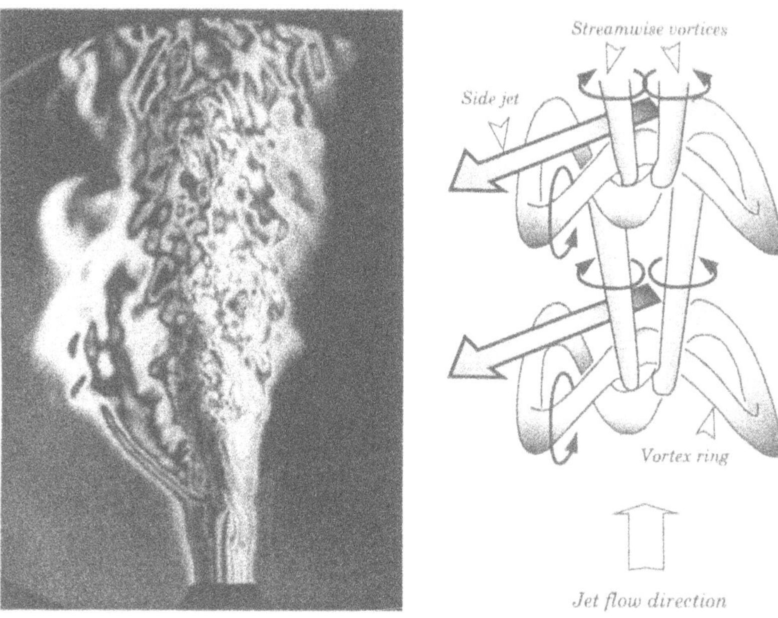


*Figure 2.2.* Radial ejections in an isothermal helium round jet (Doc. IMFT-TELET left). Sketch of the side jets generation process, according to Brancher *et al.*[55] (right).

mode oscillations lead to strong radial ejection of the fluid, leading to the formation of what is also called "side jets", that drastically changes the boundary of the main jet, enhances its mixing and increases its spreading rate. Such features were first observed in hot jets, but are actually density effects, since they are also present in isothermal jet flows with variable density resulting from changes in species composition. An illustration is given in figure 2.2 (left). Using a Schlieren visualization technique, a radial ejection can be observed in a pure helium round jet discharging into a quiescent atmosphere. The exit velocity is 8.24 m/s and the Reynolds number 717. The anemometer probe, also visible on the photograph on the opposite side of the ejection, has no effect on the phenomenon, figure 2.2 (right).

Amongst the possible explanations to such radial ejections, the direct simulations by Brancher *et al.* [55], suggest that they occur from coherent streamwise vortex pairs induction in the braid region, and are not directly related to the deformation of the primary vortex ring, as sketched in 2.2 (right). This question will be discussed later on in Chapter 8.

## 2.3. Density / compressibility effects in homogeneous turbulence

In theoretical studies of incompressible fluid turbulence, the assumption of statistically homogeneous and isotropic fluctuations [35] has long proved to be a powerful concept when analyzing basic mechanisms of (i) turbulence acting on itself and (ii) the response of a turbulent field to external mean strain rates associated with suitable mean velocity gradients. The concept can be extended to increase our knowledge of variable density fluid turbulence, as recalled in the next section, for low speed and high speed flows situations as well. In both cases, the main questions to be addressed are where and how correlations between the density and velocity fields take place and how density changes affect the turbulence structure.
Low speed homogeneous turbulence can be relevant to buoyancy-generated turbulence, binary or multi-species non reactive mixing and combustion; High speed homogeneous turbulence can be applied to both compressible and compressed configurations as reviewed by Lele [290], [291] in 1993 and 1994.

### 2.3.1. THE CONCEPT OF HOMOGENEOUS TURBULENCE

Statistical isotropy i.e., independence of the statistical properties of turbulent fluctuations with respect to rotation in the flow field, was first applied to the study of decaying incompressible turbulence, Batchelor [35], Monin and Yaglom [333]. Statistical homogeneity, i.e., independence of the statistical properties of turbulent fluctuations with respect to position in the flow field, has revealed a fruitful concept in identifying basic phenomena of incompressible turbulence. As pioneered by Craya [114], an extended version of the homogeneity assumptions makes it possible to analyze the response a "turbulent material" to "external changes", the characteristcis time scales of which are imposed by (constant) mean velocity gradients. In incompressible flows, such gradients are associated with plane deformation, pure shear or body rotation.

Variable density mixing can also be analyzed within the frame of homogeneous turbulence in low speed situations. In this case, the mixing process is driven by kinetic or potential energy and takes place into a fluid which composition (Chassaing *et al.* [88]), and /or temperature (Sandoval [407]) is not initially uniform in space.

The same kind of approach can be used in compressible turbulence, but restrictions on the mean fields are now necessary to maintain spatial homogeneity. As shown by Ribner [391], Feiereisen *et al.* [160], Dang and Morchoisne [121], and Cambon *et al.* [69], for instance, the homogeneous situations allowed in compressible turbulence are (i) the pure shear, which is the only acceptable time-independent deformation, (ii) some suitable combinations of time-dependent rotation and dilatation, and (iii) the interaction of a convected homogeneous field of turbulence with a plane shock front. Consequently, all other volume-preserving mean deformations and pure rotation are excluded from the scope of homogeneous compressible situations.

### 2.3.2. HOMOGENEOUS BUOYANCY-DRIVEN TURBULENCE

As already mentioned, the motion induced in a variable density fluid, initially at rest, when subjected to a constant (or gravitational) acceleration can be analyzed within the frame of spatially homogeneous turbulence.
In the presence of acceleration, density inhomogeneity acts as a source of potential energy, being converted into kinetic energy as the fluid is set to motion. As the flow develops, the mixing occurs, density gradients are smoothed by molecular and turbulent diffusion, and the density field tends toward a constant mean level. Thus the potential driving energy is reduced, and since the kinetic energy is dissipated to heat by viscosity, the flows ultimately comes to rest. To summarize, the initial amount of potential energy is (i) eroded by diffusion, (ii) converted into kinetic energy and finally (iii) dissipated to heat.

Batchelor *et al.* [37] studied statistically homogeneous, buoyancy-driven turbulence, assuming small density fluctuations and therefore using the Boussinesq approximation to the motion equations.
In 1995, Sandoval [407] using direct numerical simulations reported new original results without the Boussinesq approximation, for the same type of situation. In figure 2.3(a), which is taken from the latter reference, the time evolution of the turbulence kinetic energy during the mixing process is given. The kinetic energy $\widetilde{k}$ is defined as:

$$\widetilde{k} = \frac{1}{2}\widetilde{u_i''u_i''} \equiv \frac{1}{2}\frac{\overline{\rho u_i''u_i''}}{\overline{\rho}} ,$$

where $u_i'' = U_i - \widetilde{U}_i$ is the Favrian velocity fluctuation with respect to the density-weighted average $\widetilde{U}_i = \overline{\rho U}_i/\overline{\rho}$.

The non dimensional time is $t^* = t(gI_{\rho 0}/l_0)^{1/2}$, where $I_{\rho 0} = \sqrt{\overline{\rho_0'^2}}/\rho_0$ is the initial turbulence density intensity, $g$ the gravitation constant, and $l_0$ a characteristic length scale associated with the wavelength at which the initial spatial spectrum of the density fluctuation $\rho'$ reaches its maximum. The presented results correspond to two different numerical simulations with the same "pseudo-Reynolds number" $l_0(gl_0I_{\rho 0})^{1/2}/\nu = 256$, and two different initial density ratios $s_0 = \rho_{max}/\rho_{min}$ of 1.105 and 4. The lower value refers to the validity range of Boussinesq's approximation. The higher initial density ratio corresponds to a higher level of potential energy.

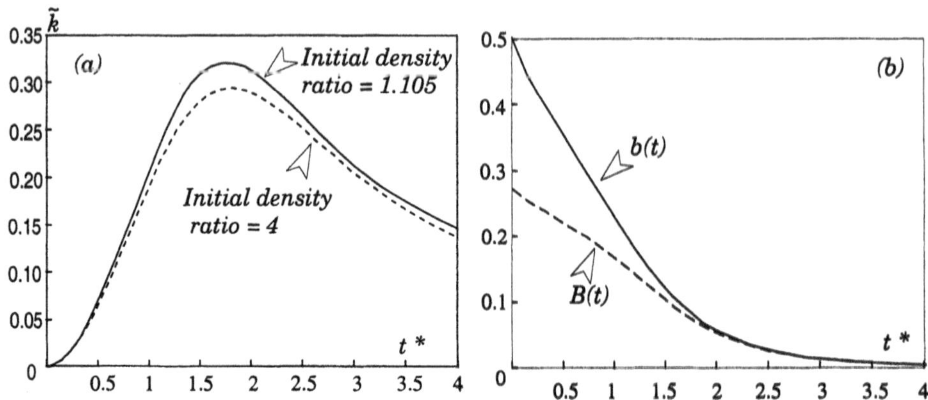


*Figure 2.3.* Buoyancy-driven homogeneous turbulence: (a) Turbulence kinetic energy (b) Departure from Boussinesq's approximation (adapted from Sandoval [407]).

Rather surprisingly, as observed in Fig.2.3(a), the effect of such an intensification of the density variation is to reduce the amount of kinetic energy produced in the turbulent flow field as compared with the situation under the Boussinesq approximation ($s_0 = 1.105$). This means that a part of the available potential energy is derived from its conversion into kinetic energy or that the efficiency of the conversion mechanism is reduced. We shall give the explanation later on and see that the reason originates from the turbulent mass flux.

The evolution of the relative density variance $B(t) = \overline{\rho'^2}/\overline{\rho}^2$ is given in Fig.2.3(b). As expected this function vanishes when the flows develops. A second parameter is also plotted in the same figure, that is the opposite correlation between the fluctuating density and the fluctuating specific volume $b(t) = -\overline{v'\rho'}$, where $v$ denotes the instantaneous specific volume ($v\rho = 1$). Hence it is clear that $b(t) = \overline{v}\,\overline{\rho} - 1$.
Since $\overline{v'\rho'} \equiv \overline{(1/\rho)'\rho'}$ and developing $1/\rho$, the following approximation to $b(t)$ can be obtained:

$$b(t) \simeq \frac{\overline{\rho'^2}}{\overline{\rho}^2} + \mathcal{O}(\overline{\rho'^3}) = B(t) + \mathcal{O}(\overline{\rho'^3}) .$$

Thus, for small density fluctuations (i.e., when the Boussinesq approximation assumption applies), $b(t) \simeq B(t)$. As deduced from Fig.2.3(b), it can be concluded that this condition is fulfilled at about $\overline{\rho'^2}/\overline{\rho}^2 = 0.04$.

### 2.3.3. COMPRESSED ISOTROPIC TURBULENCE

As sketched in Fig.2.4 (left), this situation corresponds to a body of homogeneous and isotropic turbulence (initial volume $L_0^3$) undergoing a mean deformation rate given by:

$$\frac{\partial \overline{U}_i}{\partial x_j} = \begin{pmatrix} S_1(t) & 0 & 0 \\ 0 & S_2(t) & 0 \\ 0 & 0 & S_3(t) \end{pmatrix} ,$$

with $S_\alpha(t) = V/X_\alpha(t)$, where $V$ is the compression speed and $X_\alpha(t)$ the box length at time $t$. By homogeneity, the mean velocity of the flow field $\overline{U}_i$ must be linear in the spatial coordinates, so that $\partial\overline{U}_i/\partial x_j$ and therefore $S_\alpha$ depends only on time. Now in isotropic compression — (i) in Fig.2.4 —, $S_1(t) = S_2(t) = S_3(t) \equiv S(t)$, and due to the mass conservation law, the spatial mean density value $<\rho>$ is given by:

$$<\rho> (t) X^3(t) = \rho_0 L_0^3 \, .$$

Assuming an adiabatic compression of a perfect gas, the spatial mean temperature and pressure are given by:

$$\frac{<T> (t)}{T_0} = (\frac{<\rho> (t)}{\rho_0})^{\gamma-1} \qquad \text{and} \qquad <P> (t) = R <\rho> (t) <T> (t) \, .$$

In figure 2.4(b), taken from the direct numerical simulations by Wu *et al.* [488], the time evolution of the turbulence kinetic energy $<k> = \frac{1}{2} <u_i'u_i'>$, normalized by its initial value, is plotted versus the inverse of the total strain rate dimensionless parameter:

$$\sigma = exp[\int_0^t S(t')dt'] \, .$$

Four different compression rates are applied to the same initial isotropic turbulence (initial Reynolds $q^4/\varepsilon\nu = 20.56$, where $q^2$ is twice the turbulence kinetic energy). They differ from the initial values of the ratio between the turbulence and mean strain time scales $r = |S|q^2/\varepsilon$. Four values are considered, ranging from $r = 47.04$ (the fastest) to $r = 0.10$ (the slowest). The two smallest values $r = 0.50$ and $r = 0.10$ are within the range of what is encountered during the compression stroke of an internal combustion engine (0.5 - 0.05).

As observed in the figure, the isotropic compression acts against the decaying process. When the compression rate is high ($r = 47.04$ and $r = 2.52$), the turbulent field is immediately affected by the mean strain rate, and the turbulence kinetic energy increases immediately.
When the compression is not strong enough ($r = 0.10$, for instance), the turbulence kinetic energy first decreases during an initial period, then stabilizes, reflecting the action of the mean strain rate. For $r = 0.50$, the

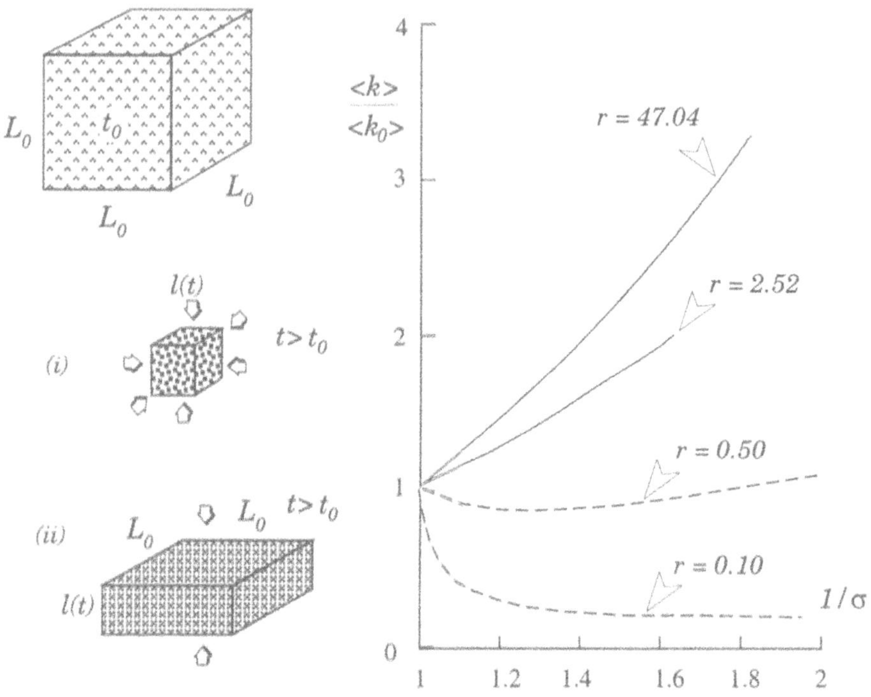


*Figure 2.4.* Compressed turbulence: Sketch of the homogeneous compression (left); Evolution of the turbulence kinetic energy for four isotropic compression rates, (right) adapted from Wu *et al.* [488].

decay and mean compression effects are roughly balanced over the period which is considered, but in all cases, the turbulent kinetic energy should increase, provided the strain is applied during a long enough period.

### 2.3.4. COMPRESSIBLE ISOTROPIC TURBULENCE

In incompressible fluid turbulence when "'homogeneous and isotropic[3]" conditions apply, the turbulence kinetic energy is evolving both in magnitude and spectrum:

- in the absence of production, the initial amount of turbulence kinetic energy is continuously reduced due to viscous dissipation;
- according to the Kolmogorov cascade, the energy spectrum is shifted in time towards low wave numbers.

When the solenoidal condition is fulfilled by the velocity fluctuations, the pressure-velocity correlation is identically zero, and the previous features

[3]The quotation marks mean that, strictly speaking, both designations are mathematically redundant since invariance by rotation (isotropy) implies invariance by translation (homogeneity).

are entirely determined by the initial energy spectrum (initial amount of turbulence kinetic energy and characteristic length scales).

This is no longer the case in compressible isotropic turbulence. Density and pressure fluctuations are now present that influence the kinetic energy evolution, as shown by the first direct numerical simulations of Passot & Pouquet [359] in 1987, for instance. As recalled by Zeman [497], the turbulence evolution is found to depend on (i) the given initial rms (or turbulence) Mach number ($M_t$), (ii) the level of initial pressure or density fluctuations and (iii) the ratio of compressible to total kinetic energy ($\chi$). Splitting the instantaneous velocity ($u'_i$) into an incompressible, solenoidal part ($u'^I_i$) and a compressible part ($u'^C_i$) — see Chapter 3 —, the latter parameter is defined as:

$$\chi = \frac{< u'^C_i u'^C_i >}{< u'^C_i u'^C_i > + < u'^I_i u'^I_i >} , \tag{2.2}$$

where the angle brackets denote an ensemble average (space average in statistical homogeneous turbulence). The turbulence Mach number is defined as usually, viz.

$$M_t =< u'_i u'_i >^{1/2} / < \gamma RT >^{1/2} .$$

***Turbulence Mach number.*** The strong role of the turbulence Mach number in two-dimensional calculations was demonstrated by Passot and Pouquet [359] in 1987. From the three-dimensional simulations available at the time of his review, Lele [290], [291] concluded that the decay rate of turbulence kinetic energy slightly increases with $M_t$. This result is confirmed by the more recent simulations by Cai *et al.* [67], as shown in figure 2.5, where three evolutions are given, corresponding to the same initial energy ratio $\chi$=0.6, near statistical equipartition of energy in vortical and compressible modes. The non dimensional time is based on the initial values of the Taylor micro-scale and the rms of the velocity fluctuations.

***Thermodynamic fluctuations.*** The role of the fluctuations of the three thermodynamic variables ($p'$, $\rho'$ and $\theta'$ for pressure, density and temperature respectively) in decaying compressible turbulence is more complex to analyze. Without consideration of heat conduction in the energy equation, and according to Sarkar *et al.* [416], the behavior of pressure fluctuations can be obtained by considering the incompressible ($p'^I$) and compressible ($p'^C$) parts of the pressure fluctuation associated with the corresponding parts of the velocity fluctuation (see Chapter 3):
$p'^I$ and $u'^I$ are governed by the incompressible equations (Poisson's equation

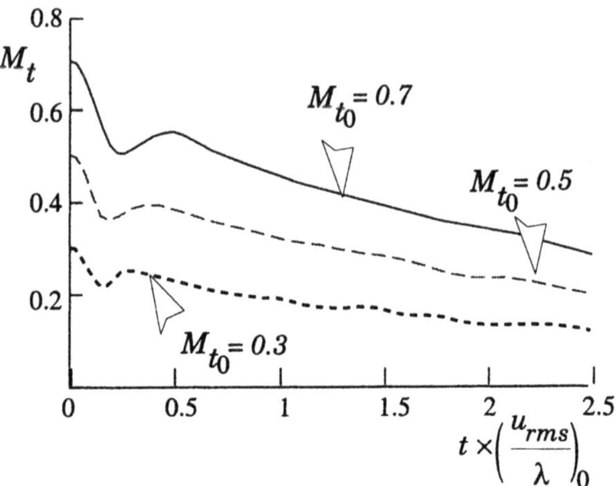


*Figure 2.5.* Time evolution of the turbulence Mach number in decaying compressible turbulence, adapted from Cai *et al.* [67].

for the pressure fluctuation),
$p'^C$ and $u'^C$ satisfy the wave equation on the acoustic time scale. Obviously, $p'^I$ is of the order of $M_t^2$, and it was found that:

$$< (p'^C)^2 > \simeq \chi\gamma^2 M_t^2 \,, \tag{2.3}$$

where, as usual, $\gamma$ stands for the polytropic coefficient. Thus, provided that heat conduction can be neglected (temperature fluctuations are initially negligible with respect to the pressure ones), eq.(2.3) predicts that the compressible pressure fluctuation depends strongly on $\chi$, at any given turbulence Mach number.
The question now arises as to how this result applies to situations with initially dominant temperature fluctuations. Following Cai *et al.* [67], there exist three distinct types of turbulence, associated with different range of $\chi$ and corresponding to different scaling of density fluctuations:

(a) nearly incompressible turbulence, dominated by vorticity ($\chi \ll 1$). Under the nearly incompressible approximation detailed in Chapter 3, (Zank and Matthaeus [494], Bayly *et al.* [40]), $\theta' = \mathcal{O}(M_t)$, $p' = \mathcal{O}(M_t^2)$ and density and temperature fluctuations are anti-correlated, i.e., $\rho'/\rho = -\theta'/T$;
(b) statistical equipartition of energy between vortical and compressive contributions ($\chi \simeq 0.5$);
(c) nearly pure acoustic turbulence, dominated by compressive modes ($\chi \simeq 1$); The density and pressure fluctuations are linked by $p'/P = \gamma\rho'/\rho$ and the density-temperature correlation is zero.

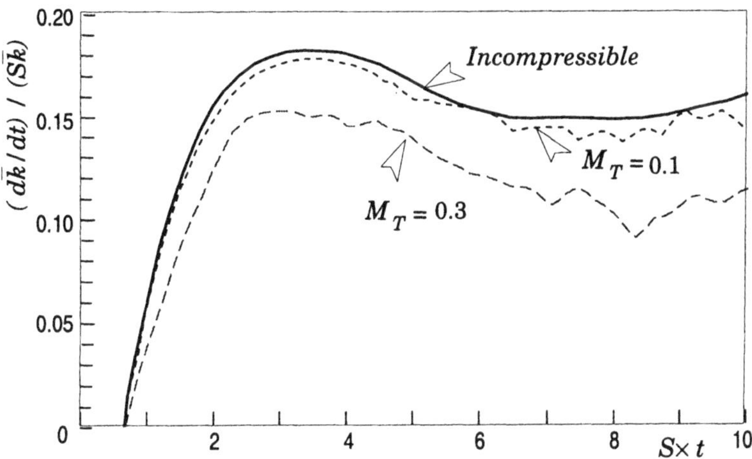


*Figure 2.6.* Compressibility effect on the growth rate of turbulent kinetic energy in homogeneous shear flow, adapted from Hamba [203].

According to the numerical simulations of Cai *et al.* [67], the asymptotic states of (a) nearly incompressible and (c) acoustic turbulence are approximately achieved when the initial values of the turbulence Mach number and the energy ratio are respectively $M_{t0} < 0.3$ with $\chi_0 = 0$ and $M_{t0} = 0.7$ with $\chi_0 = 1$. For the latter case, density and temperature fluctuations, which are initially perfectly anti-correlated, become uncorrelated at two eddy turnover times.
Finally, eq.(2.3) can be extended to accurately predict the compressible part of the pressure, within the following ranges:

- $\chi_0 = 0$ and $M_{t0} < 0.5$;
- $\chi_0 = 0.6$ and $0.3 < M_{t0} < 0.7$;
- $\chi_0 = 1$ and $0.3 < M_{t0} < 0.7$.

### 2.3.5. COMPRESSIBLE HOMOGENEOUS SHEAR FLOW

***Turbulence kinetic energy.*** Direct numerical simulations of compressible homogeneous turbulence submitted to a uniform mean shear have been carried out, among others, by Sarkar *et al.* [415] in 1991, Blaisdell *et al.* [48], [47] in 1991 and 1993 and more recently by Hamba [203] in 1999. A consistent trend in such simulations is the reduction of kinetic energy growth rate with increasing compressibility, as illustrated for example in figure 2.6 adapted from the last reference. Time normalization is based on the constant mean shear rate scale $S = d\bar{U}_1/dx_2 \equiv C^t$.

The influence of the turbulence Mach number $M_t = \sqrt{2\bar{k}}/\bar{c}$ on the history of the non-dimensional growth rate of the turbulence kinetic energy

can be clearly observed in the figure. After a transient stage, until about $S{\times}t = 6$, compressibility acts as a stabilizing factor, reducing the turbulence kinetic energy by 30%, as compared with the incompressible case. In 1993, it was argued (Blaisdell *et al.* [47]) that the principal reason for this reduction was the dissipation rate due to dilatation ($\overline{\epsilon}_d = \frac{4}{3}\overline{\mu}\overline{\vartheta'^2}/\overline{\rho}$) and the pressure-dilatation correlation ($\overline{p'\vartheta'}$) where $\vartheta'$ denotes the divergence of the fluctuating velocity ($\partial u_i'/\partial x_i$).
Zeman [495] introduced the *dilatation* dissipation concept in 1990 from phenomenological considerations, based on the presence of eddy shocklets (see Chapter 6, section 6.5).

A rather similar concept, called *compressible* dissipationwas introduced independently by Sarkar et al. [416] in 1991 from a totally different approach. These authors developed an asymptotic analysis of compressible turbulence for the scaling of the compressible and solenoidal parts of the turbulent dissipation, and used direct numerical simulations of homogeneous turbulence to validate the following decomposition of the dissipation rate $\overline{\varepsilon}$ (see Chapter 6, section 6.4.4) :

$$\overline{\varepsilon} = \overline{\varepsilon}_s + \overline{\varepsilon}_d \,, \tag{2.4}$$

where: $\overline{\varepsilon}_s \equiv \overline{\rho}\,\overline{\epsilon}_s = \overline{\mu}\,\overline{\omega_k'\omega_k'}$ and $\overline{\varepsilon}_d \equiv \overline{\rho}\,\overline{\epsilon}_d = \dfrac{4}{3}\overline{\mu}\,\overline{\vartheta'^2}$ .

- The first component $\overline{\epsilon}_s$ is associated with the vortical component of the fluctuating velocity field, and is therefore called the *solenoidal* dissipation. It is the only contribution present in constant density flows;
- The second component $\overline{\epsilon}_d$, associated with the dilatational characteristic of the fluctuating velocity field, increases the solenoidal dissipation in variable density fluid motions. It will be called the *dilatation* dissipation, according to Zeman's original denomination.

In 1992, Sarkar [408] found that the pressure-dilatation covariance was more important than the dilatation dissipation. This conclusion is in agreement with the more recent analysis by Ristorcelli [394] from which it results that the importance of the pressure-dilatation depends on the type of flow.

Later literature on the topic, (Sarkar [409] in 1995) showed that the reduction in the kinetic energy growth rate was primarily the consequence of an important implicit effect of compressibility on the reduced production level, as we shall see now.

***Turbulent shear stress anisotropy.*** The DNS results of Sarkar [409] show that in compressible homogeneous shear flows, the turbulence eventually evolves to approximately constant values of the Reynolds shear stress anisotropy: $b_{12} = \widetilde{u''v''}/2\widetilde{k}$. Here $u_i'' = U_i - \widetilde{U}_i$ where the tilde denotes a

density-weighted or Favre average. $\widetilde{k} = \frac{1}{2}\widetilde{u''_i u''_i} \equiv \frac{1}{2}\overline{\rho u''_i u''_i}/\overline{\rho}$ is the turbulence kinetic energy based on the Favrian velocity fluctuations.
The long time values ($S \times t = 20$) obtained by Sarkar [409] are given in the next table at a constant turbulence Mach number, $(\sqrt{\widetilde{k}}/c)_0 = 0.4$. Hence, the initial value of the gradient Mach number $M_{g0} = Sl_0/c_0$ accounts for compressibility effects. $l_0$ is the initial value of a representative integral length scale of the turbulence in the transverse direction of shear.
As shown in table 2.1, the increase in the initial gradient Mach number results in an important reduction of the turbulent shear stress anisotropy. Now, in such a flow, the production term is simply $P_{rod} = -S\,\widetilde{u''v''}$, so that the normalized production rate is just $P_{rod}/(\widetilde{k}S) = -2b_{12}$, the decrease of which is a direct consequence of the reduction in the shear stress anisotropy which, in turn, is achieved by the reduction of the pressure strain term $\overline{p(\partial u'_i/\partial x_j + \partial u'_j/\partial x_i)}$ according to Hamba [203].
As demonstrated in Sarkar [409], this reduction in the turbulence production level is predominantly responsible for the reduced growth rate of the turbulence kinetic energy. Since in Sarkar's simulations the mean density is constant and does not change with flow evolution, the inhibition of turbulence level and anisotropy by compressibility in homogeneous shear flow is definitively not a mean density effect.

TABLE 2.1. Long time values of the turbulent shear stress anisotropy from DNS results by Sarkar [409].

| Gradient Mach number $M_{g0}$ | Turbulent shear stress anisotropy $-b_{12}$ |
|---|---|
| 0.22 | 0.29 |
| 0.44 | 0.24 |
| 0.66 | 0.18 |
| 1.32 | 0.12 |

### 2.3.6. SHOCK/HOMOGENEOUS TURBULENCE INTERACTION

The shock/turbulence interaction has received considerable attention since the pioneering work of Ribner [391], [390], almost some fifty years ago, initiating the analytical study of sound generated by the interaction of a single vortex with a shock, and introducing a linear analysis in which vorticity, entropy and acoustic modes are considered separately, according to Kovasznay's modal decomposition [259].
This "linear interaction analysis" (LIA) is able to predict the post-shock

properties (intensities and spectra) of a given homogeneous pre-shock turbulence at a Mach number ranging from about 1 to 10. The general principle of the method, which consists in isolating the various physical mechanisms underlying shock wave interaction, is still a useful basis to give predictions which can be directly compared with RDT results, DNS and measurements, as reviewed by Andreopoulos *et al.* [3].

Forty years later, the first direct numerical simulations of shock/turbulence interaction were published in 1991 and 1993 by Lee, Lele & Moin [283], [285] and Hannapel & Friedrich [206]. By the same time, new experimental studies on isotropic turbulence/shock wave interaction were carried out by Keller & Merzkirch [244], the Andreopoulos group [58], [217], Barre *et al.* [31] and Jacquin *et al.* [230], for instance.

Incidence of a shock wave is known to strongly affect both the mean flow and the turbulent fluctuations, including magnitude, scale and spectrum modifications, as shown by the following examples.

***Turbulence kinetic energy.*** One of the well known major features of shock-turbulence interaction is the amplification turbulence kinetic energy when passing across a shock wave. Making use of the LIA assumptions, and according to Kovasznay's analysis [259] (see Chapter 3), small amplitudes fluctuations in compressible turbulence are decomposed into three modes of mutually independent waves: vortical (or shear), acoustic (or sound) and entropy (or temperature).
When any one of such elementary waves encounters a shock, the interaction generates a triad of all three modes downstream of the shock, the amplitudes of which can be related to the upstream values through so-called transfer functions, which can be derived with a LIA approach, for instance, as long as all fluctuations upstream of the shock are small, so that linearization applies and the shock front is not substantially distorted. In this case, the upstream turbulence is represented as a superposition of plane Fourier waves, each of them interacting independently with the shock. By integrating over all the incident waves, the solution obtained by solving the linearized Euler equations for each individual incident wave, one can obtain ([313], [228], [231]) the statistical properties of the after shock turbulence. As computed by Jamme [231] with this approach, the turbulence kinetic energy amplification (as defined by the ratio of the turbulence kinetic energy after and before shock) is plotted versus the incident Mach number in figure 2.7. Three different types of incident turbulence are considered:

- solenoidal turbulence: the pre-shock velocity fluctuations are purely rotational (vortical mode);
- acoustic turbulence: the pre-shock thermodynamic fluctuations are isentropic, so that: $p'/P = \gamma\rho'/\rho$ and $\theta'/T = (\gamma - 1)\rho'/\rho$;

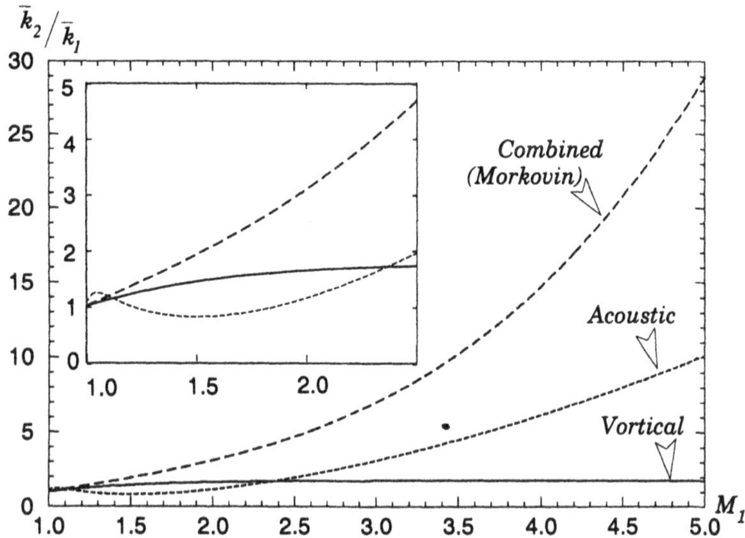


*Figure 2.7.* Turbulence kinetic energy amplification for different types of incident turbulence passing through a shock, (courtesy of S. Jamme [231]).

– combined turbulence: the pre-shock agitation includes both vortical and entropy modes. They are coupled according to the Morkovin or "strong Reynolds analogy" (SRA) assumption [339] (see Chapter 3), that is, velocity, temperature and density fluctuations are linked with:

$$\frac{\theta'}{\overline{\overline{T}}} = -\frac{\rho'}{\overline{\rho}} = -(\gamma - 1)M^2\frac{u'}{\overline{\overline{U}}}, \tag{2.5}$$

where $M$ is the local Mach number.

As observed in the figure, the lowest amplification rate is obtained for the pure vortical incident mode. With an incident Mach number $M_1 = 5$, for instance, the amplification encountered by a pure acoustic turbulence is more than five times the one of a pure vortical turbulence. On the other hand, the strongest amplification is obtained for entropy/vorticity modes coupled by the SRA relation.

As pointed out by Mahesh *et al.* [313], the role of the upstream entropy fluctuations strongly depends upon the correlation between velocity and temperature (density) fluctuations: negative upstream velocity/temperature correlations enhance the amplification of the turbulence kinetic energy, while positive correlation has a suppressing effect. The SRA, which yields negatively correlated velocity/temperature fluctuations, is actually responsible for the strong increase in the amplification level.

Some of the previous considerations could give a plausible explanation to experimental observations. Indeed, in experiments of grid turbulence

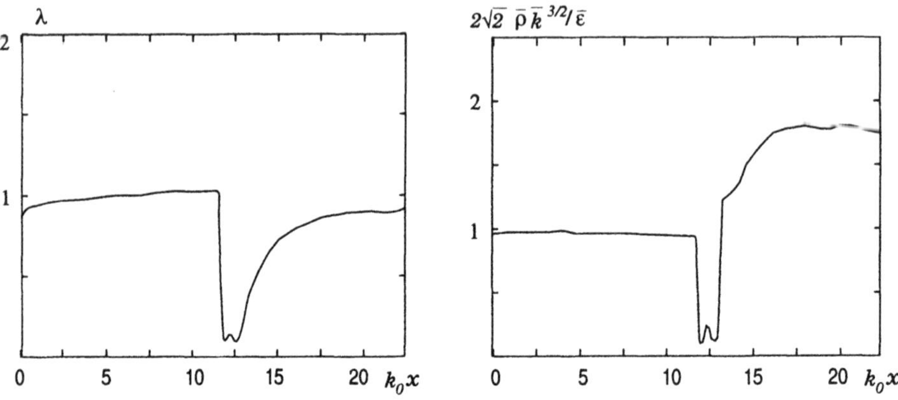


*Figure 2.8.* Turbulence length scales variation through a shock at an incident Mach number 1.5: Taylor longitudinal micro-scale (left) and Batchelor dissipation scale (right), adapted from Jamme [231].

interaction with a normal shock-wave, an amplification of turbulent velocity and vorticity fluctuations is generally observed. However, the amplification level is found to depend on the shock wave strength and the characteristics of the upstream turbulent flow (incoming compressibility level and grid generation). As reported by Andreopoulos *et al.* [3], the available experimental data indicate that the interaction is very sensitive to the upstream conditions of the flow. For instance, no amplification is evident downstream the shock when the upstream turbulence is generated by fine grids into a moderate compressible incoming flow.

***Length scales.*** A controversy has long existed between experimental data and theoretical results about the type of changes of the characteristic length scales of a turbulence passing through a shock. This point could be now elucidated thanks to direct numerical simulations by Lee *et al.* [281] and recent measurements by Barre *et al.* [31] showing that most usual turbulence length scales (longitudinal integral scales, Taylor micro-scales) decrease across the shock. A noticeable exception is the Batchelor dissipation length scale $\bar{\rho}\bar{k}^{3/2}/\bar{\epsilon}$, which increases slightly for shock waves with $M_1 < 1.65$, as shown by Lee *et al.* [281]. The numerical simulations of Jamme [231], with $M_1 = 1.5$, are used to illustrate these results in figure 2.8. The distance normal to the shock ($x$) is normalized by the wave number corresponding to the peak in the initial energy spectrum. Both length scales are normalized by their values just before the shock. The incident turbulence is purely solenoidal.

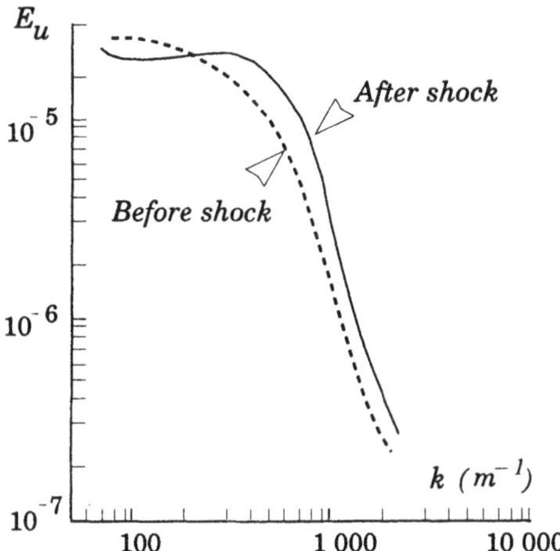


*Figure 2.9.* Shock effect on homogeneous turbulence spectrum (adapted from Barre et al.[32]).

***Energy spectrum.*** As expected, the changes observed on the characteristic length scales of a turbulence passing across a shock wave, have their counterpart on the spectra which can be predicted by LIA, and are confirmed by direct numerical simulations and measurements. As an example to illustrate the spectrum modification, the experimental result by Barre *et al.* [31] have been chosen. As noticed by Ribner, the product 'velocity times wave number' preserves invariance of frequency in a shock-fixed frame. Thus, the wave number of the post-shock spectra is reexpressed to the pre-shock one, by division by the mean flow velocity ratio.

It can then be observed that:

– the post-shock spectrum area is about 1.5 times the pre-shock one, in agreement with the theory,

– the small scales are more amplified than the large scales, the post-shock spectrum lying slightly above the pre-shock one, at wave numbers $k > 200\,m^{-1}$. As quoted by Barre *et al.* [31], [32], the maximum amplification ratio is equal to 3.5 and obtained for wave numbers at about $1\,300\,m^{-1}$. From these results, one can infer that the turbulent micro-scale decreases across the shock, as already mentioned.

## 2.4. Density / compressibility effects in fully developed turbulent shear flows

We turn now to kinematically inhomogeneous flows in which the mean velocity gradient is no longer spatially constant throughout the flow field. Accordingly, compressibility/density effects are now interesting both mean and fluctuating values as resulting from:

- **mean** variations of density associated with mean temperature, concentration or pressure fields;
- density **fluctuations**.

As we shall see later on, mean velocity profiles in compressible wall bounded flows (boundary layers and channel flows) can be derived from the Van Driest transformation of their incompressible counterparts: this is an example of **mean** variation effect.

On the other hand, as shown in section 2.3.5, the growth rate of turbulence kinetic energy is reduced in homogeneous compressible shear flows, or, as we shall see in section 2.4.2, the growth rate of thickness of high speed mixing layers decreases when the Mach number increases: these are examples of compressibility effects related to turbulent **fluctuations**. We shall detail both types of effects in the following sections, devoted to simple shear flows, but including high an low speed situations, wall bounded and free shear layers as well.

### 2.4.1. COMPRESSIBLE BOUNDARY LAYER

Several reviews and textbooks have been devoted to this important topic by Cebeci and Smith [72], Bradshaw [52], Cousteix [106], Dussauge *et al.* [142], Eléna and Gaviglio [145], Smits and Dussauge [434], among many others. Some major arguments and results, taken from these references are reviewed and discussed here.

***Mean velocity logarithmic profile.*** Starting from the incompressible regime, one of the most important results applying to equilibrium turbulent boundary layers is the famous logarithmic profile in the inertial sublayer:

$$\frac{\overline{U}}{u_w} = \frac{1}{\kappa} \ln y^+ + C \,. \tag{2.6}$$

$u_w = \sqrt{\tau_w/\rho_w}$ is the friction velocity, $\tau_w$ is the friction at the wall and $\rho_w$ the density at the wall and $y^+$ the distance to the wall, normalized by the "inner" scale $y^+ = yu_w/\nu$, where $\nu$ is the kinematic viscosity of the fluid. The so-called von Kármán constant is $\kappa = 0.41$ and good agreement with experimental data is achieved with $C = 5$.

The validity of eq.(2.6) comes from the existence of an overlapping region, in a multiple scales analysis of the viscous effects at high Reynolds number (see Tennekes and Lumley [459], for instance). Introducing the "inner scaling" $y^+$ and the "outer scaling" $\eta = y/\delta$, where $\delta$ denotes the boundary layer thickness, a region can exist where the two conditions $y^+ \gg 1$ and $\eta \ll 1$ are simultaneously satisfied, provided the Reynolds number is sufficiently

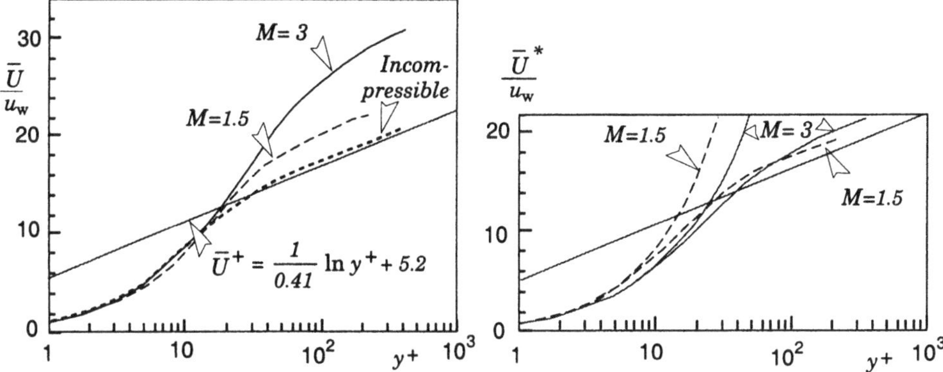


*Figure 2.10.* Mean velocity profiles and the log law in supersonic wall channel flows without (left) and with (right) Van Driest transformed velocity, adapted from Huang and Coleman [220].

high.
Hence the question directly arises about the validity of such scaling considerations in compressible boundary layers, since the Reynolds number is no longer the only significant parameter in this case.

In actual fact, eq.(2.6) does not apply to compressible flows, as clearly shown by Huang and Coleman [220], for instance, using results from direct numerical simulations of supersonic, isothermal, wall channel flow. In figure 2.10, adapted from the previous reference, the departure from the incompressible case is all the more important than the Mach number, based on bulk velocity and wall sound speed is high ($M=1.5$, with $T_c/T_w=1.38$ and $M=3$, with $T_c/T_w=2.47$ where $T_c/T_w$ is the center-line to wall temperature ratio).

As a consequence of the high Mach number effect, the viscous dissipation is no longer negligible in the enthalpy equation for high speed boundary layer flows. Indeed, due to the stagnation condition of the fluid at the wall, high levels of dissipation are generated near the wall, developing strong temperature gradients, which, in turn, are responsible for low-density and high viscosity effects close to the wall. To appreciate this dissipation heating, the ratio of wall temperature to the free stream temperature in a boundary layer on an adiabatic wall is 1.9 at a mean flow Mach number $M_\infty = 2.2$, 4.7 at $M_\infty = 4.5$ and nearly 20 at $M_\infty = 10$ (Lele [290]). As observed experimentally, the consequence is a thicker boundary layer and the extension of the viscous dominant zone, as compared with the incompressible situation at the same Reynolds number.

Now, as far as the scaling issue is concerned, it is clear that the "inner length scale" $\nu/u_w$ is affected by any change in viscosity. As discussed in

Smits and Dussauge [434], the temperature dependence of the viscosity is taken into account in the scaling of the viscous sublayer only, so that the modification of the scaling in the full-turbulent sublayer is purely density (or temperature) dependent.
In other words, based on the mean velocity gradient, a unique "inner" time scale is used for incompressible flows $\mathcal{T} = (\partial \overline{U}/\partial y)^{-1} \simeq y^+/u_w$ while a double scaling is needed for compressible boundary layers:

$$\mathcal{T}^{(\nu)} \simeq y^+/u_w \times \mu^*(T^*) \quad \text{and} \quad \mathcal{T}^{(\rho)} \simeq y^+/u_w \times \rho^*(T^*) \,,$$

where $\mu^*(T^*)$ and $\rho^*(T^*)$ stand for two non-dimensional functions of a non-dimensional temperature accounting for viscosity and density effects separately.
To summarize, and *recalling that the analysis are based on different arguments*[4], the following "inner" scaling for incompressible and compressible boundary layers are as given in Table 2.2.

TABLE 2.2. Comparison of the inner scaling for incompressible and compressible boundary layers.

| Regime | $y^+$ | viscous sublayer | Log. region |
|---|---|---|---|
| Incompressible | $\frac{y\,u_w}{\nu}$ | $\frac{\partial \overline{U}^+}{\partial y^+} = 1$ | $\frac{\partial \overline{U}^+}{\partial y^+} = \frac{1}{\kappa y^+}$ |
| Compressible | $\frac{y\,u_w}{\nu_w}$ | $\frac{\mu}{\mu_w}\frac{\partial \overline{U}^+}{\partial y^+} = 1$ | $\sqrt{\frac{\overline{\rho}}{\rho_w}}\frac{\partial \overline{U}^+}{\partial y^+} = \frac{1}{\kappa y^+}$ |

As proposed by Van Driest in 1951 [468], a logarithmic distribution in the compressible regime can be obtained by integrating the corresponding scaling for the "transformed velocity":

$$\overline{U}^* = \int_0^{\overline{U}} \sqrt{\frac{\overline{\rho}}{\rho_w}}\, dU \,, \tag{2.7}$$

so that it formally results:

$$\frac{\overline{U}^*}{u_w} = \frac{1}{\kappa} \ln y^+ + B \,. \tag{2.8}$$

[4]The scaling of the fully-turbulent part of the inner region in compressible boundary layers with zero pressure gradient can be simply obtained with the assumption of a constant turbulent shear stress equals to the friction at the wall: $-\overline{\rho}\,\overline{u'v'} = \tau_w$. Adopting an eddy viscosity concept with a mixing length formulation, this assumption yields: $\mu_t \frac{\partial \overline{U}}{\partial y} = \overline{\rho}\,\kappa^2\, y^2 (\frac{\partial \overline{U}}{\partial y})^2 = \tau_w$, which is equivalent to the expression given in Table 2.2

To complete the transformation — eq.(2.7) —, the mean density must be explicited as a function of the mean velocity. It is usually obtained by applying Crocco's integral (see [72], for instance):

$$\frac{\rho_w}{\overline{\rho}} = \frac{\overline{T}}{T_w} = 1 + [(1 + \frac{\gamma - 1}{2} M_e^2) \frac{T_e}{T_w} - 1] \frac{\overline{U}}{U_e} - \frac{\gamma - 1}{2} M_e^2 \frac{T_e}{T_w} (\frac{\overline{U}}{U_e})^2 ,$$

where the subscript $e$ refers to external conditions.
Thus, if transformation (2.7) is relevant, the compressibility and heat transfer effects on the mean velocity profile in the logarithmic region can be accounted for by a square root density-weighted integration of the mean velocity, and only slight variations of the constant $B$ in eq.(2.8) with the Mach number and the heat flux at the wall are to be expected. This point is rather well supported by the direct numerical simulations by Huang and Coleman [220], as shown in Fig. 2.10 (left). In addition, as reviewed in [434], such adaptation of the low speed analysis is effective for compressible boundary layers at a Mach number $M_\delta$ ranging between 3.5 and 7.2, with $\kappa = 0.4$ and $B = 5.1$.

Slightly different conclusions are reached by So *et al.* [435] based on an extension of the dimensional similarity arguments of the incompressible case to the compressible one. It is shown that the existence of an overlap between the wall layer and the defect law layers leads to logarithmic expressions of the mean velocity profiles for both "inner" and "outer" scaling formulations:

$$Inner\ scaling: \quad \overline{U}^+ = \frac{1}{\kappa(M_\tau, B_q, \gamma}) \ln y_w^+ + B(M_w, B_q, \gamma, P_{r_w}) , \qquad (2.9)$$

$$Outer\ scaling: \quad \frac{U_\infty - \overline{U}}{u_w} = -\frac{1}{\kappa(M_w, B_q, \gamma}) \ln \eta + A(M_\tau, B_q, \beta, \gamma) . \qquad (2.10)$$

In the previous relations, $\overline{U}^+ = \overline{U}/u_w$, where $u_w = \sqrt{\tau_w \overline{\rho}_w}$, $U_\infty$ is the free stream velocity, $y^+ = y u_w/\nu_w$, $\eta = y/\delta$, where $\delta$ is the boundary-layer thickness.
The "constants" $\kappa$, $A$ and $B$ are now parametric in the friction Mach number $M_w = u_w/a_w$ where $a_w$ is the speed of sound evaluated at the wall conditions. The dimensionless wall heating is $B_q = Q_{tot}/(\rho_w C_p u_w T_w)$ where $Q_{tot}$ is the total heat flux. The dimensionless pressure gradient is $\beta = (\delta dP/dx) / \tau_w$. The Prandtl number at the wall is $P_{r_w} = \mu_w Cp/k_w$, and the ratio of the specific heats evaluated at the wall conditions is $\gamma = (C_p/C_v)_w$.

From comparisons with measurements, So *et al.* [435] establish in particular that:

- the von Kármán parameter $\kappa$ remains constant at approximately 0.41 for adiabatic compressible boundary layers on a flat plate with a free-stream Mach number ranging from 0 to 4.5;
- $\kappa$ rises sharply as the free stream Mach number increases beyond 5;
- $\kappa$ decreases (increases) as the wall to free stream temperature ratio decreases (increases).

***Reynolds stresses.*** As noted, for instance, by Dussauge *et al.* [142], the measurement of turbulent quantities, Reynolds stresses, rms values of temperature and density fluctuations in supersonic boundary layers is exceedingly difficult. In fact, the longitudinal normal stress component $\overline{u'^2}$, where a prime denotes a centered fluctuation, currently seems the only one to be rather well-documented.

When the free stream Mach number ($M_\infty$) increases, smaller values of the turbulence intensity $\sqrt{\overline{u'^2}}/u_w$, where $u_w = \sqrt{\tau_w/\rho_w}$ are observed ([434], page 216). For instance, in the boundary layer on a flat plate, this intensity is about 1.8 for the incompressible regime, at a distance normal to the wall $y/\delta = 0.2$. It is reduced to about 1.4 ($M_\infty = 1.7$) and 1.0 ($M_\infty = 4.7$), based on the experimental data reported in Schlichting [422].

Now, if a square root density weighting is applied to the normal stresses, i.e., if the following expressions $\sqrt{\overline{\rho}/\rho_w}\sqrt{\overline{u'^2}}/u_w$ and $\sqrt{\overline{\rho}/\rho_w}\sqrt{\overline{v'^2}}/u_w$ are considered, the compressible profiles are close to those measured in the incompressible boundary layer, as shown in the next figure adapted from Eléna and Gaviglio [145], an observation which emerges even more clearly from the data compiled by Smits and Dussauge [434].

A similar conclusion also holds for the turbulent shear stress $-\overline{\rho}\overline{u'v'}/\tau_w$ and the correlation coefficient $-\overline{u'v'}/(\sqrt{\overline{u'^2}}\sqrt{\overline{v'^2}})$.

***Velocity-Density (temperature) correlation.*** As reported in ref. [72], pages 70-72, experimental evidence has been given that in supersonic, adiabatic, turbulent boundary layers:

(a) the maximum fluctuations of the total temperature $T + U^2/2C_p$ are less than 5% at $M_\infty = 4.67$. Thus the mean total enthalpy is nearly constant and the fluctuating total enthalpy can be neglected;

(b) the temperature fluctuations are essentially isobaric when Mach numbers are less than 5.

As we shall see in Chapter 3, these observations provide experimental evidence to support Morkovin's strong Reynolds analogy [339]. For the present discussion, a simpler derivation of the corresponding relations is given here.

- Introducing $H = C_pT + U^2/2$, the first observation (a) yields $\overline{H} = C^t$ and

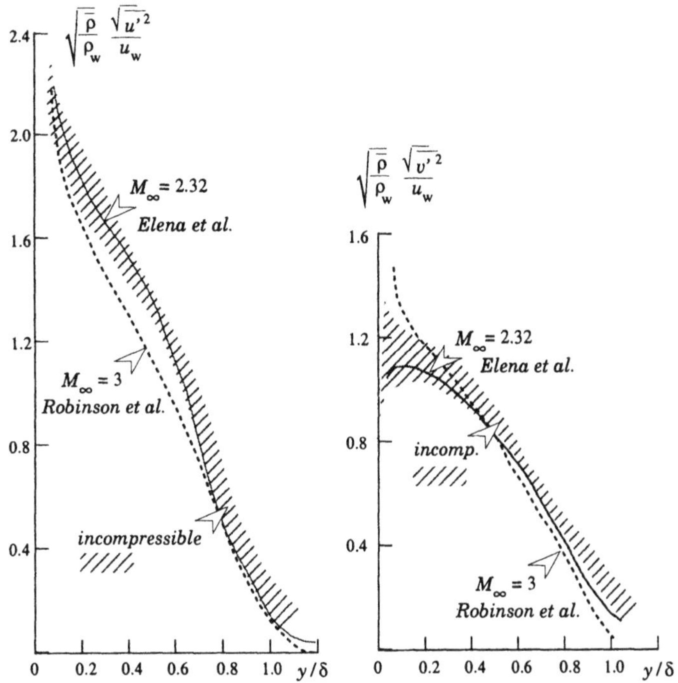


*Figure 2.11.* Density normalized turbulence intensities in compressible boundary layers, adapted from Eléna and Gaviglio [145].

$h'(\equiv H - \overline{H}) = 0$, that is, with a linear approximation to the square of the velocity, $C_p\overline{T} + \frac{\overline{U}^2}{2} = C^t$ and $C_p\theta' = -\overline{U}u'$, where $\theta' = T - \overline{T}$. Thus, it can be easily deduced that:

$$\frac{\theta'}{\overline{\overline{T}}} = -(\gamma - 1)M^2\frac{u'}{\overline{\overline{U}}}, \tag{2.11}$$

where $M = \overline{U}/\sqrt{\gamma R\overline{T}}$ is the local Mach number based on the mean values of velocity and temperature within the boundary layer.

– Now, according to second comment (b) on the isobaric temperature fluctuations, the linearized fluctuating equation of state $p'/\overline{P} = \rho'/\overline{\rho} + \theta'/\overline{T}$ reduces to:

$$\frac{\theta'}{\overline{\overline{T}}} = -\frac{\rho'}{\overline{\rho}}. \tag{2.12}$$

Combining eqs.(2.11) and (2.12), it is deduced that:

$$-\frac{\overline{\theta' u'}}{\overline{\overline{T}}\,\overline{\overline{U}}} = \frac{\overline{\rho' u'}}{\overline{\rho}\,\overline{\overline{U}}} = (\gamma - 1)\overline{M}^2\frac{\overline{u'^2}}{\overline{\overline{U}}^2} \qquad -\frac{\overline{\rho'\theta'}}{\overline{\rho}\,\overline{\overline{T}}} = \frac{\overline{\rho'^2}}{\overline{\rho}^2} = -\frac{\overline{\theta'^2}}{\overline{\overline{T}}^2} = (\gamma - 1)^2\overline{M}^4\frac{\overline{u'^2}}{\overline{\overline{U}}^2} \tag{2.13}$$

so that, for $\gamma = 1.4$ density-temperature fluctuations and temperature-velocity fluctuations are negatively correlated, while density-velocity fluctuations are positively correlated.
As shown by Eléna and Gaviglio [145], the previous relations are found to be approximately satisfied in an adiabatic supersonic boundary layer with $M_\infty = 2.32$ where, for instance, the following values are measured for $0.2 < y/\delta < 1$:

$$\frac{\sqrt{\overline{\theta'^2}}/\overline{T}}{(\gamma-1)\overline{M}^2\,(\sqrt{\overline{u'^2}}/\overline{U})} \simeq 0.95 \ \text{(expected 1.0)} ,$$

$$-\frac{\overline{\theta' u'}}{\sqrt{\overline{\theta'^2}}\sqrt{\overline{u'^2}}} \simeq 0.8 \ \text{(expected 1.0)} ,$$

$$-\frac{\sqrt{\overline{\theta'^2}}}{\overline{T}} = A(y/\delta)(\gamma-1)\overline{M}^2\frac{\sqrt{\overline{u'^2}}}{\overline{U}} \ \text{(expected } A = 1.0) ,$$

where $A(y)$ is a slightly motonously increasing function of the non dimensional distance to the wall ($A(0)=1$ and $A(0.8)\simeq 1.2$ according to [434]). From the previous values, quantitative effects of density changes driven by temperature variations can be appreciated by comparison with the constant density situation. For example, when temperature acts as a passive contaminant, the correlation between velocity and temperature fluctuations is lower, close to 0.5, as reported in [145].

To summarize, in equilibrium supersonic boundary layers up to $M_\infty < 5$ with adiabatic wall conditions, pressure fluctuations are usually small at such supersonic speeds, and density effects mainly result in mean temperature induced variations.
Within the scope of this traditional description of turbulence in supersonic boundary layers, the departure in the one-point statistical properties of the mean ($\overline{U}_i$) and turbulent ($\overline{u'_i u'_j}$) velocity fields between incompressible and compressible regimes, can be accounted for by *mean* density variations. Accordingly, this means in particular that:

- Van Driest transformed mean velocity profiles have a same log-law with the same slope as in both flow regimes;
- turbulence velocity intensities and correlations in the compressible regime can be recovered from the incompressible situation by an appropriate scaling in mean density.

This leads to talk about *mean-density* driven effects in this case, as opposed to *compressibility* driven effects, as typically encountered in mixing layers (see next section).

However, there exists experimental evidence showing that the traditional picture neither *strictly* applies nor applies to *all* properties of the turbulent boundary layer. New results reported by Smits and Dussauge [434], for instance, show that the normal stresses $\overline{v'^2}$ and $\overline{w'^2}$ appear to increase slightly when increasing the free stream Mach number, contrasting with the streamwise stress component $\overline{u'^2}$. Therefore, when using the mean density scaling applied to the normal stresses profiles, the anisotropy ratios $\overline{v'^2}/\overline{u'^2}$, $\overline{w'^2}/\overline{u'^2}$ will not be recovered from the incompressible situation. Other major differences, with respect to the free stream Mach number, are observed on the large scale structures and the geometry of the turbulent/non-turbulent interface: the intermittency level, for instance, decreases as the Mach number increases, even moderately ($M_\infty = 2.5$ [434]). Of course, such departures from the incompressible case are all the more important than a substantial heat flux existing at the wall or the Mach number being largely increased, such as in hypersonic boundary layers (Bradshaw [52]).

Thus, when present, strong mean temperature and velocity gradients lead to significant temperature and velocity fluctuations, ***and*** mean transverse density gradients $(\partial\overline{\rho}/\partial y)$ which are expected to alter the larger eddies mainly responsible for the expansion, entrainment and mixing of free stream fluid. In other words, as pointed out by Lele [290], variable *inertia* effects and *compressible* effects are imbedded in high Mach number boundary layers. Due to the rather lack of available data, the identification of both effects will not be discussed here in this flow configuration. On the other hand, they will be addressed separately in two different free flows: the compressible mixing layer and the low speed jet.

### 2.4.2. VARIABLE DENSITY EFFECTS IN MIXING LAYERS AND JETS

Incompressible free simple shear flows, such as two dimensional mixing layers and jets, are basically governed by a time scales equilibrium, between advection (or convection) and diffusion. Denoting by $U_c$ a characteristic velocity scale of convection / advection, the time scale of the convective/advective transport along a distance $x$ is simply $T_c \propto x/U_c$, while the time scale of the diffusion transfer is $T_d \propto \delta/\Delta U$, where $\delta$ denotes a characteristic length scale of the diffusion or spreading of the flow associated with a velocity variation $\Delta U$ across the flow field (see Chassaing [84], for instance). From the basic time-scales equilibrium, it simply results that:

$$\frac{\delta}{x} = \beta \frac{\Delta U}{U_c} , \tag{2.14}$$

where $\beta$ is a constant, for a given type of shear flow. Thus the spreading rates of such flows are directly linked with the streamwise variations of two characteristics velocity scales.

Now, if a similar basic equilibrium ($T_c \propto T_d$) still prevails in variable density flows, it can be inferred from experimental data, that $\beta$ is no longer a constant but depends upon some characteristic parameters associated with the fluid density variation, such as a given density ratio $s$ or/and Mach number $M$. Accordingly, eq.(2.14) can be formally rewritten as:

$$\frac{\delta}{x} = \beta(s, \text{or/and}\, M)\frac{\Delta U}{U_c} \,. \tag{2.15}$$

The previous analysis is restricted to fully developed turbulent flows, so that the main diffusion process is due to continuous motions, and molecular effects can be discarded. This explains why $\beta$ does not depend on the Reynolds number.

The most important conclusions concerning the spreading rate of such free shear flows are as follows:

- *low speed situation*: heavier-than-ambient jets spread slower than constant density and lighter-than-ambient jets;
- *high speed situation*: (i) the thickness growth rate of a compressible, single species, shear layer decreases with increasing Mach number, and (ii) the influence of the density ratio, separate from the Mach number, as investigated from DNS by Sarkar and Pantano [414] for a temporally evolving binary free shear layer consists in a decrease of the momentum thickness growth rate with increasing the density ratio at a given Mach number.

### 2.4.3. THE VARIABLE DENSITY MIXING LAYER

***Mean flow expansion.*** Variable density effects due to differences in temperature and/or fluid composition between the two incoming flows of a turbulent, *low-speed* mixing layer were reported in particular by Brown and Rohsko [61] in 1974. In order to account for such effects, two parameters are to be considered, the density ratio $s = \rho_2/\rho_1$ and the velocity ratio $r = U_2/U_1$, where subscripts 1 and 2 indicate the high speed and low-speed side respectively. As shown by Papamoschou and Roshko [358], a fairly good agreement with the measurements is obtained with the following expression of the low-speed mixing layer growth rate:

$$\frac{d\delta_{inc}}{dx} = C_\delta \, \frac{(1-r)(1+\sqrt{s})}{1+r\sqrt{s}} \,, \tag{2.16}$$

where $\delta(x)$ denotes some characteristic length scale of the spanwise expansion of the mixing layer, the subscript '*inc*' referring to the *low-speed* ("incompressible") situation. The numerical value of the constant $C_\delta$ depends on the definition adopted for $\delta_{inc}$ (see Chapter 9 for more details).

A strong effect of the Mach number on the spreading rate of compressible mixing layers was observed independently very early, in 1968 by Siriex and Solignac [432] in France, for instance. Indeed, the spreading rate was found to decrease sharply with the Mach number, as compared with the incompressible situation at the same free-stream velocity and density ratios. This stabilizing trend of compressibility by the thinning of the shear zone is remarkably evident when considering the "*normalized*" spreading rate, i.e., the ratio of the compressible spreading rate $\dot{\delta} = d\delta/dx$ to the incompressible one $\dot{\delta}_{inc}$, where $\delta(x)$ denotes some characteristic length scale of the spanwise expansion of the mixing layer.

Thus, one can question oneself about the identification of *the suitable* compressibility parameter able to correlate such a reduction in the growth rate of similar and non-similar gases mixing layers.
The first significant contribution to this issue is due to Bogdanoff [49] in 1983. His proposal, introducing a so-called "convective Mach number" ($M_c$), was confirmed by Papamoschou and Roshko [358] in 1988. Referring to the high speed zone of the mixing layer (subscript 1) for instance, it can be defined as the Mach number in a frame of reference convecting with the velocity of the dominant structures $U_c$:

$$M_c = \frac{U_1 - U_c}{a_1} ,$$

where $U_1$ is the velocity of the outer flow and $a_1$ the speed of sound in the same flow. Assuming an identical value of the polytropic coefficient in both streams ($\gamma_1 = \gamma_2$), the convective velocity can be approximated by ([358]):

$$U_c = U_1 \frac{1 + r\sqrt{s}}{1 + \sqrt{s}} = \frac{a_2 U_1 + a_1 U_2}{a_1 + a_2} , \tag{2.17}$$

from which it results that:

$$M_c = \frac{U_1 - U_2}{a_1 + a_2} .$$

As shown by Bogdanoff [49], Papamoschou *et al.* [358] and Barre *et al.* [30] among others, the *convective* Mach number was actually found to be *a* pertinent parameter accounting for compressibility effects on the *normalized* growth rate of heterogeneous mixing layers, i.e., yielding a fair correlation in the form:

$$\frac{\dot{\delta}}{\dot{\delta}_{inc}} = f(M_c) .$$

The consensus adopted for the function $f(M_c)$, based on the experimental data available before 1972, is known as the "Langley curve". A remarkable feature of this curve is the strong reduction of the normalized growth rate

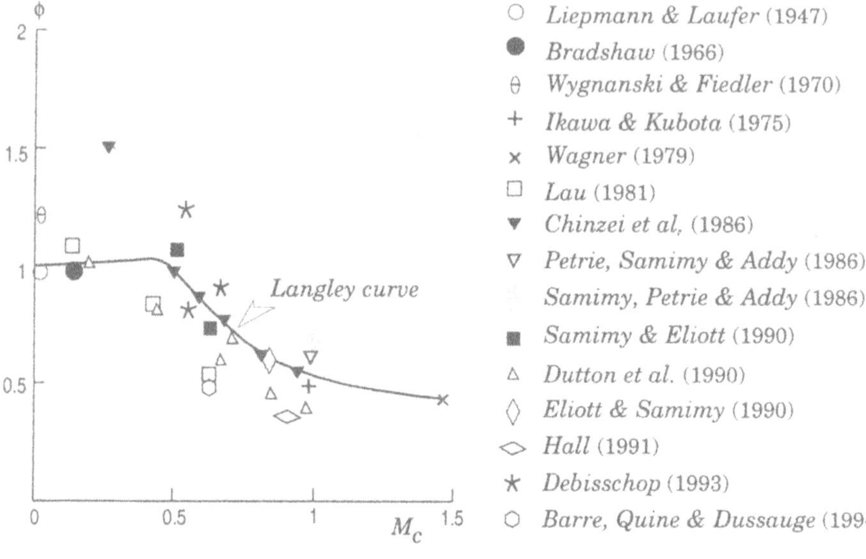


*Figure 2.12.* Compressibility effect on the spreading rate of a mixing layer, adapted from Barre et al.[30].

from 1 to about 0.2 over the range $1 < M_c < 0.8$, and a rather small reduction beyond $M_c = 0.8$. It is plotted in fig.2.12 according to the expression of Barre, Quine & Dussauge [30]:

$$\frac{d\delta}{dx} = (\frac{d\delta}{dx})_{inc} \frac{U_1 - U_2}{2U_c} \times \phi(M_c) \times f(s, M_c) \; , \tag{2.18}$$

where $f(s, M_c)$ is a correction factor, close to unity, introduced to correlate results for both temporally and spatially evolving mixing layers.

The comparison with the various experimental data compiled by Barre *et al.* [30], shows that, even reduced, departures from a unique curve $\phi(M_c)$ are still present.

A large amount of the scientific literature in the field has been devoted to the explanation to such intrinsic compressibility effects in both homogeneous and heterogeneous mixing layers. For the former situation, a very early popular explanation, as recalled, for instance, by Debiève *et al.* [125], was the existence of shocklets, or small shock waves. Due to the presence of such structures, one can expect a reduction of turbulence kinetic energy, both by viscous and thermal dissipation and acoustic radiation. However, experimental evidence of the importance of such a reduction has not been definitely found so far.

An other possible explanation can be found in the analysis of explicit compressibility terms which emerge in the turbulence kinetic energy balance. Such terms as dilatation dissipation and pressure-dilatation correlation were taken for a moment as the plausible candidates for such an explanation (Zeman [495], Sarkar [408], Sarkar *et al.* [415]).

However, later literature on the topic, such as Sarkar [409], Vreman *et al.* [477] tends to demonstrate that the production term actually plays the dominant role. It seems that there is now a wide agreement to consider that the effect of *intrinsic* compressibility (i.e., resulting from the non zero divergence condition) is to reduce, as compared with an "equivalent" pure solenoidal situation, the amount of kinetic turbulence energy produced by mean shear. A direct consequence is of course the reduction of turbulent mixing. The reduced production level could be the consequence of pressure-strain terms Vreman *et al.* [477], which could be the in depth origin of such a well known characteristic of compressible flows.

However, as noticed by Cambon and Simone [68], this single point explicit correlation analysis poorly reflects other compressibility effects on the structure of the pressure field and the anisotropic structure of the velocity field for instance.

At last, as far as heterogeneous compressible mixing layers are concerned, the recent DNS results obtained by Sarkar and Pantano [414] support the idea that the density ratio is to be considered as an independent parameter, in addition to the convective Mach number, to accounting for variable density effects in such flows. For instance, with $M_c = C^t = 0.7$, the momentum thickness is about three times lower when the density ratio changes from 1 to 8.

***Turbulence characteristics.*** The reduction of the spreading rate can be hardly conceived without considering that turbulence characteristics should be significantly modified, as compared with the constant density situation. Actually, both experimental investigations and direct numerical simulations provide evidence on statistical and structural changes in the turbulent properties of the compressible mixing layer (see, for instance, Elliott *et al.* [146], [147], [405], Vreman *et al.* [477]).

The maximum values of the turbulence intensities and shear stress profiles, normalized by the difference of the free stream velocities $(\Delta\overline{U})$ are plotted in figure 2.13 (left), as measured by Elliott and Samimy [146]. The significant components of the Reynolds stress anisotropy tensor $b_{ij}$ as deduced from the previous results (assuming $\overline{v'^2}=\overline{w'^2}$) and those obtained by Vreman *et al.* [477] are given in the same figure (right). The Reynolds stress anisotropy tensor $b_{ij}$ is defined as:

$$b_{ij} = \frac{\overline{u'_i u'_j}}{2\overline{k}} - \frac{1}{3}\delta_{ij} \ ,$$

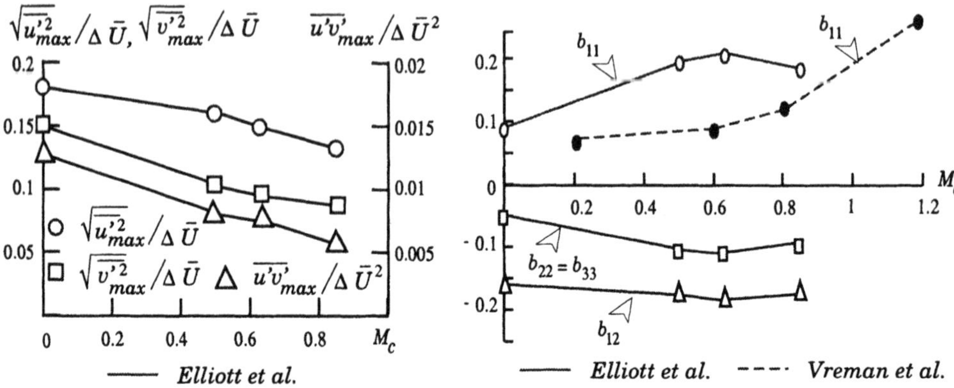


*Figure 2.13.* Reynolds stress components and anisotropy in compressible mixing layers, adapted from Elliott et al. [146] and Vreman et al. [477].

where $\overline{k} = \overline{u'_i u'_i}/2$.

Two main observations can be drawn from these results about compressibility effects on turbulence in such flows:

- The maximum levels of the turbulence fluctuations *decrease* (almost linearly) with increasing convective Mach number. Indeed, the maximum Reynolds stress values measured by Elliott *et al.* [146] at $M_c = 0.51$ are about 40% lower than the subsonic results;
- The reduction in the diagonal terms is associated with an *increasing* anisotropy in the normal stresses. According to Vreman *et al.* [477], for instance, $b_{11}$ at $M_c = 1.2$ is about five times greater than the incompressible value.

Modifications induced by compressibility on the large scale structures in the mixing layer, as first observed by flow visualizations [61], [357], [358] have been confirmed by measurements, using space-time statistics, for instance, [147], [136], and direct numerical simulations [288]. This point is illustrated here by comparing energy spectra in subsonic incompressible and supersonic mixing layers, as shown in the next figure, adapted from [125].

In both cases, the Reynolds number $\Delta U \times \delta/\nu$ is about $10^5$ and the convective Mach number of the supersonic mixing layer is 0.62. The spectra $E(k)$ are normalized to unity and multiplied by the wave number $k$ to highlight the production zone where $E(k) \propto k^{-1}$. The wave number is normalized by the layer thickness $\delta$. From figure 2.14, it appears that the production scales extend over a wider range in the supersonic layer, centered at greater non-dimensional wave numbers. This suggests that the energetic scales are greater in an incompressible mixing layer than in a compressible

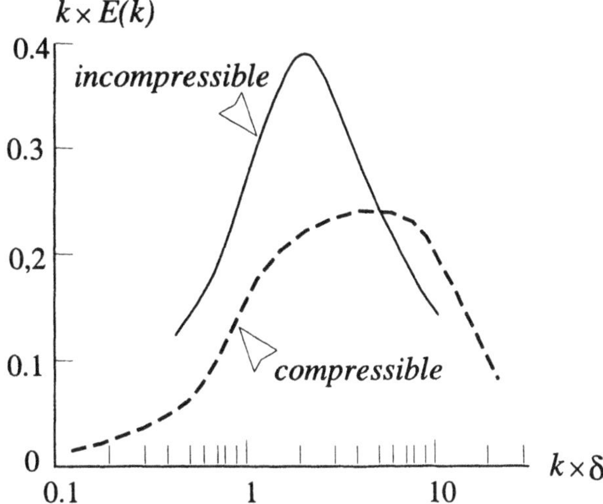


*Figure 2.14.* Compressibility effect on turbulence spectra in mixing layers, adapted from Debiève et al. [125].

one. Actually, the integral scales deduced from two point correlation measurements by Dupont *et al.* [136] are respectively equal to $0.44\delta$ and $0.20\delta$.

It should be noticed that these conclusions could be dependent upon the convective Mach number range. As shown by Samimy *et al.* [405] from space-time correlations, the large structures in a compressible mixing layer at $M_c = 0.51$ are similar to those in the incompressible case, but less organized. On the other hand, at $M_c = 0.86$, the structures are highly three-dimensional, with a good spatial organization, but a poor temporal one.

### 2.4.4. THE LOW SPEED JET

The last example of density effects presented here is devoted to fully turbulent free jets. Like the mixing layer, density variation can be associated with compressibility and/or non-uniformity in temperature or concentration at the exit. When acting separately, these effects lead to two distinct situations (i) high velocity jets and (ii) inhomogeneous low-speed jets, the only one to be considered in the following.

Strong variations in temperature and/or composition in low-speed turbulent jets, associated with a density ratio at the exit $s_0 = (\rho_{jet}/\rho_\infty)_0$ which significantly departs from unity, were found to alter the flow field, even far downstream, where the density variation has been lowered. Global characteristics (mixing and spreading of the jet), mean and turbulent profiles, center-lines evolutions... are changed, as compared with an equivalent[5] constant density situation. Experimental evidence was first reported in the literature by many authors, Ricou and Spalding [393] in 1961, Chassaing [80], [87] in 1979, Sautet [417] in 1992, Richards and Pitts [392] in 1993, Panchapakesan and Lumley [352] in 1993, Djeridane [132] in 1994, for instance. Many other references can be found in the reviews by Schefer *et al.* [420] in 1986 and Chassaing *et al.* [86] in 1994.

Turning back to the time scale analysis previously introduced (see section 2.4.2), it is clear that, for such a low-speed variable density jet discharging into a quiescent atmosphere, the velocity scales are such that $\Delta U \propto U_c \propto U_{\mathbb{C}}$, where $U_{\mathbb{C}}$ is the mean velocity along the jet axis. Thus the fundamental time scale equilibrium (2.15) reduces to:

$$\frac{\delta}{x} = \beta(s_0) \ , \tag{2.19}$$

so that the linear expansion of the flow depends now on the density ratio at the exit.

As a direct consequence of eq.(2.19), the decreasing rates of the centerline velocity and temperature or concentration also depend on the density ratio. Unlike the mixing layer, such centerline properties decrease in the far field jet, according to *exactly* hyperbolic laws in the self-preserving region of a constant density round jet developing into a quiescent atmosphere. When the density is not uniform, such hyperbolic expressions are still usually adopted as *approximated* formulations to the centerline decrease:

$$\frac{U_{\mathbb{C}}}{U_0} = \frac{A(s_0)}{(x - x^*)/D_0} \quad \text{and} \quad \frac{F_{\mathbb{C}}}{F_0} = \frac{B(s_0)}{(x - x^*)/D_0} \ , \tag{2.20}$$

where the subscript $_0$ refers to exit conditions, and $F$ stands for either mean concentration or mean temperature difference. The downstream location is normalized by the diameter at the exit $D_0$ and $x^*$ denotes a virtual origin.

A compilation of experimental data taken from ref. [86] is given in figure 2.15. The abscissa range is $20 \leq (x - x^*)/D_0 \leq 100$, where approximated eqs.(2.20) can be used.

From the previous compilation, a strong effect of the density variation can be observed when compared with the constant density situation ($s_0 = 1$). For instance, when increasing the density ratio by 50%, the velocity coefficient $A$ is multiplied by 1.3 (1.2 for $B$) while $A$ and $B$ are both roughly multiplied by 0.65, when reducing the density ratio by 50%. As discussed in [86], correlations with density fluctuations play an important part to the explanation of such effects.

[5]Both kinematic and dynamic equivalences have been considered for such comparisons. In the first case, the same velocity at the exit is adopted, in the second one, the momentum is conserved for the constant and variable density situations respectively.

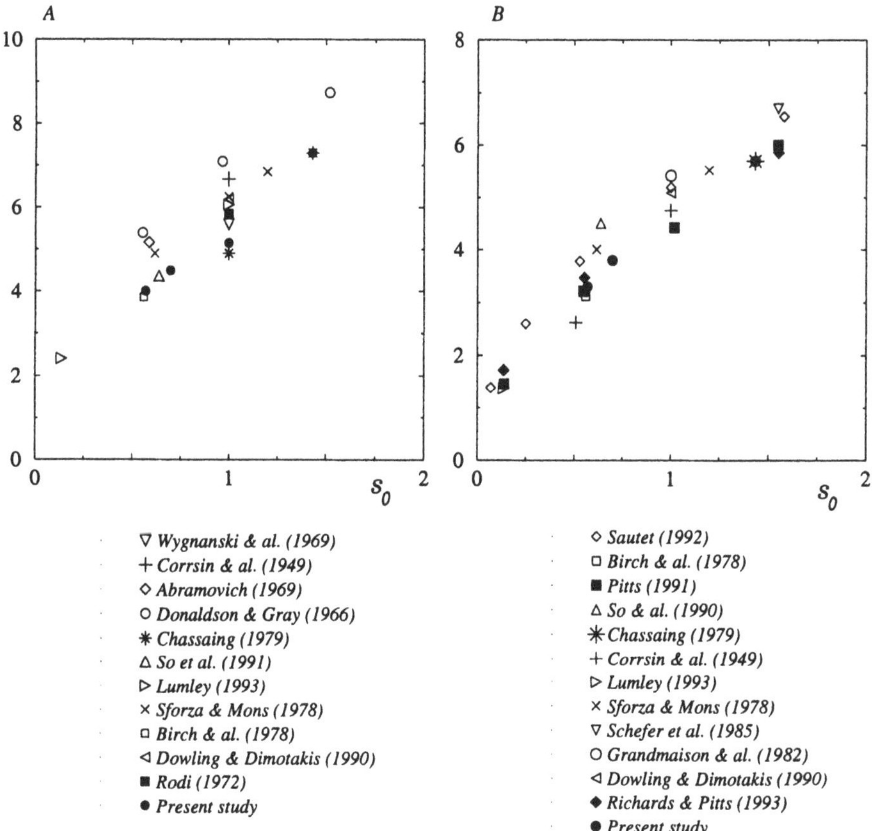


*Figure 2.15.* Density effect on the decreasing rates of low speed jets, from Chassaing *et al.* [86].

*CHAPTER 3*

# APPROXIMATE MODELS FOR VARIABLE DENSITY FLUID MOTIONS

*This chapter aims at giving a comparative overview of some of the various models which, derived from the general Navier-Stokes equations, account for density variations according to several types of approximations. The role of the pressure is first examined. The Helmholtz decomposition is introduced and the linear analysis of Kovasznay compressible modes is presented. Then several models are discussed, referring to Boussinesq's approximation, and other approximations which are concerned with (i) filtering acoustic effects, (ii) incorporating density variations in pseudo-incompressible formulations and (iii) deriving weakly compressible limits to the general compressible equations.*

## 3.1. Introduction

All sources driving density variations are not always present, in the same orders of magnitude, in any given flow configuration. Consequently, using the general Navier-Stokes model is not necessarily the most appropriate choice to capture pertinent density variation effects, without too much complexity. For example, in many flows, the density variations play no important part in the dynamics. This is the case for most liquid motions and low speed, constant composition and temperature gas flows. In aerodynamics applications for instance, when velocities are lower than 60 m/s, the air can be considered as a fairly incompressible gas, insofar as the departure with results derived from compressible formulae is not exceeding 1%.
In other circumstances, the constant density assumption can be a bad approximation, even for low speed motions, as for instance in convection driven flows. However in this case too, one can get rid of the full Navier-Stokes equations, using Boussinesq's or anelastic approximations.
Another situation, in which constant density approximation is obviously inappropriate for the fluctuations, concerns sound propagation. In this case, *acoustic* density fluctuations are specifically associated with pressure fluctuations and cannot be described by any incompressible approximation. Thus in several cases, some simplified, but approximate, forms of the Navier-

Stokes equations can be derived, according to the dominant effects of the density variation under consideration. This is the question to be addressed in the present chapter.
Some other more theoretical aspects should also be considered. They concern the mathematical conditions under which the solution $(\overrightarrow{V}^c, P^c)$ of the compressible Navier-Stokes equations converges to the solution of the incompressible model $(\overrightarrow{V}^I, P^I)$ as the Mach number goes to zero. Such aspects will not be presented here. Some important results can be found in the paper by Klainerman and Majda [251] including a rigorous justification of the use of linearized acoustic as a uniformly valid principal correction in the deviation of $\overrightarrow{V}^c - \overrightarrow{V}^I$ as $M \to 0$.

In order to discuss some approximated models, we shall first go back to the role of the pressure as a prelude to the decomposition of the pressure field and the so-called "Helmholtz" decomposition of the velocity vector field.

The last part of the chapter will be more directly concerned with approximations in weakly compressible flows, starting from the linear small perturbation analysis of Kovasznay.

## 3.2. The role of the pressure in compressible turbulence

### 3.2.1. PRESSURE IN INCOMPRESSIBLE FLOWS: POISSON'S EQUATION

In order to clarify the role of the pressure in variable density fluid motion, we shall first recall classical results for incompressible $(\rho = \rho_0 = C^t)$ and isovolume $(\partial U_i/\partial x_i = 0)$ evolution. In this case the momentum equation reduces to:

$$\rho_0(\frac{\partial U_i}{\partial t} + U_j\frac{\partial U_i}{\partial x_j}) = \rho_0 F_i - \frac{\partial P}{\partial x_i} + \frac{\partial \tau_{ij}^{(I)}}{\partial x_j},$$

where $F_i$ stands for the external body forces and $\tau_{ij}^{(I)}$ is the restriction of the viscous stress tensor to its incompressible expression.
Taking the divergence of the momentum equation and using the solenoidal condition, one obtains,

$$-\frac{\partial^2 P}{\partial x_i \partial x_i} = \rho_0 \frac{\partial}{\partial x_i}(U_j\frac{\partial U_i}{\partial x_j} - F_i - \frac{1}{\rho_0}\frac{\partial \tau_{ij}^{(I)}}{\partial x_j}) \equiv \rho_0 \frac{\partial U_j}{\partial x_i}\frac{\partial U_i}{\partial x_j} - \rho_0\frac{\partial F_i}{\partial x_i} - \frac{\partial^2 \tau_{ij}^{(I)}}{\partial x_i \partial x_j} \tag{3.1}$$

Thus, the pressure field of an incompressible fluid motion obeys a Poisson's equation the source terms of which include advection or transport, external body forces and external viscous stresses.

As a direct consequence in incompressible turbulence associated with a solenoidal random velocity field, the pressure fluctuation is entirely determined by the mean and fluctuating velocity fields, due to the non linearity of the source term in eq.(3.1). Temperature and/or concentration fluctuations, if present, are regarded as "passive contaminant", in the sense that they are not introducing any dynamic effect on the turbulence itself.

### 3.2.2. PRESSURE IN COMPRESSIBLE FLOW: THE GENERALIZED WAVE EQUATION

When the restriction of incompressibility is removed, sound radiation from turbulent zones arises. This is a new aspect of the turbulence problem specific to variable density fluctuations. In compressible fluid, any disturbance from a local source will propagate at a **finite** speed and will influence a **finite** domain of the flow field at a given time. In other words, in order to know the flow properties at a given instant, one has to know the behavior of the disturbance source earlier in time. This is a direct consequence of a generalized wave equation for the pressure.
To derive this equation, we shall adopt the formulation by Laufer *et al.* [272] leading to a slightly different form as the one first obtained by Phillips [363] in 1956. The analysis is concerned with the mass conservative motion of a compressible, viscous, heat-conducting an perfect gas, using a velocity-pressure-entropy formulation. In addition to the equation of state ($P = \rho RT$), the transport equations can be written in this case as follows:

$$\textit{Continuity}: \quad \frac{1}{\rho}\frac{d\rho}{dt} + \frac{\partial U_i}{\partial x_i} = 0 \, ,$$

$$\textit{Momentum}: \quad \frac{\partial U_i}{\partial t} + U_j \frac{\partial U_i}{\partial x_j} = F_i - \frac{1}{\rho}\frac{\partial P}{\partial x_i} + \frac{1}{\rho}\frac{\partial \tau_{ij}}{\partial x_i} \, ,$$

$$\textit{Energy}: \quad \frac{1}{C_p}\frac{dS}{dt} = \frac{1}{\gamma P}\frac{dP}{dt} - \frac{1}{\rho}\frac{d\rho}{dt} \, .$$

In the energy equation (second law of thermodynamics), the specific entropy $S$ is taken as a function of the absolute temperature and the specific volume or the density: $T\delta S = C_v \delta T + P\delta v \equiv C_v \delta T - (P/\rho^2)\delta\rho$, where $C_v$ is the specific heat at constant volume.
Taking the divergence of the momentum equation, and introducing the isentropic speed of sound $a^2 = \gamma P/\rho$ one obtains:

$$-\frac{\partial}{\partial x_i}\left(\frac{a^2}{\gamma P}\frac{\partial P}{\partial x_i}\right) = \frac{\partial}{\partial x_i}\left(\frac{\partial U_i}{\partial t} + U_j \frac{\partial U_i}{\partial x_j} - F_i - \frac{1}{\rho}\frac{\partial \tau_{ij}}{\partial x_j}\right) . \tag{3.2}$$

Combining the continuity and energy equations, it is easy to deduce that:

$$\frac{1}{\gamma P}\frac{dP}{dt} = \frac{1}{C_p}\frac{dS}{dt} - \frac{\partial U_i}{\partial x_i} \, .$$

Taking the material derivative of this relation and adding it to eq.(3.2) we finally obtain ($\gamma = C^t$):

$$\frac{d}{dt}(\frac{1}{P}\frac{dP}{dt}) - \frac{\partial}{\partial x_i}(\frac{a^2}{P}\frac{\partial P}{\partial x_i}) = \gamma\frac{\partial}{\partial x_i}(\frac{\partial U_i}{\partial t} + U_j\frac{\partial U_i}{\partial x_j} - F_i - \frac{1}{\rho}\frac{\partial \tau_{ij}}{\partial x_j}) + \gamma\frac{d}{dt}(\frac{1}{C_p}\frac{dS}{dt} - \frac{\partial U_i}{\partial x_i}) \,. \tag{3.3}$$

Applied to small pressure perturbation $p' = P - P_0$, this equation can be rewritten in a more usual linearized form:

$$\frac{1}{P_0}(\frac{d^2p'}{dt^2} - \frac{\partial}{\partial x_i}a^2\frac{\partial p'}{\partial x_i}) = \gamma\frac{\partial}{\partial x_i}(\frac{\partial U_i}{\partial t} + U_j\frac{\partial U_i}{\partial x_j} - F_i - \frac{1}{\rho}\frac{\partial \tau_{ij}}{\partial x_j}) + \gamma\frac{d}{dt}(\frac{1}{C_p}\frac{dS}{dt} - \frac{\partial U_i}{\partial x_i}) \,. \tag{3.4}$$

Two main comments can be drawn from this result:

**a** The pressure fluctuation is governed by a generalized wave equation, in which the usual wave operator in the left-hand-side is replaced by a more general one, since the *material* derivative of the pressure fluctuation is substituted to the usual *partial* time derivative;

**b** Turning now to the right-hand-side of eq.(3.4), various sources of noise generation can be identified, as produced by perturbations through transport, body forces, viscous forces, dilatation ($\partial U_i/\partial x_i$) and entropy.

If the right-hand-side terms in eq.(3.4) are known, this equation provides the starting point of the calculation of sound radiation. However in general, these terms are implicit functions of the pressure. Therefore, additional assumptions have to be introduced in order to decouple these terms from the radiated pressure field. Some of them will be given in the next sections.

### 3.2.3. LIGHTHILL'S ACOUSTIC ANALOGY

Lighthill was the first to succeed in formulating the aerodynamic noise problem [295] and identifying turbulence as a source of sound [296].
In Lighthill's formulation, a rather seemingly simpler rearrangement of the equations of motion than before is introduced. It is motivated by the original seeking for a physical analogy between the acoustic phenomenon in a uniform medium at rest and the aerodynamic radiation in a (turbulent) moving fluid.

The starting equations are the continuity and momentum equations in conservative formulation and without body forces:

$$\frac{\partial \rho}{\partial t} + \frac{\partial \rho U_i}{\partial x_i} = 0$$

$$\frac{\partial \rho U_i}{\partial t} + \frac{\partial \rho U_i U_j}{\partial x_j} = -\frac{\partial P}{\partial x_i} + \frac{\tau_{ij}}{\partial x_j} .$$

Taking the partial time derivative of the continuity equation and subtracting the divergence of the momentum equation yields:

$$\frac{\partial^2 \rho}{\partial t^2} - \frac{\partial^2 P}{\partial x_i \partial x_i} = \frac{\partial^2 (\rho U_i U_j)}{\partial x_i \partial x_j} - \frac{\partial^2 \tau_{ij}}{\partial x_i \partial x_j} .$$

In a purely formal manner, this equation can be equally rewritten as:

$$\frac{\partial^2 (\rho - P/a_0^2)}{\partial t^2} + \frac{1}{a_0^2}\frac{\partial^2 P}{\partial t^2} - \frac{\partial^2 P}{\partial x_i \partial x_i} = \frac{\partial^2 (\rho U_i U_j)}{\partial x_i \partial x_j} - \frac{\partial^2 \tau_{ij}}{\partial x_i \partial x_j} , \tag{3.5}$$

or

$$\frac{\partial^2 \rho}{\partial t^2} - \frac{\partial^2 (P - \rho a_0^2)}{\partial x_i \partial x_i} - a_0^2 \frac{\partial^2 \rho}{\partial x_i \partial x_i} = \frac{\partial^2 (\rho U_i U_j)}{\partial x_i \partial x_j} - \frac{\partial^2 \tau_{ij}}{\partial x_i \partial x_j} , \tag{3.6}$$

where $a_0$ simply stands for a constant velocity.

Lighthill's approximation for low Mach number flows consists in the following assumptions:
- the dominant pressure and density fluctuations $p' = P - P_0$ and $\rho' = \rho - \rho_0$ are linked with, i.e., $p' \simeq a_0^2 \rho'$ where $a_0$ denotes now the reference speed of sound,
- mean temperature or density gradients in the flow field can be neglected and therefore $\rho \simeq \rho_0$ in the right-hand-side source term.

With these assumptions, equations (3.5) and (3.6), when applied to the pressure and density fluctuations, yield:

$$\frac{1}{a_0^2}\frac{\partial^2 p'}{\partial t^2} - \frac{\partial^2 p'}{\partial x_i \partial x_i} = \frac{\partial^2 (\rho_0 U_i U_j - \tau_{ij})}{\partial x_i \partial x_j} ,$$

$$\frac{\partial^2 \rho'}{\partial t^2} - a_0^2 \frac{\partial^2 \rho'}{\partial x_i \partial x_i} = \frac{\partial^2 (\rho_0 U_i U_j - \tau_{ij})}{\partial x_i \partial x_j} .$$

These equations are quite similar to the inhomogeneous wave equations governing density and pressure fluctuations in a uniform medium at rest, leading to the Lighthill acoustic analogy:

*The density and pressure fluctuations in a low Mach number moving fluid may be identified with those occurring in a uniform acoustic medium at rest produced by the source term $\partial^2 \mathbf{T}_{ij}/\partial x_i \partial x_j$. The effective forcing tensor $\mathbf{T}_{ij}$ is given by:*

$$\mathbf{T}_{ij} = \rho_0 U_i U_j - \tau_{ij} \simeq \rho_o U_i U_j . \tag{3.7}$$

The last approximation is a consequence to the low Mach number assumption, as pointed out by Laufer *et al.* [272].

## 3.3. Decomposition of velocity and pressure fields

As shown by the previous analysis — eq.(3.4) —, various ways of generating sound exist in fluid motions with external fluctuating forces and heat sources. When focusing on aerodynamic sound generation in turbulence, one is first concerned with the generation of sound by a *fluctuating vorticity* field. This explains why it can be interesting to be able to make a distinction in the vortical part of the velocity field.

### 3.3.1. HELMHOLTZ DECOMPOSITION

As quoted by Batchelor [35], page 87, the velocity field $\vec{U}$ which is consistent with specified values of the rate of expansion $\vartheta \equiv \mathrm{div}\vec{U}$ and the vorticity $\vec{\Omega} \equiv \overrightarrow{curl}\,\vec{U}$ at each point of a flow field is given by:

$$\vec{U} = \vec{U}^I + \vec{U}^C + \vec{V} \,, \tag{3.8}$$

where $\vec{V}$ is both solenoidal and irrotational, i.e.,

$$\mathrm{div}\,\vec{V} = 0 \quad \text{and} \quad \overrightarrow{\mathrm{curl}}\,\vec{V} = \vec{0} \,. \tag{3.9}$$

$\vec{V}$ is determined by the boundary conditions to be satisfied by the flow, and is unique for an unbounded domain of fluid at rest at infinity.
Now, in eq.(3.8):

- $\vec{U}^I$ stands for the *solenoidal* or *isovolume* contribution such as:

$$\mathrm{div}\,\vec{U}^I = 0 \,, \tag{3.10}$$

  which can be derived from a vector function $\vec{\Psi}$, according to $\vec{U}^I = \overrightarrow{curl}\,\vec{\Psi}$. This function reduces to the usual stream function in two-dimensional situations;

- $\vec{U}^C$ accounts for variable density effects, and is therefore called the *dilatational* or *compressible* part, with:

$$\overrightarrow{\mathrm{curl}}\,\vec{U}^C = \vec{0} \,. \tag{3.11}$$

  It can be derived from a velocity potential (scalar function) $\phi$, according to $\vec{U}^C = \overrightarrow{\mathrm{grad}}\,\phi$. Consequently, $\vec{U}^C$ is also called the *potential* contribution.

When a statistical averaging is introduced (see Chapter 5), this decomposition can be applied to velocity *fluctuations.* Dividing the instantaneous velocity component $U_i$ into the sum of a mean ($\overline{U}_i$) and a centered fluctuation ($u'_i$, with $\overline{u'_i} = 0$), the Helmholtz decomposition leads, uniquely in homogeneous flows for instance, to:

$$u'_i = u'^I_i + u'^C_i \ , \tag{3.12}$$

with:
$$\frac{\partial u'^I_i}{\partial x_i} = 0 \quad \text{and} \quad \epsilon_{ijk}\frac{\partial u'^C_k}{\partial x_j} = 0 \ , \tag{3.13}$$

where $\epsilon_{ijk}$ is the third-order rank alternate tensor.
Based on this decomposition, one can quantify the degree of compressibility of the kinetic energy of turbulence as the ratio of the dilatational mode (or irrotational component) to the total kinetic energy [150], [66]:

$$\chi_e = \overline{u'^C_i u'^C_i}/\overline{u'_i u'_i} \ . \tag{3.14}$$

As shown by Cai *et al.* [66], the coefficient $\chi_e$ can also be used, in homogeneous compressible turbulence, to get the ratio[1] of dilatational scalar flux to solenoidal scalar flux $r = \overline{\rho u'^C_i y''}/\overline{\rho u'^I_i y''}$, where $y'' \equiv Y - \widetilde{Y}$ denotes the fluctuation of the mass fraction ($Y$) associated with a density weighted mean value or Favre-average ($\widetilde{Y} = \overline{\rho Y}/\overline{\rho}$, see Chapter 5).

### 3.3.2. DECOMPOSITION OF THE PRESSURE FIELD

As opposed to the velocity field, the decomposition of the pressure field is based upon physical rather than mathematical arguments.
A simple distinction can be introduced, using scaling arguments. It consists in considering an "incompressible" or "hydrodynamic" pressure $\pi$, scaling as $\rho U^2$ where $U$ is the local fluid velocity, and a "compressible" or "thermodynamic" pressure $P$, scaling as $P/\rho = RT = a^2/\gamma$ where $a$ is the local speed of sound. The comparison of both types of scaling introduces the Mach number of the fluid motion $M = U/a$, since $\pi/P = \gamma M^2$. Depending upon the order of magnitude of the Mach number, various approximations to the full set of Navier-Stokes equations can be derived. This issue is addressed in the next Chapter.

Another possibility to analyzing the various roles of pressure in variable density fluid motions refers to the equation governing this function, as discussed now.
The imposition of a constant density constraint $\rho = C^t (\equiv \rho_0)$ in a solenoidal

[1]The expression is $r = \frac{\chi_e}{1-\chi_e}\frac{\Lambda^C}{\Lambda^I}$ where $\Lambda^C$ and $\Lambda^I$ are the Lagrangian integral time scales associated with the dilatational and solenoidal components respectively.

fluid motion leads to the "incompressible" pressure $P^I$ which satisfies the incompressible Poisson's equation eq.(3.1).
On the other hand, a purely acoustic pressure field $P^A$ propagating in a medium at rest satisfies a simple sound wave equation — a crude restriction of the general equation eq.(3.4) —, viz.

$$\frac{\partial^2 P^A}{\partial t^2} - \frac{\gamma P^A}{\rho^A}\Delta P^A \simeq \frac{\partial^2 P^A}{\partial t^2} - a^2 \Delta P^A = 0 \,. \tag{3.15}$$

In this equation, $a = \sqrt{\gamma P^A/\rho^A}$ is the celerity of propagation.

Hence, in compressible fluid motions, it may seen natural to consider that the pressure field includes an "*acoustic*" or "*compressible*" pressure which propagates and an "*advective*" or "*incompressible*" or "*pseudo*" pressure which is associated with the convective motions in the fluid. These different denominations result in several definitions which have been introduced by various authors. As reported in Ristorcelli [394], it seems that the term "pseudo-pressure" was first coined by Blokhintsev in 1956.
Formally, any type of such decomposition reads:

$$P = P^I + P^C \,, \tag{3.16}$$

denoting by $P^I$ and $P^C$ respectively, some kind of incompressible and compressible contributions to the pressure field that are to be specifically defined.
The distinction between these two contributions can be done in slightly different ways, as shown by various proposals in the literature (see [408] and [203] for example). However, in any case, the incompressible pressure field is recovered from a velocity field by using the general Poisson's equation for the pressure, as recalled now.

Taking the divergence of the momentum equation[2] and using the mass conservation equation, the extension of eq.(3.1) to the variable density fluid motion is obtained as:

$$\Delta P = \frac{\partial^2 \rho}{\partial t^2} - \frac{\partial^2 (\rho U_i U_j)}{\partial x_i \partial x_j} + \frac{\partial^2 \tau_{ij}}{\partial x_i \partial x_j} \,, \tag{3.17}$$

where $\tau_{ij}$ is now the actual viscous stress tensor in *variable* density fluid motions.
Within the *adiabatic approximation*, this equation can be rewritten as:

$$\Delta P - \frac{1}{a_\infty^2}\frac{\partial^2 P}{\partial t^2} = -\frac{\partial^2 (\rho U_i U_j)}{\partial x_i \partial x_j} + \frac{\partial^2 \tau_{ij}}{\partial x_i \partial x_j} \,, \tag{3.18}$$

[2]The external body forces are temporarily discarded from the present discussion. They will be reintroduced in the modeling issue (see Chapter 11).

where $a_\infty^2 = \gamma P_\infty / \rho_\infty$.
Assuming for instance that, *by definition*, the incompressible pressure satisfies:

$$\Delta P^I = -\frac{\partial^2(\rho U_i U_j)}{\partial x_i \partial x_j} + \frac{\partial^2 \tau_{ij}}{\partial x_i \partial x_j} , \tag{3.19}$$

it results from eq.(3.18) that the compressible pressure is governed by:

$$\Delta P^C - \frac{1}{a_\infty^2}\frac{\partial^2 P^C}{\partial t^2} = \frac{1}{a_\infty^2}\frac{\partial^2 P^I}{\partial t^2} . \tag{3.20}$$

We shall see in the next section how such a decomposition of the pressure field can be applied in a turbulent flow.

### 3.3.3. PRESSURE DECOMPOSITION IN TURBULENT FLOW

In turbulent, variable density fluid motions, pressure *fluctuations* associated with both types of pressure fields can be simultaneously encountered, giving rise to different situations depending upon which one may be considered as the dominating process. This question will be addressed in more details in the next Chapter. Here, we focus on the coupling between mean and fluctuating motions which results in contributions from non linear terms.

To be applied to turbulent motion (see Chapter 5), each instantaneous function can be decomposed into a mean value and a *centered* fluctuation, respectively denoted by an overbar and a superscript $'$:

$$U_i = \overline{U}_i + u_i' , \quad P = \overline{P} + p' , \quad \rho = \overline{\rho} + \rho' , \quad \tau_{ij} = \overline{\tau}_{ij} + \tau_{ij}' ,$$

with $\overline{u'}_i = \overline{p'} = \overline{\rho'} = \overline{\tau'}_{ij} = 0$.
Substituting these decompositions into eq.(3.17), and averaging gives:

$$\Delta \overline{P} = \frac{\partial^2 \overline{\rho}}{\partial t^2} - \frac{\partial^2}{\partial x_i \partial x_j}(\overline{\rho}\overline{U}_i\overline{U}_j + \overline{\rho}\,\overline{u_i' u'_j}) + \frac{\partial^2 \overline{\tau}_{ij}}{\partial x_i \partial x_j}$$
$$- \frac{\partial^2}{\partial x_i \partial x_j}(\overline{\rho u'_i}\,\overline{U}_j + \overline{\rho u'_j}\,\overline{U}_i + \overline{\rho' u_i' u_j'}) . \tag{3.21}$$

In the right-hand-side of eq.(3.21), the first two terms depend on the *mean* density. They consist in (i) the second time derivative of the mean density, (ii) the contribution of the velocity field which includes both the mean flow and turbulent (Reynolds stresses) contributions. Thus, in incompressible flow, Poisson's equation for the mean pressure field reduces to the three terms in the first line of eq.(3.21), where the viscous contribution (third term in the RHS) is restricted to its incompressible formulation eq.(3.1).

Accordingly, in variable density turbulence, the density *fluctuation* introduces in the mean pressure equation two additional contributions, due to the turbulent mass flux ($\overline{\rho u_i'}$, second-order correlation) and the third-order moment $\overline{\rho' u_i' u_j'}$.

By subtracting eq.(3.21) from eq.(3.17), we obtain the resulting equation governing the pressure fluctuation. After simple algebraic manipulations, it reads:

$$\begin{aligned}\Delta p' = & -2\frac{\partial^2(\overline{\rho}\overline{U}_j u_i')}{\partial x_i \partial x_j} - \frac{\partial^2(\overline{\rho} u_i' u_j' - \overline{\rho}\overline{u_i' u'_j})}{\partial x_i \partial x_j} \\ & + \frac{\partial^2 \rho'}{\partial t^2} \\ & -2\frac{\partial^2[(\rho' u_i' - \overline{\rho' u'_i})\overline{U}_j]}{\partial x_i \partial x_j} - \frac{\partial^2(\rho' u_i' u_j' - \overline{\rho' u_i' u'_j})}{\partial x_i \partial x_j} - \frac{\partial^2(\rho' \overline{U}_i \overline{U}_j)}{\partial x_i \partial x_j} + \frac{\partial^2 \tau_{ij}'}{\partial x_i \partial x_j} .\end{aligned} \tag{3.22}$$

As shown in eq.(3.22), both *mean* and *fluctuating* values of the density contribute to the source terms of the equation governing the "*instantaneous*" pressure fluctuation[3]:

- The terms in the first line of the right-hand-side of eq.(3.22) depend explicitly on the *mean* density. They are the only ones present in the incompressible case, introducing the classical linear and quadratic contributions;
- In contrast with all the other terms in eq.(3.22), which are based on *spatial* derivative operators, the one in the second line refers to the *time* rate of change of the density fluctuation. Using the methodology in Tennekes and Lumley [459], it is estimated by Sarkar [408] to give the dominant contribution in the pressure fluctuation equation for compressible flows at small turbulence Mach number;
- In addition to the viscous fluctuating contribution, the terms in the third line result from spatial variation of the density fluctuation, with and without coupling with velocity fluctuations.

When applied to the turbulent motion, eq.(3.16) leads to splitting the pressure fluctuation into an incompressible $p'^I$ and compressible $p'^C$ part, viz.

$$p' = p'^I + p'^C . \tag{3.23}$$

Based on the Helmholtz decomposition of the velocity field (see section 3.3.1), a first possibility consists in defining the incompressible pressure as

[3]The present formulation is based on the *time* derivative of density fluctuations. An equivalent expression, using the *material* derivative of this function, can be found in Sarkar [408]. As noticed by this author, the use of such a derivative makes clearer the Galilean invariance of the formulation.

the pressure associated with the conservative equations for the *solenoidal* part of the velocity field. This is the choice adopted in Hamba [203] for the analysis of the effects of pressure fluctuations on turbulence growth in compressible homogeneous shear. However, as shown from eq.(3.17), the pressure cannot be obtained from the velocity field independently from the density, and, in a turbulent flow, the pressure fluctuation is governed by equation (3.22) in which both mean and fluctuating values of density are involved. Thus, in [203], only the mean density source terms are taken into account.

In Sarkar [408], a distinction which does not make use of the Helmholtz decomposition is introduced: the incompressible part of the pressure fluctuation is simply given as the solution of eq.(3.22) when restricting the source terms to the mean density contributions:

$$\Delta p'^{I} = -2\frac{\partial^2(\overline{\rho}\overline{U}_j u_i')}{\partial x_i \partial x_j} - \frac{\partial^2(\overline{\rho}u_i'u_j' - \overline{\rho}\overline{u_i'u_j'})}{\partial x_i \partial x_j} . \tag{3.24}$$

As argued by this author, there are three main reasons for employing such a decomposition:
- the leading-order density dependence in $p'^{I}$ is through the mean density;
- the incompressible limit for $p'^{C}$ is zero, since this term is only due to density fluctuations;
- the compressible contribution $p'^{C}$ is expected to contain the fast-time-scale oscillations, as a consequence of its leading-order term $\partial^2\rho'/\partial t^2$.

## 3.4. The Kovasznay modes decomposition

In general, variable density fluid turbulence can be considered as the situation where density fluctuations result from velocity, pressure and temperature (or entropy) fluctuations. In other words, the density variance $\overline{\rho'^2}$ can be considered as a function of three main variances:
- the velocity variance, i.e. twice the kinematic kinetic energy of turbulence $\overline{u_i'u_i'}$,
- the pressure variance $\overline{p'^2}$,
- the temperature $\overline{\theta'^2}$ or entropy $\overline{s'^2}$ variance.

The high level of nonlinearity, which characterizes variable density turbulence, results in a mathematical coupling of the governing equations for these variances and a strong physical interaction of the associated effects.

In 1953, Kovasznay [259] proposed to separate the *small* turbulent fluctuations in a supersonic flow into three modes, called vorticity, acoustic and entropy modes. The velocity, pressure and entropy variances are not exact

representatives[4] for these modes due to non-linear interactions which are discarded from the first-order Kovasznay's analysis.
The procedure will be briefly recalled hereafter, in order to point out the main assumptions, without going through the analytical details.

The analysis is developed for small fluctuations in a space-time domain of the flow field, where averaged values in this domain are taken as constant references (The coordinate system is taken in such a way that the averaged reference velocity is zero). The small perturbation assumption means that (i) pressure, density and absolute temperature fluctuations are small compared with their respective reference values and (ii) velocity fluctuations are small compared with the average speed of sound. Finally, the analysis applies to a perfect gas medium, $(P/\rho = [(\gamma - 1)/\gamma]C_pT)$, with constant viscosity, heat conduction and specific heats, governed by the following conservation equations:

$$Mass: \quad \frac{\partial \rho}{\partial t} + \mathrm{div}(\rho \vec{V}) = 0 \,,$$

$$\begin{aligned} Momentum: \quad \rho\frac{\partial \vec{V}}{\partial t} &= -\overrightarrow{\mathrm{grad}}P - \rho\overrightarrow{\mathrm{grad}}(\frac{\vec{V}.\vec{V}}{2}) + \rho\,\overrightarrow{\mathrm{curl}}\vec{V}\otimes\vec{V} \\ &+\mu(\Delta\vec{V} + \frac{1}{3}\overrightarrow{\mathrm{grad}}(\mathrm{div}\vec{V})\,,) \end{aligned}$$

$$Energy: \quad \rho C_p\frac{\partial T}{\partial t} = -\rho C_p\vec{V}.\overrightarrow{\mathrm{grad}}T + \frac{\partial P}{\partial t} + \vec{V}.\overrightarrow{\mathrm{grad}}P + \lambda\Delta T + \Phi \,,$$

where $\Phi$ is the viscous dissipation and $C_p \equiv \gamma C_v$ the specific heat at constant pressure.

Introducing $P = P_0 + p'$, $\rho = \rho_0 + \rho'$, $T = T_0 + T'$ and $\vec{V} = 0 + \vec{v'}$, where it is recalled that the prime denotes centered fluctuations, the linearized form of the previous equations are:

$$Mass: \quad \mathrm{div}\,\vec{v'} = -\frac{1}{\rho_0}\frac{\partial \rho'}{\partial t} \equiv -\frac{1}{\gamma P_0}\frac{\partial p'}{\partial t} + \frac{1}{C_p}\frac{\partial s'}{\partial t} \,. \tag{3.25}$$

The last expression is obtained by substituting to the density fluctuation $\rho'$ the entropy fluctuation $s'$, according to $s'/C_p = p'/\gamma P_0 - \rho'/\rho_0$, as a direct consequence of the Gibbs relation $T\delta S = C_v\delta T - (P/\rho^2)\delta\rho$.

$$Momentum \quad \frac{\partial \vec{v'}}{\partial t} = -\frac{a^2}{\gamma P_0}\overrightarrow{\mathrm{grad}}\,p' + \nu(\Delta\vec{v'} + \frac{1}{3}\overrightarrow{\mathrm{grad}}(\mathrm{div}\,\vec{v'}))\,. \tag{3.26}$$

[4]For example, according to Helmholtz decomposition, the velocity variance includes a compressible part $\overline{u_i'^C u_i'^C}$ that corresponds to the acoustic mode.

where the speed of sound is defined as $a^2 = \gamma P_0/\rho_0$ and the kinematic viscosity by $\nu = \mu/\rho_0$.

$$\textit{Energy} \qquad \frac{\partial s'}{\partial t} - \frac{\nu}{P_r}\Delta s' = C_p \frac{\nu}{P_r}\frac{\gamma-1}{\gamma}\frac{\Delta p'}{P_0} \,. \tag{3.27}$$

Here again, temperature fluctuation has been transformed into entropy fluctuation, according to $T'/T_0 = s'/C_p + \frac{\gamma-1}{\gamma}p'/P_0$, and $P_r = \mu C_p/\lambda$ is the Prandtl number, which is 3/4 in Kovasznay's analysis.

From the momentum equation, the two following results can be derived by taking the curl and the divergence of this equation (3.26).

- first, by taking the curl of equation (3.26), one obtains:

$$\frac{\partial \vec{\omega}'}{\partial t} = \nu \Delta \vec{\omega}' \,. \tag{3.28}$$

  Thus the equation governing the vorticity fluctuation $\vec{\omega}' = \vec{\text{curl}}\,\vec{v}'$ is decoupled from those for temperature and pressure fluctuations, indicating that vorticity has no first-order interaction with the pressure and entropy (temperature) fields;
- now, by taking the divergence of the momentum equation, it results that:

$$\frac{\partial \vartheta'}{\partial t} = -\frac{a^2}{\gamma P_0}\Delta p' + \frac{4}{3}\nu \Delta \vartheta' \,, \tag{3.29}$$

  where $\vartheta' = \text{div}\,\vec{v}'$. Making use of the continuity and energy equations, eq.(3.29) can be explicited to give the equation governing the pressure fluctuation:

$$\frac{\partial^2 p'}{\partial t^2} - a^2 \Delta p' = \frac{4}{3}\gamma\nu \frac{\partial}{\partial t}(\Delta p') \,. \tag{3.30}$$

  Hence, the pressure fluctuation obeys a wave-like equation. The departure from the classical form — eq.(3.15) — is due to an additional source term which accounts for sound absorption at high frequencies.

The last approximate equation comes directly from eq.(3.27) where the pressure term is discarded, so that:

$$\text{div}\,\vec{v}' = \frac{1}{C_p}\frac{\partial s'}{\partial t} = \frac{\nu}{C_p P_r}\Delta s' \quad \text{or} \quad \vec{v}' = \frac{\nu}{C_p P_r}\,\vec{\text{grad}}\, s' \,. \tag{3.31}$$

Thus entropy gradients generate velocity fluctuations or, in other words, entropy (temperature) inhomogeneity in the field induce a motion which is a potential flow.

Referring to the Helmholtz decomposition of the velocity field (see 3.3.1), we are now able to explicit the generation of velocity fluctuations according to Kovasznay's analysis. As sketched in Fig.3.1, this generation consists in three modes, the so-called *vorticity*, *acoustic* (pressure-sound) and *entropy* (temperature) modes.

1. The *vorticity mode* ($\omega' \neq 0,\ p' = s' = \rho' = 0$) is the only source of *solenoidal* velocity fluctuations. It is governed by the vorticity fluctuation equation (3.28);
2. The *acoustic or sound wave mode* ($\omega' = 0,\ p' \neq 0,\ s' = 0,\ \rho' \neq 0$) is the **first** type of generation of velocity fluctuations due to density variations. It corresponds to *compressible* effects where the pressure fluctuations are governed by an almost wave equation at the speed of sound in the reference state (3.30);
3. The *entropy mode* ($\omega' = p' = 0,\ s' \neq 0,\ \rho' \neq 0$) is the **second** type of generation of velocity fluctuations due to density variations. In this case, velocity fluctuations are due to variations of temperature (entropy) by heat conduction which change the density and cause a (non-isovolume) motion (3.31).

To sum up briefly, the first mode has no incidence on the solenoidal (isovolume) condition, while the last two modes are directly introducing density fluctuations since they are linked with the source terms of the equation for the divergence of the velocity fluctuation vector.

Within the "weak-field" turbulence limit (small linearized fluctuations about a flow state of uniform mean velocity and constant mean thermodynamic properties), the three modes are dynamically decoupled from each other to first-order in the fluctuation amplitudes. In other words, as stressed by Kovasznay [260], (i) there is no energy transfer from one wavelength to another, and (ii) the three modes do not interact among themselves, except through the boundary conditions on solid walls, because they are imposed on the velocity and temperature, not on the separate modes.

Further developments on non linear interactions between modes have been introduced by Kovasznay and coworkers. A table of such interactions can be found in ref. [434], page 84. Such non-linear extensions of the theory are not detailed here, where the consideration of the exact set of equations (Chapters 4 and 5) is preferred.

## 3.5. Flows in bounded domains

As recalled in the introduction to this chapter, all possible sources of density variations, if present, do not necessarily contribute, at to the level, to the actual density fluctuations in a given flow. Thus it can be appropriate to

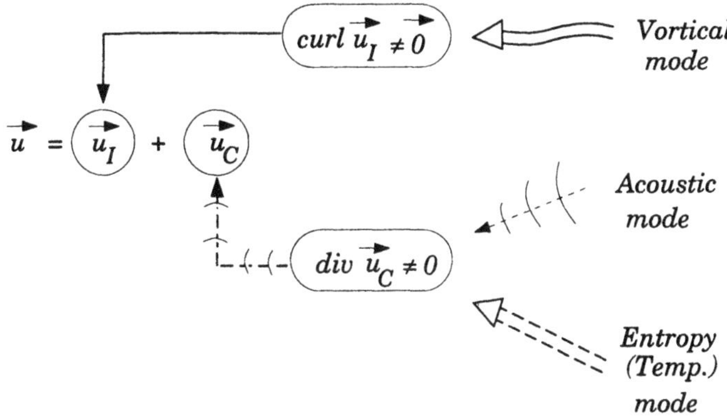


*Figure 3.1.* Kovasznay modes according to Helmholtz decomposition.

"simplify" the general Navier-Stokes equations to some restricted form, more pertinent to the origin of the density variation under consideration. Several approximate models of this type can be derived, according to different types of assumptions.

As a first example of such "restricted" models, we shall consider two generalizations of the usual incompressible flow equations to those governing compressible, *low* Mach number, inviscid flows, as derived by Mansour and Lundgren [314] in 1990 and Fedorchenko [159] in 1997. The common feature of both models is that the leading-order term for the pressure reference is *purely* time-dependent, i.e., constant throughout the flow field at any given time. Strictly speaking, this assumption does not restrict the validity of the models to flows in bounded domains, although they have been used only in such situations.

### 3.5.1. MANSOUR-LUNDGREN'S MODEL

The starting equations are the compressible Euler equations for an isentropic flow:

$$\frac{\partial \rho}{\partial t} + \frac{\partial}{\partial x_j}(\rho U_j) = 0 \ ,$$

$$\rho(\frac{\partial U_i}{\partial t} + U_j \frac{\partial U_i}{\partial x_j}) = -\frac{\partial P}{\partial x_i} \ ,$$

along with $P/\rho^\gamma = C^t$.

Denoting by the subscript $_0$ a constant reference state for the scaling of any variable, the following non dimensional quantities can be introduced:

$$\rho^* = \rho/\rho_0 \ , \quad x_i^* = x_i/L_0 \ , \quad t^* = t/t_0 \ , \quad P^* = P/P_0 \quad \text{and} \ , \quad U_i^* = U_i/a_0$$

where $a_0^2 = \gamma P_0/\rho_0$ is the reference sound speed. Thus, the non dimensional Euler equations can be written as:

$$M_0 \frac{\partial \rho^*}{\partial t^*} + \frac{\partial}{\partial x_j^*}(\rho^* U_j^*) = 0 , \qquad (3.32)$$

$$\rho^*(M_0 \frac{\partial U_i^*}{\partial t^*} + U_j^* \frac{\partial U_i^*}{\partial x_j^*}) = -\frac{1}{\gamma}\frac{\partial P^*}{\partial x_i^*} , \qquad (3.33)$$

where $M_0 = L_0/(t_0 a_0)$ is the characteristic Mach number based on the reference conditions.

For the compressed turbulence considered in ref. [314] which corresponds to the situation encountered in internal combustion engines where the characteristic Mach number is low (typically $M_0 < 0.15$), the density, pressure and velocity fields can be expanded in power series in $M_0$:

$$\rho^* = \rho^{(0)} + M_0^2\, \rho^{(1)} + \dots ,$$

$$P^* \equiv (\rho^*)^\gamma = P^{(0)} + M_0^2\, P^{(1)} + \dots ,$$

$$U_i^* = M_0\, U_i{}^{(0)} + \dots .$$

Substituting these expansions in eqs.(3.32) and (3.33), and equating the coefficients in power of Mach number to first-order in eq.(3.33) and second-order in eq.(3.32) gives:

$$\frac{\partial \rho^{(0)}}{\partial t^*} + \frac{\partial}{\partial x_j^*}(\rho^{(0)} U_j^{(0)}) = 0 ,$$

$$0 = -\frac{1}{\gamma}\frac{\partial P^{(0)}}{\partial x_i^*} \quad \text{and} \quad \rho^{(0)}(\frac{\partial U_i^{(0)}}{\partial t^*} + U_j^{(0)} \frac{\partial U_i^{(0)}}{\partial x_j^*}) = -\frac{1}{\gamma}\frac{\partial P^{(1)}}{\partial x_i^*} .$$

The second relation in the previous equations shows that $P^{(0)}$ is *independent of the space coordinates*, so that $P^{(0)} \equiv P^{(0)}(t)$ and similarly $\rho^{(0)} \equiv \rho^{(0)}(t)$, due to the isentropic linkage. Thus, returning to the dimensional equations, the simplified model is:

$$\frac{\partial \rho(t)}{\partial t} + \rho(t)\frac{\partial U_j}{\partial x_j} = 0 , \qquad (3.34)$$

$$\rho(t)(\frac{\partial U_i}{\partial t} + U_j \frac{\partial U_i}{\partial x_j}) = -\frac{\partial P}{\partial x_i} . \qquad (3.35)$$

These equations are Mansour-Lundgren's generalization of the usual incompressible equations for low Mach number, inviscid, isentropic flow.

### 3.5.2. FEDORCHENKO'S NON LOCAL MODEL

In 1997, Fedorchenko [159] proposed an original mathematical model for unsteady, subsonic flows in bounded domains in the presence of continuously distributed volume sources of mass, heat and momentum. This non-local model, where div $\vec{V} \neq 0$, is more general than the incompressible one, but less general than the full compressible Navier-Stokes equations, since all acoustic effects are excluded. Large density variations of the medium in the bounded domain are thus allowed due to considerable time changes in the average pressure or temperature, for example. But the existence of any sound wave is totally precluded.
The main features of this model are briefly summarize hereafter. For sake of clarity, the presentation is restricted to inviscid, non thermally conductive perfect gas, with no mass source in the domain. The general form is available in the original paper.

A first basic assumption in Fedorchenko's proposal consists in the following splitting of the instantaneous static pressure $P(x_i, t)$:

$$P(x_i, t) = P^*(t) + p(x_i, t) \ . \tag{3.36}$$

It should be emphasized that eq.(3.36) is not the usual decomposition of the pressure as the sum of a mean value and a small perturbation. Here, $P^*(t)$ is a certain global pressure, spatially constant in the flow domain, which scales as $\rho_0 a^2$, where $\rho_0$ is a reference value of the density and $a$ the adiabatic speed of sound throughout the domain $D$. It is estimated that $\|p/P^*\|_D \ll 1$, as a consequence of the low speed motion assumption without acoustic effects.

The global pressure $P^*$ is obtained from an integro-differential equation which is derived from the continuity and energy equations:

$$\frac{\partial \rho}{\partial t} + \frac{\partial(\rho U_i)}{\partial x_i} = 0 \qquad \frac{\mathrm{d}s}{\mathrm{d}t} = q \ , \tag{3.37}$$

where $s$ is the specific entropy and $q$ the entropy source per unit mass due to heat release (inviscid fluid).
Now, for a perfect gas, $s = C_v \ln(P/\rho^\gamma)$, so that $\partial s/\partial P = C_v/P$ and $\partial s/\partial \rho = -C_p/\rho$.

Hence:
$$\frac{\mathrm{d}s}{\mathrm{d}t} = \frac{\partial s}{\partial P}\frac{\mathrm{d}P}{\mathrm{d}t} + \frac{\partial s}{\partial \rho}\frac{\mathrm{d}\rho}{\mathrm{d}t} = \frac{C_v}{P}\frac{\mathrm{d}P}{\mathrm{d}t} - \frac{C_p}{\rho}\frac{\mathrm{d}\rho}{\mathrm{d}t} \ ,$$

or:
$$\frac{q}{C_p} - \frac{1}{\gamma P}\frac{\mathrm{d}P}{\mathrm{d}t} = -\frac{1}{\rho}\frac{\mathrm{d}\rho}{\mathrm{d}t} \equiv \frac{\partial U_i}{\partial x_i} \ .$$

Introducing $P = P^* + p$, an approximation to this equation is derived by Fedorchenko, assuming that the characteristic time scale of the substantial variation of the small pressure $p$ is much greater than the convective time scale governing the velocity divergence. Hence, the pressure term in the approximate expression of the velocity divergence is only concerned with $P^*$:

$$\frac{q}{C_p} - \frac{1}{\gamma P^*}\frac{\mathrm{d}P^*}{\mathrm{d}t} = -\frac{1}{\rho}\frac{\mathrm{d}\rho}{\mathrm{d}t} \equiv \frac{\partial U_i}{\partial x_i} \,. \tag{3.38}$$

By integration of this equation over the confined domain $D$, which has a moving boundary $\Gamma(t)$ and accordingly a changing volume $\mathcal{V}(t)$, the following integro-differential equation for the global pressure (constant in the domain at any time) is obtained, with the use of Gauss' theorem:

$$\frac{\mathcal{V}}{\gamma P^*}\frac{\mathrm{d}P^*}{\mathrm{d}t} = \int_D \frac{q}{C_p}\mathrm{d}v - \int_\Gamma \mathrm{U}_n \mathrm{d}\sigma \,, \tag{3.39}$$

where $U_n = \vec{U}.\vec{n}$ denotes the normal velocity component on the boundary ($\vec{n}$ is the outward unit normal vector on $\Gamma(t)$).

In addition to eq.(3.38) or eq.(3.39), the model governing the seven scalar functions $P^*$, $p$, $U_i$, $\rho$ and $S$ consists in the continuity and entropy equations — eqs.(3.37) — along with the following simplified momentum equation:

$$\frac{\partial}{\partial t}(\rho U_i) + \frac{\partial}{\partial x_i}(\rho U_i U_j) = \rho F_i - \frac{\partial p}{\partial x_i} \,, \tag{3.40}$$

in which only the small pressure contribution $p(x_i, t)$ must be taken into account in the spatial derivative.

As noted by the author, the second assumption introduced when deriving eq.(3.38) is equivalent to a thermodynamic postulate of the equation of state as:

$$f(P^*\rho, s) = 0 \,, \tag{3.41}$$

where the global pressure $P^*$ is used instead of the actual one. Since the pressure field $p$ is omitted from this equation, there is no explicit connection between local variations of $p$ and $\rho$. This results in the absence of sound wave propagation in the flow domain. However, a significant aspect of the model is that it makes possible to "estimate" a value of the speed of sound as $a^2 = \gamma P^*/\rho$, which is not exactly local, due to $P^* = P^*(t)$. Based on this value, the Mach number can be estimated in order to check whether the condition assumed in the model, $p/P^* \simeq M^2 \ll 1$, is actually satisfied.

## 3.6. Approximated models for low speed free buoyant flows

We are concerned here with another class of variable density fluid motions where density variations result mainly from temperature variations introduced by some process independent of the flow dynamics. Thus, density variations arising from adiabatic compression/expansion or from viscous dissipation are not relevant to the present situation, which basically refers to low speed convective flows, with main applications to geofluid flows (oceanography, micrometeorology) engineering, human activity and environment (fires flows) problems.

### 3.6.1. BOUSSINESQ'S APPROXIMATIONS

For such applications, the most widely used approximate equations have been introduced by Boussinesq[5] [51] in 1903. The so-called Boussinesq's approximations can be viewed as a simplified version of the *anelastic approximations*, which, as it will be discussed in section (3.7.1), are based on a *linearization* of the initial set of equations with respect to the *thermodynamic variables.*

The initial set of equations consists in the following relations:

$$Continuity: \quad \frac{\partial \rho}{\partial t} + \frac{\partial}{\partial x_j}(\rho U_j) = 0 \ ,$$

$$Momentum: \quad \rho(\frac{\partial U_i}{\partial t} + U_j \frac{\partial U_i}{\partial x_j}) = -\rho g \delta_{i3} - \frac{\partial P}{\partial x_i} + \frac{\partial \tau_{ij}}{\partial x_j} \ ,$$

$$Energy: \quad \rho C_p(\frac{\partial T}{\partial t} + U_j \frac{\partial T}{\partial x_j}) = -\frac{dP}{dt} + \lambda \frac{\partial^2 T}{\partial x_j \partial x_j} + \Phi \ ,$$

$$Equation\, of\ state: \quad P = R\rho T \ .$$

The coordinate system is taken so that the gravity acceleration (modulus $g$) is aligned with the coordinate axis $x_3$; $\tau_{ij}$ is the viscous stress tensor, $\lambda$ the heat conductivity and $\Phi$ the viscous dissipation. For sake of clarity, extra terms in momentum and energy equations, such as Coriolis acceleration and radiative heat transfer for instance, have been omitted.

A reference state is introduced (see De Moor [124] for instance), which corresponds to a steady horizontal motion, solution of the previous equations under adiabatic and non viscous conditions (the effects of radiation, heat conduction, viscous dissipation are neglected). Denoting the corresponding reference values by the subscript $_0$, the reference state, which satisfies such a thermodynamic hydrostatic adiabatic equilibrium, is defined as:

$$U_{0i} = U_0(z)\delta_{i1} \ , \qquad \frac{\partial P_0}{\partial x_i} + \rho_0 g \delta_{i3} = 0 \ , \qquad \frac{\partial T_0}{\partial x_i} = -\frac{g}{C_p}\delta_{i3} \ .$$

[5] As quoted in Dutton and Fichtl, [143] it seems that quite similar equations were introduced by Oberbeck in 1857. The reference is given in [143].

The question is now to find out simplified forms of the general equations governing the *deviation* motion from this reference state, in which the contribution of the thermodynamic variables is linear.
Thus we proceed in introducing the following perturbations:

$$U_i = U_{0i} + u_i' \,, \qquad P = P_0 + p' \,, \qquad T = T_0 + T' \,, \qquad \rho = \rho_0 + \rho' \,,$$

into the previous equations.

- *Equation of state*: by taking the logarithmic derivative of the equation of state at the reference conditions and taking local variations as small perturbations, it becomes:

$$\frac{\rho'}{\rho_0} = \frac{p'}{P_0} - \frac{T'}{T_0} \,. \tag{3.42}$$

Now, a dynamic estimation of the pressure variations is: $p' \sim \rho_0 u'^2 = \rho_0 M'^2 a_0^2$, where $M' = u'/a_0'$ is the Mach number of the velocity perturbation based on the speed of sound in the reference state: $a_0 = \sqrt{\gamma P_0/\rho_0}$. Thus:

$$\frac{p'}{P_0} \simeq \frac{\rho_0}{P_0} M'^2 a_0^2 = \gamma M'^2 \,. \tag{3.43}$$

By substituting eq.(3.43) into eq.(3.42), it comes:

$$\frac{\rho'}{\rho_0} \simeq \gamma M'^2 - \frac{T'}{T_0} \,,$$

which reduces to:

$$\frac{\rho'}{\rho_0} \simeq -\frac{T'}{T_0} \,, \tag{3.44}$$

for $M' \ll 1$.
This result corresponds to negatively correlated density-temperature fluctuations. It is exactly satisfied in pure convective flows. In this case, density variations come from temperature variations only, so that $\rho'/\rho_0 = -\beta T'$, where $\beta$ is the thermal expansion coefficient of the gas. Since, for a perfect gas, $\beta = 1/T$, the relative density fluctuation is now exactly given by eq.(3.44).

- *Momentum equation*: when dividing both sides of the momentum equation by the density, the resulting pressure term reads:

$$-\frac{1}{\rho}\frac{\partial P}{\partial x_i} = -\frac{1}{\rho}\frac{\partial P_0}{\partial x_i} - \frac{1}{\rho}\frac{\partial p'}{\partial x_i} \,.$$

The corresponding linearized form in density and pressure fluctuations is:

$$-\frac{1}{\rho}\frac{\partial P}{\partial x_i} \simeq -\frac{1}{\rho}\frac{\partial P_0}{\partial x_i} - \frac{1}{\rho_0}\frac{\partial p'}{\partial x_i} \,. \tag{3.45}$$

From the hydrostatic relation applying to the reference state ($\partial P_0/\partial x_i + \rho_0 g \delta_{i3} = 0$) it can be easily deduced that:

$$-\frac{1}{\rho}\frac{\partial P_0}{\partial x_i} = \frac{\rho_0}{\rho} g \delta_{i3} \simeq (1 - \frac{\rho'}{\rho}) g \delta_{i3} \,.$$

Substituting this expression in eq.(3.45), the linear form of the pressure term is obtained:

$$-\frac{1}{\rho}\frac{\partial P}{\partial x_i} \simeq (1 - \frac{\rho'}{\rho}) g \delta_{i3} - \frac{1}{\rho_0}\frac{\partial p'}{\partial x_i} \,.$$

With this expression — assuming that $|\rho'/\rho_0| \ll 1$ —, the approximate form of the momentum balance equation of the perturbation field which is linear in the thermodynamic variables, is:

$$(\frac{\partial u_i'}{\partial t} + u_j' \frac{\partial u_i'}{\partial x_j}) = -\frac{\rho'}{\rho_0} g \delta_{i3} - \frac{1}{\rho_0}\frac{\partial p'}{\partial x_i} + \nu_0 (\frac{\partial^2 u_i'}{\partial x_j \partial x_j} + \frac{1}{3}\frac{\partial \vartheta'}{\partial x_i}) \,, \tag{3.46}$$

where $\nu_0 = \mu/\rho_0$ is the constant kinematic viscosity and $\vartheta' = \partial u_i'/\partial x_i$.

- *Energy equation*: a similar procedure can be applied to the energy equation. Details can be found in Dutton and Fichtl [143] for instance. The approximate form is:

$$C_p \frac{d}{dt}\left(\frac{T'}{T_0}\right) + \frac{C_p w'}{T_0}(\frac{\partial T_0}{\partial z} + \frac{g}{C_p}) - \frac{P_0}{\rho_0 T_0}\frac{d}{dt}\left(\frac{p'}{P_0}\right) = \frac{\lambda_0}{\rho_0}\frac{\partial^2 T'}{\partial x_i \partial x_i} \,. \tag{3.47}$$

- *Continuity equation*: the continuity equation can be rewritten as follows:

$$\frac{\partial U_i}{\partial x_i} = -\frac{1}{\rho/\rho_0}\frac{d(\rho/\rho_0)}{dt} \equiv -\frac{d}{dt}\ln(\rho/\rho_0) \,. \tag{3.48}$$

Applied to the perturbed field, the first-order approximation of the continuity equation is:

$$\frac{\partial u_i'}{\partial x_i} = -\frac{d}{dt}(\rho'/\rho_0) \,, \tag{3.49}$$

so that further simplification of eq.(3.49) needs additional assumptions. One of them could be, for instance, the *adiabatic* condition for a perfect gas motion with $M' \ll 1$. The first law of thermodynamics implies that if an infinitesimal amount of heat is communicated to a gas, one part serves to increase the internal enthalpy ($C_p T$), and the other does work against the external pressure, which, for a perfect gas is expressed as:

$$\delta Q = C_p dT - \frac{\gamma - 1}{\gamma} C_p T \frac{dP}{P} \,.$$

Applying this expression to the perturbed field with the pressure estimate given by eq.(3.43) yields:

$$\delta Q \simeq C_p T' - (\gamma - 1) C_p T_0 M'^2 \, .$$

Thus, for $M' \ll 1$:

$$\frac{\rho'}{\rho} \simeq -\frac{T'}{T_0} \simeq -\frac{\delta Q}{C_p T_0} \, .$$

With this result, it can be concluded from eq.(3.49) that any *adiabatic* motion of this kind is solenoidal.

A similar conclusion also holds for *shallow convection* in the atmosphere, as shown by Dutton and Fichtl [143]. When the vertical scale of the motion is much less than the scale height (typically 500 m and 8 km respectively), eq.(3.49) can be reduced to the isovolume condition $\partial u_i'/\partial x_i = 0$. This leads again to usual Boussinesq's approximations which consist, from a practical point of view, in ignoring variations of all thermodynamic fluid properties, excepting only the density, insofar as its variation gives rise to a gravitational body force.

### 3.6.2. REHM-BAUM'S MODEL

As shown in the previous section, the isovolume condition under Boussinesq's approximations is not fulfilled in low speed motions with large heat release. The most illustrative examples of such situations are fire flows. In 1978, Rehm & Baum [379] derived a system of equations governing such flows, where the fluid motions are driven by the heat source $Q_s$ in the gravity field, so that buoyant forces are important. Rehm and Baum introduced an appropriate scaling of such flow fields, considering that the characteristic time scale of the energy variation is the same magnitude as the one of the source term. They derived the following equations as the leading-order approximations to the full system of variable density, inviscid, fluid flows at low Mach number:

$$\frac{\partial \rho}{\partial t} + \frac{\partial}{\partial x_j}(\rho U_j) = 0 \, ,$$

$$\rho \frac{\mathrm{d} U_i}{\mathrm{dt}} \left(\equiv \rho \frac{\partial U_i}{\partial t} + \rho U_j \frac{\partial U_i}{\partial x_j}\right) = \rho g_i - \frac{\partial P*}{\partial x_i} \, , \tag{3.50}$$

$$\rho C_p \frac{\mathrm{d} T}{\mathrm{dt}} = \frac{\mathrm{d} P_{th}}{\mathrm{dt}} + Q_s \, . \tag{3.51}$$

It is worth noticing that the changes in the model concern the momentum and energy equations,(3.50) and (3.51), in which two different pressures are present: the dynamic pressure $P*$ and the thermal pressure $P_{th} = \rho RT$, for a perfect gas. The actual pressure is obtained as:

$$P = P * + P_{th} \; .$$

As in Fedorchenko's formulation (section 3.5.2), it is required that the thermal pressure is spatially constant, and so is $\rho T$, from the equation of state. Thus, from eq.(3.51) it can be deduced that:

$$\frac{\partial U_j}{\partial x_j} = -\frac{1}{\gamma}\frac{\partial(\rho T)}{\partial t} + \frac{Q_s}{C_P} \, , \tag{3.52}$$

in which the departure from the isovolume condition results in the non stationarity of the thermal pressure field and the heat source.

## 3.7. Isovolume approximations to the continuity equation

As shown in the previous section, the key point in the formulation of approximations to the isovolume (solenoidal) condition is the identification of the source terms in the expression of the divergence of the velocity field which are pertinent to the physical situation under consideration. A more general discussion of this question is now presented.

### 3.7.1. ANELASTIC APPROXIMATION

When studying small-scale atmospheric circulations, one is frequently interested in approximating the full continuity equation by a formulation which represents *non-acoustic* modes with complete confidence. One class of such formulations refers to "anelastic" versions of the continuity equation. The name*anelastic*[6] was coined by Ogura and Phillips [351] in 1962. In steady flows, the anelastic approximation consists in adopting the following form of the continuity equation:

$$\mathrm{div}(\widetilde{\rho}\vec{V}) = 0 \, , \tag{3.53}$$

where a *characteristic* value of the density, say $\widetilde{\rho}$, has been substituted to the actual value of the density $\rho$. More generally, in unsteady flows, the

[6]The external energy density of small *linearized* perturbations on a resting stratified atmosphere consists in kinetic energy, "thermobaric" energy, and "elastic" energy, which is proportional to the square of the pressure perturbation. When the same linearization is applied to the set of equations derived by Ogura and Phillips [351], the elastic contribution is not present, justifying the use of the word anelastic. The "sound-proof" denomination might also be appropriate, as we shall see from the physical interpretation of the anelastic assumptions.

anelastic approximation is associated with the following formulation of the continuity equation:

$$\operatorname{div} \vec{V} = -\frac{1}{\tilde{\rho}} \frac{\mathrm{d}\tilde{\rho}}{\mathrm{dt}} . \tag{3.54}$$

The anelastic formulation of the continuity equation usually applies to fluid motions which slightly depart from a given reference state. To clearly discern the actual assumptions validating such an approximation, let us introduce the density decomposition $\rho = \tilde{\rho} + \rho'$ into the exact continuity equation. To second-order terms in density fluctuations, it is deduced that:

$$\operatorname{div} \vec{V} = -\frac{1}{\rho} \frac{\mathrm{d}\rho}{\mathrm{dt}} \simeq -\frac{1}{\tilde{\rho}} \frac{\mathrm{d}\tilde{\rho}}{\mathrm{dt}} + \frac{1}{\tilde{\rho}} (\frac{\rho'}{\tilde{\rho}}) \frac{\mathrm{d}\tilde{\rho}}{\mathrm{dt}} - \frac{1}{\tilde{\rho}} \frac{\mathrm{d}\rho'}{\mathrm{dt}} + \mathcal{O}(\rho'^2) . \tag{3.55}$$

Hence, to be reduced to eq.(3.54), which only retains the first term in the right of hand side of eq.(3.55), the last two terms are to be discarded in this first-order development. Since they refer in a different way to the material derivatives of both $\tilde{\rho}$ and $\rho'$ it can be concluded that two distinct assumptions are required. They are explicited here for the usual convective motion in the atmosphere.

As noticed by Durran [140], different definitions of the density reference $\tilde{\rho}$ have been proposed, which are not equivalent and result from different versions of the anelastic approximation. We are here concerned with the originate version of Batchelor [33], dealing with fluid motions which slightly depart from a given reference state assimilated to an atmosphere in adiabatic equilibrium. The corresponding vertical density profile $\rho_a(z)$ is linked with the pressure and temperature profiles:

$$\left(\frac{\rho_a(z)}{\rho_{a0}}\right)^{\gamma-1} = \left(\frac{P_a(z)}{P_{a0}}\right)^{\frac{\gamma-1}{\gamma}} = \frac{T_a(z)}{T_{a0}} = 1 - \frac{\gamma-1}{\gamma} \frac{\rho_{a0} g}{P_{a0}} (z - z_0) ,$$

and the anelastic expression of eq.(3.55), as derived by Batchelor [33] is:

$$\operatorname{div} \vec{V} = -\frac{1}{\rho_a} \frac{\mathrm{d}\rho_a}{\mathrm{dt}} (\equiv -\frac{1}{\rho_a} W \frac{\mathrm{d}\rho_a}{\mathrm{dz}} = W \frac{\rho_a}{\gamma P_a} g = W \frac{g}{a_a^2}) , \tag{3.56}$$

where $W$ is the vertical component of the velocity and $a_a$ the speed of sound for the adiabatic reference state (last equivalent expressions in eq.(3.56) apply to inviscid fluid motions).

As pointed out by Ogura and Phillips [351], such an anelastic approximation, when compared with the first-order development — eq.(3.55)—, requires that:

– (i) the relative density variations are small:

$$\frac{\rho - \rho_a}{\rho_a} \ll 1 ,$$

so that the second term in the right-hand-side of eq.(3.55) can be neglected, as compared to the first one;

- (ii) the frequencies of the density fluctuations are limited by the convective frequency of the flow, say $\mathcal{U}/\mathcal{L}$, where $\mathcal{U}$ and $\mathcal{L}$ are the characteristic velocity and length scales of the convective motion under consideration. Hence, the third term in the right-hand-side of eq.(3.55) can also be discarded. Since acoustic waves in general have high frequencies, the physical meaning of this second assumption is that the anelastic approximation eq.(3.56) acts as a "low-pass" filter and does not support sound waves.

The original version of the anelastic continuity equation, Batchelor [33], Ogura and Phillips [351], has been improved in several ways. In 1969 for instance, Dutton and Fichtl [143] presented alternative "soundproof" equations, in which the actual mean state values of the thermodynamic variables are substituted to the ones associated with the adiabatic reference state. They reached the conclusion that, in the momentum equation, the perturbations of pressure appear in several places for deep atmospheric convection, in which the vertical scale of motion is about the same order of magnitude as the scale height. On the other hand, when the vertical scale motion is much less than the scale height, the perturbation pressure appears only in the pressure gradient term, as it is the case with the "usual" Boussinesq's formulation.

### 3.7.2. PSEUDO-INCOMPRESSIBLE APPROXIMATION

The physical approximation associated with the use of the anelastic continuity equation is that *the perturbation density field has no influence on the mass balance.* This is obviously apparent from the comparison of the exact expression of the continuity equation of the perturbed field:

$$\left(\frac{1}{\rho}\frac{d\rho}{dt}\right)' + \frac{\partial u'}{\partial x} + \frac{\partial v'}{\partial y} + \frac{\partial w'}{\partial z} = 0 ,$$

with the anelastic approximation:

$$\frac{w'}{\rho_a(z)}\frac{d\rho_a(z)}{dz} + \frac{\partial u'}{\partial x} + \frac{\partial v'}{\partial y} + \frac{\partial w'}{\partial z} = 0 .$$

In 1989, Durran [140] demonstrated that, for the study of small-scale atmospheric circulations, sound waves can be filtered from the governing equations with a less strict assumption, based on the so-called "pseudo-incompressible" form of the continuity equation. The physical assumption, when adopting such a formulation, is now that *the influence of perturbation*

*pressure on perturbation density can be neglected in the mass balance.* Here again, sound waves are filtered, but in a different way.
The "pseudo-incompressible" form of the continuity equation is now:

$$\operatorname{div} \vec{V} = -\frac{1}{\widetilde{\rho}} \frac{\mathrm{d}\widetilde{\rho}}{\mathrm{dt}} \qquad \text{with} \qquad \widetilde{\rho} = \overline{\rho} \frac{\overline{\theta}}{\theta} . \tag{3.57}$$

In eq.(3.57), $\theta = T(P_0/P)^{(\gamma-1)/\gamma}$ is the instantaneous value of the potential temperature, and an overbar denotes a *vertically varying* mean state of the thermodynamic fields. Hence, with the "pseudo-incompressible" approximation, the response of $\widetilde{\rho}$ to compression or expansion is limited to that forced by changes in the mean-state pressure and only the density fluctuations that arise through fluctuations in the temperature field are figured into the mass balance equation, while those arising in response to fluctuations in the pressure field are neglected.

## 3.8. Incompressible limits to the general equations

The equations of compressible, ideal fluid flow take the form of a hyperbolic system which is distinctly different from the Euler equations of incompressible fluid flow. However, one expects that, under appropriate conditions on the initial and boundary data, the solution of the compressible system "converge" to that of the incompressible one when a typical parameter, characterizing the compressibility influence on the fluid motion (namely a Mach number) tends to zero. A mathematical contribution to the solution of this problem can be found in Klainerman & Majda [251] for an ideal, barotropic fluid ($P(\rho) = A\rho^\gamma + B$).

From a physical point of view, the only one to be adopted here, the key point in deriving any incompressible limit of the general compressible equations is to be able to distinguish between sound and vortical fluctuations, in order to separate pressure, temperature and density fluctuations in the appropriate way.
Schematically, it looks reasonable to identify the "rapid" motions in the fluid as acoustic or essentially compressible, and "slow" motions as vortical or essentially incompressible.

In the previous sections, various *incompressible* approximations have been presented in which acoustic effects are excluded by *filtering out* or *averaging over* fast acoustic motions.
In the present section, we are interested in examining how incorporating acoustic effects yielding *pseudo-compressible* approximations.

Such pseudo-compressible approximations can be understood as some incompressible limits to the general compressible motion equations including various effects of density variations. Two kinds of them will be presen-

ted here, namely the *nearly incompressible* and the *weakly compressible* flows, as proposed by Zank & Matthaeus [494], and Bayly *et al.* [40]. In 1991, Zank and Matthaeus [494] proposed a unified analysis delineating the conditions relating incompressible and compressible equations of a perfect gas. One year later Bayly *et al.* [40] elucidated the nature of incompressible approximations to the compressible flow equations for a fluid with a general equation of state and produced direct numerical simulations of the compressible Navier-Stokes equations to verify the consistency of the different asymptotic models.
Based on some kind of asymptotic expansion, the pseudo-compressible limits are concerned with equations for *correction* of the leading-order equations, which generally refer to the constant density and isovolume situation. Thus weakly compressible equations govern first-order corrections for density, pressure,.. variations.

The way the various models incorporate acoustic modes depends on the assumptions used to derive the equations. As shown by Zank & Matthaeus [494], two parameters are necessary to the distinction of acoustic effects. This can be easily seen from the comparison of the advective and acoustic time scales in a moving fluid. Indeed the advective time scale of a given body of fluid is $\tau = L/U_0$, denoting by $L$ and $U_0$ the characteristic length and velocity scales, whereas the acoustic time scale of a propagating wave of typical wavelength $\lambda$ is $\tau_a = \lambda/a_0$, where $a_0$ is the typical speed of sound. Hence:

$$\tau_a = \frac{\lambda}{L} M_0 \, \tau \, , \tag{3.58}$$

so that the Mach number $M_0$ is not the only pertinent parameter.
In Zank & Matthaeus [494], a mixed fast-time, long-wavelength multiple scales analysis is developed and two different situations are investigated in which heat fluctuations either (i) *dominate* or (ii) *modify* the dynamics of the compressible fluid.
In Bayly *et al.* [40], such a double expansion is avoided by assuming that the two significant (small) parameters $\epsilon$ and $\delta$ are related. Here again two situations are depicted within the limit $\epsilon \to 0$ depending whether $\delta \simeq \mathcal{O}(\epsilon)$ or $\delta \simeq \mathcal{O}(\epsilon^2)$, where $\delta$ measures the scale of the thermal fluctuations relative to the absolute temperature of the reference state. The definitions of $\epsilon$ according to [40] and [494] are given hereafter.

Basically, two expressions of the expansion of the density (and the temperature) are associated with the two investigated situations:

$$\rho = \rho^{(0)} + \epsilon \, \rho^{(1)} + ... \quad \text{or} \quad \rho = \rho^{(0)} + \epsilon^2 \, \rho^{(1)} + ... \, , \tag{3.59}$$

where $\epsilon \ll 1$ stands for $\sqrt{\gamma}\, M_0$ (Zank *et al.* [494]) and $\sqrt{\gamma - 1}\, M_0$ (Bayly *et al.* [40] for a perfect gas). In both cases, the pressure fluctuations remain $\mathcal{O}(\epsilon^2)$ and the velocity fluctuations $\mathcal{O}(\epsilon)$.

***Nearly incompressible, heat-fluctuation-dominated flows*** This situation corresponds to the first one (i) depicted in Zank *et al.* [494], when temperature and density fluctuations are much more significant than pressure fluctuations. According to Bayly *et al.* [40], it is obtained for $\delta \simeq \mathcal{O}(\epsilon)$ and is equivalent to assuming that the external heating mechanism in the energy balance equation is much stronger than both the viscous heating and the pressure induced temperature fluctuations. It represents situations in which large temperature and density fluctuations can persist for long times, as long as they are in appropriate pressure equilibrium.

According to Bayly *et al.* [40], the lowest-order non trivial equations which govern the leading-order terms $U_i^{(0)}$, $P^{(1)}$, $\rho^{(1)}$ and $T^{(1)}$ that determine the non dimensional fluctuations are:

$$\frac{\partial U_i^{(0)}}{\partial x_i} = 0\,,$$

$$\frac{\partial U_i^{(0)}}{\partial t} + U_j^{(0)} \frac{\partial U_i^{(0)}}{\partial x_j} = f_i - \frac{\partial P^{(1)}}{\partial x_i} + \frac{1}{R_e} \frac{\partial^2 U_i^{(0)}}{\partial x_j \partial x_j}\,,$$

$$\frac{\partial T^{(1)}}{\partial t} + U_j^{(0)} \frac{\partial T^{(1)}}{\partial x_j} = \frac{1}{P_r\, R_e} \frac{\partial^2 T^{(1)}}{\partial x_j \partial x_j}\,,$$

along with the following equation of state for a perfect gas:

$$\rho^{(1)} + T^{(1)} = 0\,.$$

$R_e$ and $P_r$ are the reference Reynolds and Prandtl numbers.

Thus, in this model, similarly with Boussinesq's approximations:

- the leading-order velocity is solenoidal;
- the nonvanishing pressure fluctuations at order $\epsilon^2$ obey a Poisson equation involving the solenoidal velocity:

$$\frac{\partial^2 P^{(1)}}{\partial x_i \partial x_i} = \frac{\partial U_j^{(0)}}{\partial x_i} \frac{\partial U_i^{(0)}}{\partial x_j}\,;$$

- density and temperature fluctuations are anti-correlated ( $\rho' \propto -T'$);
- density fluctuations do not admit fast scale variations and fluctuate on incompressible or *low frequency* time scales only;
- acoustic effects (*high frequency*) can only be incorporated by considering a further order in the perturbation expansion.

***Weakly compressible approximations*** We are concerned now with flows where acoustic and other compressible effects are included in terms of the core incompressible fluid equations using extensions of the *pseudo-sound approximation*. In this situation — (ii), referring to Zank *et al.* [494] —, no one of the pressure, density or temperature fluctuation dominates the others. It corresponds to an expansion of the density field which is *quadratic* in the small parameter $\epsilon$, — second relation in eq.(3.59).
In Bayly *et al.* [40], a weakly compressible model is derived which is consistent with the presence of acoustic modes of order $\epsilon^2$, when $\delta \simeq \mathcal{O}(\epsilon^2)$. As compared with the nearly incompressible model, the only changes in the lowest-order non dimensional equations which govern the leading-order terms $U_i^{(0)}$, $P^{(1)}$, $\rho^{(1)}$ and $T^{(1)}$ concern the energy equation and the equation of state, that are now for a perfect gas:

$$\frac{\partial T^{(1)}}{\partial t} + U_j^{(0)} \frac{\partial T^{(1)}}{\partial x_j} = \frac{1}{P_r\, R_e} \frac{\partial^2 T^{(1)}}{\partial x_i \partial x_i} + \frac{\partial P^{(1)}}{\partial t} + U_j^{(0)} \frac{\partial P^{(1)}}{\partial x_j} + \frac{1}{2\, R_e} \Sigma_{ij}^0 \Sigma_{ij}^0$$

and

$$T^{(1)} = \frac{\gamma - 1}{\gamma} (\rho^{(1)} + T^{(1)}) \,,$$

where, in the energy equation, the operator:

$$\Sigma_{ij} = \frac{\partial U_i}{\partial x_j} + \frac{\partial U_j}{\partial x_i} - \frac{2}{3} \frac{\partial U_k}{\partial x_k} \delta_{ij} \,,$$

is applied to the solenoidal velocity field.
Thus, for weakly-compressible flows:

- temperature fluctuations no longer behave like passive scalar;
- an approximate form of the heat transfer equation, *modified* by acoustic effects can be adopted;
- *fast scale* effects along with viscous heating by mechanical dissipation (entropy creation) are present in the modified thermal equation;
- density fluctuations are no longer anti-correlated with temperature fluctuations, but related to both temperature *and* pressure fluctuations.

*CHAPTER 4*

# GENERAL EQUATIONS AND CLASSIFICATION OF VARIABLE DENSITY FLUID MOTIONS

*The first goal of this chapter is to recall the "general, instantaneous, local" equations governing variable density fluid motions, according to the classical approach of continuum thermo-mechanics and using the local equilibrium hypothesis. On this basis, the classical numbers associated with density changes in a fluid motion are introduced and the departure from the solenoidal condition is discussed. The final part of the chapter is devoted to the presentation of a general time-scale analysis which is relevant to distinguishing and classifying several turbulent flows of variable density fluid.*

## 4.1. General assumptions and equations

In the present part of the monograph, devoted to the presentation of basic theoretical grounds, the analysis is restricted to single phase (mainly gaseous) non reactive fluid motions, within the framework of perfect gases mixtures[1]. Thus, in the turbulent flows under consideration in the present chapter, density *variations* are supposed to occur from various origins as depicted in Chapter 2 and encountered in compressed turbulence, compressible turbulence, dilatational turbulence, inhomogeneous mass or/ and heat mixing..., with no production/destruction of the fluid species. Accordingly, turbulent density *fluctuations* can be related only to temperature and/or concentration *fluctuations* in low-speed situations. When compressibility effects become important, as it is the case in high-speed flows, pressure *fluctuations* are also contributing to the density ones. As opposed to the previous chapter, no *a priori* considerations about the dominant contributions will be introduced when deriving the equations.

[1]For sake of simplicity and clarity of the formalism, the equations are mainly detailed for the case of two different non-reactive species (binary mixtures); the extension to multi species mixing is straightforward, although a bit cumbersome.

### 4.1.1. THERMODYNAMICS ELEMENTS ON MIXTURES OF FLUIDS

Let us consider a fluid mixture of $n$ constituents, so that every "point" in the fluid is occupied by particles of all constituents. Hence, the balance equations of mass, momentum and energy of the *mixture* are obtained by summation, over all constituents, of the corresponding balance equations for *each* constituent. Now, according to Müller [341], the balance equations applying to a *mixture* can be obtained in the same form as those derived for a single *homogeneous fluid*, provided suitable definitions of the functions describing the mechanic and thermodynamic properties of the *mixture* are adopted. These concern primarily the density ($\rho$), the velocity ($U_i$) and the specific internal energy ($e$).
The general form of the main balance equations of a mixture is then:

$$\underbrace{\frac{\partial(\rho\phi)}{\partial t}}_{(i)} + \underbrace{\frac{\partial(\rho\phi U_k)}{\partial x_k}}_{(ii)} = \underbrace{\frac{\partial(J_k)}{\partial x_k}}_{(iii)} + \underbrace{S}_{(iv)}, \qquad (4.1)$$

where $\phi$ stands for any velocity component or any one of the previous scalar functions.
Referring to eq.(4.1), the unsteady variation ($i$) and the convective variation ($ii$), written here in the conservative (divergence) form, are balanced by molecular diffusion ($iii$) and "algebraic sources" ($iv$). The expressions corresponding to mass, momentum and energy balances are detailed in the next table.

TABLE 4.1. Detailed expressions of the fluxes and source terms in the balance equations of a mixture — eq.(4.1).

| Function | $\phi$ | $J_k$ | $S$ |
|---|---|---|---|
| Mass of the mixture | 1 | 0 | 0 |
| Momentum of the mixture | $U_i$ | $\sigma_{ik}$ | $\rho F_i$ |
| Internal energy of the mixture | $e$ | $-q_k$ | $\sigma_{ik}\partial U_i/\partial x_k$ |

As shown in Table 4.1 and due to the assumptions of the present study, the mixture is a mass-conservative system, which means that no mass is produced nor destroyed within the flow field nor supplied through the walls in case of solid boundaries. $F_i$ is the specific external body force applied to the fluid mixture particle due to gravitation (say). $\sigma_{ik}$ is the stress tensor and $q_k$ denotes the convective flux of internal energy per unit of time. Finally, as shown in the last line of the table, any supply of internal energy into the volume (radiation) is discarded.

Since we are dealing with variable composition mixtures, we also need mass balance equations for each species[2]. Adopting the mass fractions as the additional "internal" variables[3], namely:

$$C_\alpha = m_\alpha/m \qquad \text{with} \qquad \alpha = 1, 2, \ldots\ldots n-1 \; ,$$

where $m_\alpha$ denotes the specific mass of the constituent $\alpha$ and $m$ the total mass of the mixture.
The general form of these equations reads:

$$\frac{\partial(\rho C_\alpha)}{\partial t} + \frac{\partial(\rho C_\alpha U_k)}{\partial x_k} = -\frac{\partial(\rho J_k^\alpha)}{\partial x_k} + S^\alpha \; . \tag{4.2}$$

In eq.(4.2), $J_k^\alpha$ accounts for the mass diffusion of the constituent $\alpha$, and $S^\alpha$ is the chemical mass production rate of the same species. In the present study (non reacting mixture), this term will be taken equal to zero.

To be used as the field equations governing density, velocity, internal energy and mass fractions, eqs.(4.1) and (4.2) must be supplemented by *constitutive relations* which relate the *unknown* "fluxes" ($\sigma_{ij}$, $q_i$, $J_k^\alpha$) to *known* "forces" based on the main internal variables.
At this stage, the following assumptions, corresponding to the classical linear thermodynamic behavior will be adopted:

- Cross diffusion effects, such as baro-diffusion, temperature driven mass diffusion (Soret effect), mass driven heat transfer (Dufour effect), are ignored;
- As a direct consequence, any diffusion flux vector $\vec{J}$ only depends upon a *single* driving force in each case (momentum, mass and heat transfer);
- The molecular diffusion can be modeled according to gradient-type laws (Newton, Fick and Fourier) for momentum, mass and heat transfer respectively;
- According to Stokes' approximation, a single viscosity coefficient ($\mu$) is introduced in Newton's law (i.e., the bulk viscosity is zero);
- The molecular transfer coefficients: $\mu$, the dynamic viscosity, $\mathcal{D}$, the mass diffusivity and $\lambda$, the thermal conductivity are not necessarily *constant* physical properties of a variable composition mixture, and are to be regarded as functions of the mass fraction and temperature for a gas mixture (say).

[2]Only $(n-1)$ independent relations are required, since by summation over all constituents, the mass conservation equation of the mixture is recovered.

[3]This is not the only possible choice. One can use molar fraction as well. The corresponding relationship can be found in Bird et al.[46], page 498 for instance.

Thus, the molecular diffusion fluxes of momentum ($\sigma_{ij}$), mass ($J_i^\alpha$) and heat ($q_i$) are respectively expressed as:

$$Newton-Stokes: \quad \sigma_{ij} = -P\delta_{ij} + \tau_{ij} \quad \text{with} \quad \tau_{ij} = 2\mu S_{ij} - \frac{2}{3}\mu S\delta_{ij} \ , \tag{4.3}$$

$$Fick: \qquad J_i^\alpha = -\mathcal{D}\frac{\partial C_\alpha}{\partial x_i} \ , \tag{4.4}$$

$$Fourier: \qquad q_i = -\lambda\frac{\partial T}{\partial x_i} \ . \tag{4.5}$$

In eq.(4.3), $P$ is the pressure[4] and $S_{ij} = \frac{1}{2}(\frac{\partial U_i}{\partial x_j} + \frac{\partial U_j}{\partial x_i})$ the strain rate tensor, with $S = S_{ii} \equiv \operatorname{div} \vec{U}$.

### 4.1.2. BALANCE EQUATIONS OF A COMPRESSIBLE BINARY MIXTURE

We restrict now the formulation to a binary mixture and denote by $C \equiv C_\alpha$ the mass fraction of the constituent $\alpha$, (say), so that the mass fraction of the second constituent is simply $C \equiv C_\beta = 1 - C$.

As detailed in the next section, the perfect gas thermodynamic behavior is adopted for the mixture. In particular, the internal energy variation can be obtained from the absolute temperature according to Joule's law $de = C_v\, dT$, where $C_v$ is the specific heat at constant volume.
Hence, to characterize the evolution of such a compressible mixture, *seven* scalar functions are required for density, pressure, temperature, concentration and the three velocity components.

With the assumptions which have been previously introduced, *six* balance equations are readily obtained by substituting eqs.(4.3), (4.4), (4.5) into eqs.(4.1) and (4.2). They read:

$$Continuity: \frac{\partial \rho}{\partial t} + \frac{\partial(\rho U_i)}{\partial x_i} = 0 \ , \tag{4.6}$$

$$Momentum: \rho\frac{\mathrm{d}U_i}{\mathrm{dt}} = \rho F_i - \frac{\partial P}{\partial x_i} + \frac{\partial}{\partial x_j}[\mu(\frac{\partial U_i}{\partial x_j} + \frac{\partial U_j}{\partial x_i}) - \frac{2}{3}\mu S\delta_{ij}] \ , \tag{4.7}$$

$$Mass\, fraction: \rho\frac{\mathrm{d}C}{\mathrm{dt}} = \frac{\partial}{\partial x_j}(\rho\mathcal{D}\frac{\partial C}{\partial x_j}) \ , \tag{4.8}$$

$$Energy: \rho\frac{\mathrm{d}(C_v T)}{\mathrm{dt}} = -P\frac{\partial U_i}{\partial x_i} + \Phi_\nu + \frac{\partial}{\partial x_j}(\lambda\frac{\partial T}{\partial x_j}) \ . \tag{4.9}$$

[4]When adopting Stokes' assumption, mechanical and thermodynamical pressures are identical.

The convective form of the transport term is adopted in the previous equations, with:

$$\frac{\mathrm{d}}{\mathrm{dt}} \equiv \frac{\partial}{\partial t} + U_j \frac{\partial}{\partial x_j} \, .$$

Due to continuity equation (4.6), it can be easily verified that transport and conservative formulations are equivalent:

$$\rho \frac{\mathrm{d}F}{\mathrm{dt}} \equiv \frac{\partial(\rho F)}{\partial t} + \frac{\partial(\rho F U_j)}{\partial x_j} \, .$$

From these main balance equations, the following remarks can be drawn:

- as far as the momentum equation is concerned, any change in density is *directly* reflected by inertia and external body forces. Any density variation is also responsible for *indirect* modifications to the momentum balance through the pressure term (since $P, \rho, C, T$ are now linked) and the viscous term ($\mu$ is no longer a constant in presence of $P, C, T$ variations);
- in the energy equation (4.9), $\Phi_\nu$ stands for the viscous dissipation rate:

$$\Phi_\nu = \tau_{ij} S_{ij} = 2\mu S_{ij} S_{ij} - \frac{2}{3}\mu S^2 \, .$$

- an alternate form of the energy equation is often adopted when using the specific enthalpy, that is, for a perfect gas $h = C_p T = e + P/\rho$, where $C_p$ is the specific heat at constant pressure. It is easy to deduce from this definition that:

$$\rho \frac{\mathrm{d}h}{\mathrm{dt}} = \rho \frac{\mathrm{d}e}{\mathrm{dt}} + \frac{\mathrm{d}P}{\mathrm{dt}} - \frac{P}{\rho} \frac{\mathrm{d}\rho}{\mathrm{dt}} \equiv \rho \frac{\mathrm{d}e}{\mathrm{dt}} + \frac{\mathrm{d}P}{\mathrm{dt}} - P \frac{\partial U_i}{\partial x_i} \, ,$$

  where the last identity directly results from the continuity equation (4.6). Thus, from eq.(4.9), it is now easy to obtain the enthalpy form of the energy equation:

$$\textit{Enthalpy} : \rho \frac{\mathrm{d}(C_p T)}{\mathrm{dt}} = \frac{\mathrm{d}P}{\mathrm{dt}} + \Phi_\nu + \frac{\partial}{\partial x_j}(\lambda \frac{\partial T}{\partial x_j}) \, . \tag{4.10}$$

### 4.1.3. THE EQUATION OF STATE

To close the model, one more equation is needed for a binary mixture[5]. It is obtained from thermodynamics, taking here both species as ideal gases. Considering a volume $\mathcal{V}$ of a given mass $m = m_\alpha + m_\beta$ of the mixture, the equations of state of each species are:

[5]When dealing with the mixing of $n$ different species, $(n-1)$ equations are required.

$$P_\alpha \mathcal{V} = \frac{m_\alpha}{\mathcal{M}_\alpha} rT \equiv C \frac{m}{\mathcal{M}_\alpha} rT \quad \text{and} \quad P_\beta \mathcal{V} = \frac{m_\beta}{\mathcal{M}_\beta} rT \equiv (1-C) \frac{m}{\mathcal{M}_\beta} rT \,,$$

where $P_\alpha$ and $\mathcal{M}_\alpha$ (resp. $\beta$) are the partial pressure and molecular weight of each species, $r$ is the universal perfect gas constant and $T$ the temperature of the mixture. In the equivalent expressions, $C = m_\alpha / m$ ($\equiv C_\alpha$) is the mass fraction of one species, $\alpha$ say.
Now, according to Dalton's law for ideal gases mixing, the pressure of the mixture is $P = P_\alpha + P_\beta$, and the equation of state reads:

$$P\mathcal{V} = (\frac{C}{\mathcal{M}_\alpha} + \frac{1-C}{\mathcal{M}_\beta}) mrT \qquad \text{or} \qquad P\mathcal{V} = \frac{m}{\mathcal{M}} rT \,, \tag{4.11}$$

with $1/\mathcal{M} = C_\alpha/\mathcal{M}_\alpha + C_\beta/\mathcal{M}_\beta$.
Introducing the density of the mixture $\rho = m/\mathcal{V}$, the equation of state (4.11) can be rewritten as follows:

$$\frac{P}{\rho} = (a^* C + b^*) T \,, \tag{4.12}$$

where $a^* = r(\mathcal{M}_\beta - \mathcal{M}_\alpha)/\mathcal{M}_\alpha \mathcal{M}_\beta$ and $b^* = r/\mathcal{M}_\beta$ are two constant parameters for any mixture of two given species.

Two simplified forms of eq.(4.12) can be considered for:

- single (homogeneous) ideal gas ($C \equiv 0$) where the classical equation of state is recovered:
$$\frac{P}{\rho} = RT = \frac{\gamma - 1}{\gamma} C_p T \,,$$
with $\gamma = C_p/C_v$;
- isothermal *and* isobaric mixing, where eq.(4.12) reduces to:
$$\rho = a\rho C + b \,, \tag{4.13}$$
with $a = (\rho_\alpha - \rho_\beta)/\rho_\alpha$ and $b = \rho_\beta$, where $\rho_\alpha$ and $\rho_\beta$ are the densities of the pure species at the (constant) pressure and temperature of the mixture.

Eq.(4.13) was first introduced in the theoretical analysis of the statistical properties of binary mixtures by Chassaing [80] in 1979 who pointed out some of its consequences on second order modeling [85]. It was also adopted by Shih *et al.*[427] in 1987, Panchapakesan & Lumley [352] in 1993 and Djeridane [132] in 1994.

### 4.1.4. PHYSICAL PROPERTIES AND TRANSPORT COEFFICIENTS

***Definitions*** In classical thermodynamics, several coefficients are defined as to characterize the fluid response to the action of a given variable during a specific evolution or "transformation". Some of them, such as $C_p$ and $C_v$ were already introduced. Others will be useful in further developments. Their definitions are recalled in Table 4.2 where $s$, $h$ and $e$ stand for the specific entropy, enthalpy and internal energy respectively. In addition to the general expressions given in the table, it is recalled that, for a homogeneous perfect gas, $\chi = 1/P$ and $\beta = 1/T$.

TABLE 4.2. Some classical thermodynamic coefficients.

| Definition | Coefficient |
|---|---|
| Specific heat at constant pressure | $C_p = T(\partial s/\partial T)_{P,C} = (\partial h/\partial T)_{P,C}$ |
| Specific heat at constant volume | $C_v = T(\partial s/\partial T)_{\rho,C} = (\partial e/\partial T)_{\rho,C}$ |
| Coefficient of compressibility at constant temperature | $\chi = \rho^{-1}(\partial \rho/\partial P)_{T,C}$ |
| Coefficient of thermal expansion at constant pressure | $\beta = -\rho^{-1}(\partial \rho/\partial T)_{P,C}$ |
| Coefficient of mass expansion at constant pressure | $\alpha = -\rho^{-1}(\partial \rho/\partial C)_{P,T}$ |

***Influence of density variation*** Within a fluid motion, the thermodynamic coefficients can be considered as physical properties of the fluid, that is,they are independent of the motion, provided the range of variation of the thermodynamic variables ($P$, $T$, $C$) is not too large. Obviously, this condition is no longer satisfied when dealing with the motion of any compressible mixture. As far as the general equations of such a fluid motion are concerned, this results in introducing non linear coupling since such "physical" properties, as well as all transport coefficients (molecular diffusivities), are now local functions of the flow field variables $P$, $T$, $C$...
The derivation of expressions accounting for temperature, pressure, mass variations of such parameters is not addressed in the present monograph. It is discussed for instance in Bird *et al.* [16] where various semi-empirical formulas, such as Wilke's formula for the molecular viscosity and the Chapman-Enskog formula for the mass diffusivity can be found. They can be used with a generally acceptable accuracy. Alternately, numerical fitting of experimental data can be adopted when available (see Novotny & Irvine [347] for Carbon dioxide-air mixtures, for instance).

Now, in variable density turbulent flows, and due to their linkage with variable field functions, such coefficients are to be considered as fluctuating quantities. For most practical applications, their fluctuations can be mainly derived from fluctuations in temperature and concentration. Since the semi-empirical formulas are not necessarily linear in both temperature and concentration, they do not provide tractable expressions for such a purpose. Simpler linearized expressions are often adopted, as suggested for example by Varma *et al.*[475], for the viscosity:

$$\mu' = \overline{\mu}\,(\overline{T},\overline{C})\theta' + \overline{\mu}_\alpha \gamma' + \overline{\mu}_\beta(1-\gamma') \,. \tag{4.14}$$

In eq.(4.14), $\overline{\mu}\,(\overline{T},\overline{C})$ is the value of the viscosity at the local mean values of temperature and concentration $\overline{T}$ and $\overline{C}$, $\overline{\mu}_\alpha \equiv \overline{\mu}\,(\overline{T},1)$ and $\overline{\mu}_\beta \equiv \overline{\mu}\,(\overline{T},0)$ denote the viscosities of the pure species at the mean temperature. $\theta' = T - \overline{T}$ and $\gamma' = C - \overline{C}$ stand for the temperature and mass-fraction fluctuations respectively.

## 4.2. Non dimensional parameters in variable density motions

Classically, non dimensional parameters characterizing the influence of compressibility, and more generally, variable density effects in fluid motions are obtained from non dimensional formulations of the governing equations. This will be done now, previous to a more specific time scales analysis presented in the next section.

To proceed with non dimensional formulation of eqs.(4.6)-(4.9), let us introduce, for any dimensional quantity $f$, a non-dimensional value $f^* = f/f_0$ based on a given (fixed) reference scale $f_0$. In the following, such a formulation applies to any function of the flow field and all its arguments, i.e., the eulerian variables $x_i$ and $t$.

*Continuity:* when the non-dimensional values are introduced in the continuity equation, it comes:

$$S_t \frac{\partial \rho^*}{\partial t^*} + \frac{\partial}{\partial x_j^*}(\rho^* U_j^*) = 0 \,.$$

Only one non-dimensional parameter emerges, the Strouhal number $S_t = l_0/(t_0 U_0)$, which compares a convective time scale $(l_0/U_0)$ to a specific unsteadiness time scale $(t_0)$.

*Momentum:* when applied to the momentum equation (4.7), the same procedure yields:

$$S_t \frac{\partial U_i^*}{\partial t^*} + U_j^* \frac{\partial U_i^*}{\partial x_j^*} = \frac{1}{F_r^2} g_i^* - E_u \frac{\partial P^*}{\partial x_i^*} + \frac{1}{R_e}\left[\frac{\partial^2 U_i^*}{\partial x_j^* \partial x_j^*} + \frac{1}{3}\frac{\partial}{\partial x_j^*}\left(\frac{\partial U_i^*}{\partial x_i^*}\right)\right] .$$

To obtain this equation, it is assumed that the external body forces are only due to gravity with $F_i = g_i$, so that $F_r = U_0/\sqrt{g_0 l_0}$ is the Froude number[6]. Two additional non dimensional parameters are introduced: the Euler number $E_u = P_0/(\rho U_0^2)$ and the Reynolds number $R_e = \rho_0 U_0 l_0/\mu$ so that $E_u$ and $R_e^{-1}$ compare the pressure and viscosity forces to the inertia ones.

*Concentration:* considering the mass concentration equation (4.8), one obtains similarly:

$$\rho^*(S_t \frac{\partial C}{\partial t^*} + U_j^* \frac{\partial C}{\partial x_j^*}) = \frac{\partial}{\partial x_j^*}(\rho^* \frac{1}{S_c R_e} \frac{\partial C}{\partial x_j^*}) ,$$

where $S_c = \mu/(\rho_0 \mathcal{D})$ is the Schmidt number.

*Energy:* at last, the same analysis applied to the energy equation (4.9) yields, when considering $C_v = \gamma C_p$ as a constant:

$$\rho^*(S_t \frac{\partial T^*}{\partial t^*} + U_j^* \frac{\partial T^*}{\partial x_j^*}) = -\gamma E_u \times E_c P^* \frac{\partial U_i^*}{\partial x_i^*} + \gamma \frac{E_c}{R_e} \Phi_\nu^* + \gamma \frac{\partial}{\partial x_j^*}(\frac{1}{P_e} \frac{\partial T^*}{\partial x_j^*}) .$$

In this equation, $E_c = U_0^2/(C_p T_0)$ is the Eckert number and $P_e = U_0 l_0/k$ the Péclet number, where $k = \lambda/(\rho C_p)$ is the thermal diffusivity.

Alternately, the Prandtl number ($P_r = \nu/k$) and the Mach number ($M = U_0/a$, where $a = \sqrt{(\gamma - 1)C_p T}$ denotes the speed of sound) can be introduced according to the following identities:

$$P_e = P_r \times R_e \quad \text{and} \quad E_c = (\gamma - 1)M^2 .$$

To summarize, six independent governing non-dimensional parameters can be directly associated with the main balance equation of a compressible mixture. Some of them only depend on physical properties of the *fluid*, while others are linked with the *flow* itself, as recalled in Table 4.2.

## 4.3. Incompressibility and isovolume conditions

### 4.3.1. DEFINITONS

In order to avoid misinterpretation[7] or confusion between distinct classes of variable density fluid motions, it is worth recalling the general definitions of incompressible and isovolume evolutions.

[6] In stratified flows, $1/F_r^2$ is sometimes known as a Richardson number, see Tritton [462], page 185. Another classical interpretation of the Froude number is recalled in the next table.

[7] The incompressible situation, for instance, should not be taken as a quasi-constant pressure flow.

TABLE 4.3. Characteristic numbers of the flow (top) and fluid properties (bottom) of a compressible mixture motion

| *Flow* | | |
|---|---|---|
| Name | Expression | Interpretation |
| Strouhal | $S_t = \frac{l_0}{U_0 t_0}$ | $\frac{\text{Advective time scale}}{\text{Unsteadiness time scale}}$ |
| Froude | $F_r^2 = \frac{U_0^2}{g_0 l_0}$ | $\frac{\text{Inertia forces}}{\text{Gravity forces}}$ |
| | $F_r = \frac{U_0}{\sqrt{g_0 l_0}}$ | $\frac{\text{Flow velocity}}{\text{Gravity waves celerity}}$ |
| Euler | $E_u = \frac{P_0}{\rho_0 U_0^2}$ | $\frac{\text{Pressure forces}}{\text{Inertia forces}}$ |
| Reynolds | $R_e = \frac{\rho_0 U_0 l_0}{\mu}$ | $\frac{\text{Inertia forces}}{\text{Viscosity forces}}$ |
| Péclet | $P_e = \frac{U_0 l_0}{k}$ | $\frac{\text{Heat convection}}{\text{Heat diffusion}}$ |
| Mach | $M = \frac{U_0}{a}$ | $\frac{\text{Flow velocity}}{\text{Acoustic waves celerity}}$ |
| Eckert | $\frac{1}{2}E_e = \frac{\frac{1}{2}\rho_0 U_0^2}{\rho_0 C_p T_0} \equiv (\gamma - 1)M^2$ | $\frac{\text{Kinetic energy}}{\text{Enthalpy}}$ |

| *Fluid* | | |
|---|---|---|
| Name | Expression | Interpretation |
| Prandtl | $P_r = \frac{\nu}{k} \equiv \frac{\mu C_p}{\lambda}$ | $\frac{\text{Momentum diffusivity}}{\text{Heat diffusivity}}$ |
| Schmidt | $S_c = \frac{\mu}{\rho \mathcal{D}} \equiv \frac{\nu}{\mathcal{D}}$ | $\frac{\text{Momentum diffusivity}}{\text{Mass diffusivity}}$ |
| Lewis | $L_e = \frac{\mathcal{D}}{k} \equiv \frac{\rho C_p \mathcal{D}}{\lambda} = P_r \times S_c$ | $\frac{\text{Mass diffusivity}}{\text{Heat diffusivity}}$ |

***Incompressibility*** Strictly speaking, *static* incompressibility (fluid at rest) means that a given body of fluid exhibits no change in volume when submitted to any pressure variation. It can only be achieved within the limit $\chi \to 0$.

In a moving fluid, *dynamic* incompressibility means that density changes driven by *pressure* variations in the flow field can be taken as negligible. Thus, if any change in the density only occurs from pressure variations, dynamic incompressibility is equivalent to *constant density* fluid motion. However, and more generally, these two situations are not equivalent, and density variations can be present in incompressible fluid flow due to temperature and /or concentration variations.

***Isovolume condition*** In an isovolume evolution and by definition, the volume $\mathcal{V}$ of any fluid particle remains constant during the motion of that particle. In other words, translation, rotation and deformation associated with the particle motion are not changing the volume integral over the surface of the particle. Thus, the isovolume condition is simply:

$$\frac{1}{\mathcal{V}}\frac{\mathrm{d}\mathcal{V}}{\mathrm{dt}} = 0 ,$$

where d/dt stands for the material derivative.
Now, it is well known — see Batchelor [35], page 75 for instance —, that in any fluid motion, the rate of change of the volume of an elementary material element $\mathcal{V}$ is given by:

$$\frac{1}{\mathcal{V}}\frac{\mathrm{d}\mathcal{V}}{\mathrm{dt}} = \mathrm{div}\vec{U} ,$$

where $\vec{U}$ is the velocity of the fluid particle ($U_i$, $i = 1, 2, 3$).
Hence, the isovolume condition corresponds to *divergence free* or *solenoidal* motions.
The connection between incompressible, constant density and isovolume situations can be readily understood, since from the continuity equation (4.6):

$$\frac{1}{\mathcal{V}}\frac{\mathrm{d}\mathcal{V}}{\mathrm{dt}} \equiv \mathrm{div}\vec{U} = -\frac{1}{\rho}\frac{\mathrm{d}\rho}{\mathrm{dt}} . \tag{4.15}$$

Thus, the constant density fluid motion condition is equivalent to the isovolume condition, but the reciprocal is not true, the stationary stratified laminar flow, with density gradients normal to the streamlines is a well known example of such a situation (see [83] for example).
Equation (4.15) also provides the guideline to identify the origins of the departure from the isovolume situation. This can be more easily shown when adopting, to simplify, a linearized form of the equation of state $\rho = \rho(P, T, C)$ in density variation with respect to pressure, temperature and concentration variations:

$$\rho(T, P, C) \simeq \rho_0[1 - \beta_0(T - T_0) + \chi_0(P - P_0) + \alpha_0(C - C_0)] . \tag{4.16}$$

The subscript $_0$ denotes a given reference state and $\alpha_0$, $\beta_0$ and $\chi_0$ the coefficient of mass, thermal expansion and compressibility respectively (see Table 4.2).
Such a linearized form, eq.(4.16), was actually used for instance by Gray &

Giorgini [188] in the derivation of the Boussinesq approximation for homogeneous ($C = C^t$) liquids and gases.
Substituting in eq.(4.16) yields:

$$\mathrm{div}\vec{U} = -\beta_0 \frac{\rho_0}{\rho}\frac{dT}{dt} + \chi_0 \frac{\rho_0}{\rho}\frac{dP}{dt} + \alpha_0 \frac{\rho_0}{\rho}\frac{dC}{dt} \ . \tag{4.17}$$

This equation clearly shows how to proceed with the analysis, by substituting the material derivatives of the temperature, pressure and mass-fraction from the corresponding transport equations. This will be done in a next section, the analysis being restricted hereafter to simple conclusions on pure water and air flows.

### 4.3.2. APPROXIMATIONS TO THE ISOVOLUME CONDITION FOR AIR AND WATER FLOWS

The solenoidal condition can be achieved differently in liquids and gases. The distinction clearly results from the thermodynamic properties of both medium. To find out the general expression of the density variation in a pure medium, let us recall that an infinitesimal amount of heat $\delta Q$ which is algebraically given to a unit mass of a homogeneous fluid ($C = 1$), at temperature $T$ during a *reversible* evolution can be expressed (see Bruhat [63], page 123 for instance) as:

$$\delta Q = C_v \delta T - \frac{C_p - C_v}{\beta}\frac{\delta\rho}{\rho} \quad \text{or} \quad \delta Q = C_p \delta T - \beta \frac{T}{\rho}\delta P \ ,$$

where $C_p$ and $C_v$ are the specific heat at constant pressure and volume respectively, and $\beta$ the thermal expansion coefficient at constant pressure. By equating these two expressions of $\delta Q$, it is easily deduced that:

$$\frac{C_p - C_v}{\beta}\frac{\delta\rho}{\rho} = -(C_p - C_v)\delta T + \beta \frac{T}{\rho}\delta P \ .$$

Assuming local thermodynamic equilibrium[8], this relation can be applied to the substantial variation to give:

$$\left(\frac{C_p - C_v}{\beta}\right)\frac{1}{\rho}\frac{\mathrm{d}\rho}{\mathrm{dt}} = -(C_p - C_v)\frac{\mathrm{d}T}{\mathrm{dt}} + \beta \frac{T}{\rho}\frac{\mathrm{d}P}{\mathrm{dt}} \ . \tag{4.18}$$

From this relation, different expressions of the rate of variation of the density can be derived, depending on whether $(C_p - C_v)$ is different or

[8] This assumption states that, within the volume of a fluid particle, i.e., the elementary macroscopic domain, all sub-local physical, mechanical and thermal quantities are virtually independent of both position and time.

equal to zero. This introduces a distinction between water and air flows, since, as shown by the values of the physical coefficients for these fluids given in Table 4.4, the first condition $(C_p - C_v) \neq 0$ is achieved for air flows, while for water flows, $C_p - C_v = 0$.

TABLE 4.4. Some physical coefficients of air and pure water.

| Coefficient | $\beta\ (K^{-1})$ | $C_p\ (J\,kg^{-1}K^{-1})$ | $\gamma(=C_p/C_v)$ | $Pr$ |
|---|---|---|---|---|
| Air (*stand. cond.*) | $3.3 \times 10^{-3}$ | $1.01 \times 10^{3}$ | 1.4 | 0.7 |
| Water (15 *degC*) | $1.5 \times 10^{-4}$ | $4.18 \times 10^{3}$ | 1 | 10 |

*Water flows:* for this fluid, $C_p = C_v$, so that eq.(4.18) reduces to $\mathrm{d}P/\mathrm{dt} = 0$. The amount of heat given to a unit mass of water is obviously $\delta Q = C_v \delta T = C_p \delta T$. Hence, assuming local thermodynamic equilibrium, the material rate of change of the density can be obtained as:

$$\frac{\mathrm{d}Q}{\mathrm{dt}} = C_v \frac{\mathrm{d}T}{\mathrm{dt}} = C_v \frac{\mathrm{d}T}{\mathrm{d}\rho}\frac{\mathrm{d}\rho}{\mathrm{dt}} = -\frac{C_v}{\beta}\frac{1}{\rho}\frac{\mathrm{d}\rho}{\mathrm{dt}}\,, \tag{4.19}$$

and consequently, from eq.(4.15):

$$\mathrm{div}\vec{U} = \frac{\beta}{C_v}\frac{dQ}{dt}\,. \tag{4.20}$$

Thus a solenoidal evolution can only be strictly obtained in *adiabatic* water flows.

When heat transfer is present, the departure from the isovolume condition can be appreciated from the non-dimensional form of eq.(4.19). Introducing $\Delta T_0$ as the scaling of the temperature variation and assuming a single characteristic time scale, i.e., $t_0 = l_0/U_0$ the non-dimensional form of this equation is simply:

$$\mathrm{div}\vec{U}^* = \beta\,\Delta T_0 \frac{dQ^*}{dt^*}\,.$$

Now, provided a suitable scaling giving rise to same order non-dimensional terms, the departure from the solenoidal condition can be directly estimated from the value of the *non-dimensional* coefficient $\beta\,\Delta T_0$.

This is generally obtained by referring to a usually adopted characteristic number, the Grashof number $G_r = g_0\,\beta\,\Delta T_0\, l_0^3\,/\nu^2$, in combination with the Froude and Reynolds numbers, since:

$$\beta\,\Delta T_0 = \frac{F_r^2}{R_e^2}\,G_r\,.$$

*Air flows:* for such a gas, see Table 4.4, $(C_p - C_v) \neq 0$, so that dividing eq.(4.18) by $(C_p - C_v)$ directly yields:

$$\frac{1}{\rho}\frac{\mathrm{d}\rho}{\mathrm{dt}} = -\beta\frac{\mathrm{d}T}{\mathrm{dt}} + \frac{\gamma}{\gamma - 1}\beta^2\frac{T}{\rho C_p}\frac{\mathrm{d}P}{\mathrm{dt}} . \tag{4.21}$$

This expression can be simplified by considering that air is a perfect gas, so that $P/\rho = [(\gamma - 1)/\gamma]C_pT$ and $\beta = 1/T$. Hence:

$$\mathrm{div}\vec{U} \equiv -\frac{1}{\rho}\frac{\mathrm{d}\rho}{\mathrm{dt}} = \frac{1}{T}\frac{\mathrm{d}T}{\mathrm{dt}} - \frac{1}{P}\frac{\mathrm{d}P}{\mathrm{dt}} = \beta\frac{\mathrm{d}T}{\mathrm{dt}} - \gamma\chi_S\frac{\mathrm{d}P}{\mathrm{dt}} , \tag{4.22}$$

where $\chi_S = \gamma\chi$ is the isentropic coefficient of compressibility of the perfect gas. Of course, eq.(4.22) can be readily derived from the equation of state of a perfect gas[9].

As in the previous case for water flows, the non-dimensional form of this relation can be used to estimate the departure from the isovolume condition. Assuming again $t_0 = l_0/U_0$, it reads:

$$\mathrm{div}\vec{U}^* = \beta\,\Delta T_0\frac{\mathrm{d}T^*}{\mathrm{dt}^*} - \gamma\,\chi_S P_0\frac{\mathrm{d}P^*}{\mathrm{dt}^*} ,$$

which clearly shows that *strictly* isovolume motion of air can only be achieved when temperature *and* pressure variations are *both* negligible. The *first non-dimensional* coefficient $\beta\,\Delta T_0$ can be evaluated as before, from the values of the Grashof, Froude and Reynolds numbers. The *second* coefficient $\gamma\,\chi_S\,P_0$, can be deduced from the Euler and Mach numbers, according to:

$$\gamma\,\chi_S\,P_0 = \gamma\,E_u\,M^2 ,$$

since $\chi_S = \rho_0\,a_0^2$, where $a_0$ is a reference speed of sound.

## 4.4. Density fluctuations in turbulent flows

This section is devoted to providing some quantitative information about the level of density fluctuations in various types of turbulent flows. The turbulent intensity of density fluctuations $I_\rho$ will be considered as the relevant parameter for the present discussion:

$$I_\rho = \frac{\sqrt{\overline{\rho'^2}}}{\overline{\rho}} , \tag{4.23}$$

[9]In general (see Batchelor [35], page 27), the ratio of the two terms in the right-hand side of eq.(4.22) can be expressed as:

$$-\beta\frac{T}{\rho}\frac{\mathrm{d}P}{\mathrm{dt}} \Big/ C_p\frac{\mathrm{d}T}{\mathrm{dt}} = \frac{\gamma - 1}{\gamma}\frac{\chi}{\beta}\frac{\mathrm{d}P}{\mathrm{dt}} \Big/ \frac{\mathrm{d}T}{\mathrm{dt}}$$

where $\chi$ is the isothermal coefficient of compressibility of the fluid (see Table 4.2).

where $\rho' = \rho - \overline{\rho}$ denotes the centered density fluctuation. Maximum values of $I_\rho$ in various types of flows, as reported in the literature from both experimental measurements and direct numerical simulations, are given in Table 4.5.

TABLE 4.5. Mean and rms density variations in various flow configurations.

| Author - ref. | Flow configuration | $s_\rho = \frac{\rho_{max}}{\rho_{min}}$ | $(I_\rho)_{max}$ |
|---|---|---|---|
| Sarkar [410], 1996 | Uniformly sheared compressible flow. Mean shear rate $S = C^t$. DNS. Values at $S \times t = 3$ and for different initial gradient Mach numbers: 0.22, 0.44, 0.66 and 1.32 resp. | - | 0.08<br>0.13<br>0.16<br>0.23 |
| Sarkar & Pantano [414], 2000 | Temporally-evolving compressible shear layer at a convective Mach number $M_c = 0.7$ DNS, self-similarity state. | 1<br>8 | 0.04<br>0.36 |
| Sautet [417], 1992 | 80% H2-20% N2, low speed jet in a co-flow. Density ratio at the exit $s_{\rho 0} = 4$, Reynolds number $R_{e0} = 8\,700$. EXP. Section $x/D_0 = 15$. | 1.7 | 0.21 |
| Favre-Marinet & Camano [158], 1996 | He-air annular, low speed jet in a quiescent atmosphere. Density ratio at the exit $s_{\rho 0} = 7.25$, Reynolds number $R_{e0} = 3\,200$. EXP. Section $x/D_i = 2.5$. Values on the axis. | 5.5 | 0.30 |
| Adams [1], 2000 | Turbulent boundary layer along a compression ramp at $M = 3$. DNS, Downstream station # 10. | 2.3 | 0.2 |
| Sarkar, Erlebacher & Hussaini [415], 1991 | Compressible homogeneous shear. DNS, equilibrium regime at initial turbulence Mach number $M_{t0} = 0.4$ | - | 0.13 - 0.15 |
| Blaisdell, Mansour & Reynolds [47], 1993 | Compressible homogeneous turbulent shear flow $M_t < 0.55$. DNS, equilibrium regime at $M_t = 0.3$ | - | 0.06 |
| Jamme [231], 1998 | Homogeneous turbulence passing through a shock. Upstream Mach number $M_\infty = 1.5$ ($M_t = 0.173$). DNS, pure vortical mode amplification. | 1.86 | 1.4 *[10] |

[10] Amplification factor of turbulence density intensity when passing through the shock. This estimate is not accounting for intensification due to shock-wave oscillations.

The data gathered in Table 4.5 correspond to a wide variety of variable density flows, including both low and high speed flows, so that density fluctuations originate from all possible sources: temperature and/or mass inhomogeneity without or with compressibility.
In some cases and provided suitable assumptions, density fluctuations can be explicitly linked with temperature, mass-fraction or pressure fluctuations. Some of such relations are now detailed as a useful basis to the estimate of turbulent density intensities.

### 4.4.1. TURBULENT DENSITY INTENSITY IN LOW-SPEED FLOWS

A rather well documented situation of low-speed, variable density flow is the turbulent jet, where the density at the exit $\rho_0$ differs from the ambient surrounding $\rho_\infty$ due to differences in temperature or concentration. In 1986, Pitts [367] investigated density effects in isothermal axisymmetric jets of different gases, developing in the center of a square duct in the presence of a slow co-flow of another gas. Various binary mixtures of helium, methane, propane, carbon dioxyde, sulfur hexafluoride and air have been considered, yielding an initial range of densities ratio $s_{\rho 0}=\rho_0/\rho_\infty$ from 0.14 to 37.
Within this range, the turbulent mass fraction intensity[11] along the centerline $(\sqrt{\overline{\gamma'^2}}/\overline{C})_{\mathbb{C}}$ is approaching a common asymptote of 0.23. Only, the down-stream distance required for a given gas to reach this asymptote depends strongly on the density ratio $s_\rho$. A different conclusion is obtained when the turbulent *mole* fraction is considered, since then, the asymptotic level *also* depends on $s_\rho$.
A similar conclusion is reached by Sautet [417] and Sautet & Stepowski [418] in variable concentration hydrogen/air jets. The normalized rms mixture fraction fluctuations along the axis $(\sqrt{\overline{z'^2}}/\overline{Z})_{\mathbb{C}}$ is approaching a constant level of about 0.45 for various density ratios, ranging from 0.07 (pure hydrogen/air jet) to one. However, the turbulent density intensity along the axis at $x/D_0 = 12$ is about 0.66 for the pure hydrogen/air jet, and about 0.52 for the 80% hydrogen-20% air jet. At the same location in this flow, the mass concentration intensity $(\sqrt{\overline{\gamma'^2}}/\overline{C})_{\mathbb{C}}$ is about 0.71. For all the investigated jets, the asymptotic value of the turbulent axial velocity intensity along the axis $(\sqrt{\overline{u'^2}}/\overline{U})_{\mathbb{C}}$ is about 0.2, independently of $s_\rho$.
Similarly, in a turbulent wall jet of helium in an air-flow boundary layer, Harion [207] measured a value of about 40 % for $I_\rho$ at a downstream location where the turbulent longitudinal velocity intensity $(I_u=\sqrt{\overline{u'^2}}/\overline{U}_\infty)$ is roughly half this value (17%).

[11] unmixedness

All these results suggest that the turbulent density intensity $I_\rho$ can be markedly different from other turbulent intensities in such flows. The explanation can be derived from the equation of state, as we shall see now for the pure concentration and temperature mixing situations.

***Concentration mixing*** In this low-speed situation, which corresponds to an isothermal binary mixing of perfect gases, the instantaneous equation of state $\rho=\rho(C)$ reduces to eq.(4.13). The linearized form of this equation in density ($\rho'$) and mass-fraction ($\gamma'$) fluctuations is simply:

$$\rho' \stackrel{L}{=} a\bar{\rho}\gamma' + a\overline{C}\rho' \, .$$

From this relation, it can be easily deduced that:

$$\overline{\rho'^2} \stackrel{L}{=} a\bar{\rho}\,\overline{\rho'\gamma'} + a\overline{C}\,\overline{\rho'^2} \qquad \text{and} \qquad \overline{\rho'\gamma'} \stackrel{L}{=} a\bar{\rho}\,\overline{\gamma'^2} + a\overline{C}\,\overline{\rho'\gamma'} \, .$$

By eliminating $\overline{\rho'\gamma'}$ between these two equations, one obtains:

$$\frac{\sqrt{\overline{\rho'^2}}}{\bar{\rho}} \stackrel{L}{=} \left| \frac{a}{1-a\overline{C}} \right| \sqrt{\overline{\gamma'^2}} \, , \tag{4.24}$$

where it is recalled that the coefficient $a$ is a *constant*, the value of which only depends on the density of the pure species, $a = (\rho_\alpha - \rho_\beta/\rho_\alpha)$. Thus for a jet flow (gas $\rho_0$ discharging into an atmosphere $\rho_\infty$) or a mixing layer between two different gases ($\rho_0$ and $\rho_1$ say), this coefficient is only function of the density ratio $s_\rho = \rho_0/\rho_\infty$ or $\rho_0/\rho_1$, since $a=(s_\rho - 1)/s_\rho$.
Hence, eq.(4.24) can be rewritten as:

$$\frac{\sqrt{\overline{\rho'^2}}}{\bar{\rho}} \stackrel{L}{=} |\Gamma^S(\overline{C})| \sqrt{\overline{\gamma'^2}} \qquad \text{where} \qquad \Gamma^S(\overline{C}) = \frac{s_\rho - 1}{s_\rho + (1-s_\rho)\overline{C}} \, , \tag{4.25}$$

introducing the mass-fraction indicator function $\Gamma^S(\overline{C})$.

***Temperature mixing*** A similar analysis can be derived for the isobaric mixing which takes place in a low speed moving gas, in presence of high temperature variations. As shown in the previous Chapter, section8, the linkage between density and temperature fluctuations can be taken as:

$$\rho'/\bar{\rho} \simeq \omega\theta'/\overline{T} \, , \tag{4.26}$$

were the coefficient $\omega$ is constant and equal to $-1$ (nearly incompressible, heat fluctuation dominated flow) or $1/(\gamma-1)$ (weakly compressible approximation according to Bayly *et al.* [40]). Hence:

$$\frac{\sqrt{\overline{\rho'^2}}}{\bar{\rho}} \simeq |\omega| \frac{\sqrt{\overline{\theta'^2}}}{\overline{T}} \, . \tag{4.27}$$

Let us consider that the mixing takes place in a jet or a mixing layer so that the boundary conditions for the mean temperature profile can be defined from two constant values, $T_{min}$ and $T_{max}$ say. Then, it is worth introducing a bound, non-dimensional mean temperature $\overline{\Theta}$:

$$\overline{\Theta} = \frac{\overline{T} - T_{min}}{T_{max} - T_{min}} .$$

In a mixing layer, the temperature ratio can be readily deduced from the density ratio as $T_{min}/T_{max} = \rho_{max}/\rho_{min} \equiv s_\rho$.
For the jet flow situation, a distinction is to be done between heated ($T_0 > T_\infty$) and cold ($T_0 < T_\infty$) jets. In heated jets, $s_\rho (= \rho_0/\rho_\infty) = T_\infty/T_0$, so that $T_{min}/T_{max} = s_\rho$, with $0 \leq s_\rho \leq 1$. Conversely, $T_{min}/T_{max} = 1/s_\rho$, with $1 \leq s_\rho$ for cold jets.
Hence, the mean temperature in jets and mixing layers can be expressed as a function of $\overline{\Theta}$, which is parametric in the density ratio $s_\rho$, as defined by $T_{min}/T_{max}$ in mixing layers and $\rho_0/\rho_\infty$ in jet flows:

$$\textit{Mixing layer and hot jets :} \quad \frac{\overline{T}}{\Delta T} = \frac{s_\rho + (1 - s_\rho)\overline{\Theta}}{1 - s_\rho} ,$$

$$\textit{Cold jets :} \quad \frac{\overline{T}}{\Delta T} = \frac{1 + (s_\rho - 1)\overline{\Theta}}{s_\rho - 1} ,$$

where $\Delta T$ stands for $T_{max} - T_{min}$. After substitution in eq.(4.27), it is obtained:

$$\frac{\sqrt{\overline{\rho'^2}}}{\overline{\rho}} \simeq |\omega| \Gamma^S(\overline{\Theta}) \frac{\sqrt{\overline{\theta'^2}}}{\Delta T} , \tag{4.28}$$

where the temperature indicator function $\Gamma^S(\overline{\Theta})$ is defined as:

$$\Gamma^S(\overline{\Theta}) \equiv \Gamma_0^S(\overline{\Theta}) = \frac{1 - s_\rho}{s_\rho + (1 - s_\rho)\overline{\Theta}} \qquad \text{for } 0 \leq s_\rho \leq 1 ,$$

$$\Gamma^S(\overline{\Theta}) \equiv \Gamma_1^S(\overline{\Theta}) = \frac{s_\rho - 1}{1 + (s_\rho - 1)\overline{\Theta}} \qquad \text{for } 1 \leq s_\rho . \tag{4.29}$$

It can be noticed that, whatever be $s_\rho$ and $\overline{\Theta}$ ($0 \leq \overline{\Theta} \leq 1$), the temperature indicator function is always positive, and $\Gamma_0^{1/S} = \Gamma_1^S$.

***Discussion*** From eqs. (4.25) and (4.28), it is clear that the turbulent density intensity is proportional to the rms value of either mass-fraction or temperature fluctuations by a factor depending upon (i) the mean concentration or the normalized mean temperature difference, and (ii) the density ratio. According to the value of this factor — the indicator function $\Gamma^S$

— the turbulent density intensity can be similar to, or different from the corresponding variances. The indicator functions $\Gamma^S$ for both mass-fraction and temperature mixing situations are plotted in Fig. 4.1. The major conclusion which obviously emerges from this figure is that such *low-speed* fluid motions are by no way necessarily developing *low density fluctuations.* This can be easily seen from the two asymptotic limits of $\Gamma^S(\overline{C})$ for instance:

$$s_\rho \to 0, \;\; \Gamma^S(\overline{C}) \to -\frac{1}{\overline{\overline{C}}} \quad \text{and} \quad s_\rho \to +\infty, \;\; \Gamma^S(\overline{C}) \to \frac{1}{1-\overline{\overline{C}}} \,,$$

In actual fact, from the analytical expression of the indicator functions eqs. (4.25) and (4.28), three different classes of such flows can be introduced, according to the variation range of the turbulent density intensity. Based upon the density ratio $s_\rho$, they are associated with:

- *small* indicator function: $|\Gamma^S|_{max} < 1/2$, for $2/3 \leq s_\rho \leq 3/2$;
- *moderate* indicator function: $|\Gamma^S|_{min} > 1/3$ **and** $|\Gamma^S|_{max} < 1$, for $1/2 \leq s_\rho \leq 2/3$ **or** $3/2 \leq s_\rho \leq 2$;
- *high* indicator function: $|\Gamma^S|_{min} > 1/2$, for $s_\rho \leq 1/2$ **or** $2 \leq s_\rho$.

#### 4.4.2. TURBULENT DENSITY INTENSITY IN HIGH-SPEED FLOWS

***Isentropic behavior*** As a formal extension of eq.(4.26), density fluctuations can be related to pressure and temperature fluctuations through a polytropic coefficient $n$ as:

$$n(n-1)\frac{\rho'}{\overline{\rho}} = (n-1)\frac{p'}{\overline{\overline{P}}} = n\frac{\theta'}{\overline{\overline{T}}} \,. \tag{4.30}$$

$n = 0$ corresponds to isobaric fluctuating motion, $n = 1$ isothermal, and $n=\gamma$ isentropic.

As shown by Blaisdell *et al.*[47], the DNS results of the compressible homogeneous shear flow lead to conclude that, for a turbulence Mach number $M_t < 0.5$, density, temperature and pressure fluctuations follow a nearly isentropic process, with $n \simeq 1.36$, close to $\gamma = 1.4$. For $M_t = 0.3$, $\sqrt{\overline{p'^2}}/\overline{P}$ is close to 0.1, so that the turbulent density intensity is $I_\rho \simeq 0.06$.

In 1991, Zeman [497] proposed a model for the pressure variance in which the equilibrium value is assumed to scale with the turbulence Mach number as

$$\frac{\overline{p'^2}}{\overline{P}^2} = \gamma^2 M_t^2 \left( \frac{\alpha M_t^2 + \beta M_t^4}{1 + \alpha M_t^2 + \beta M_t^4} \right) \,,$$

where the model constants are $\alpha = 1$ and $\beta = 2$ in agreement with decaying isotropic turbulence. This model, which slightly overestimates the DNS

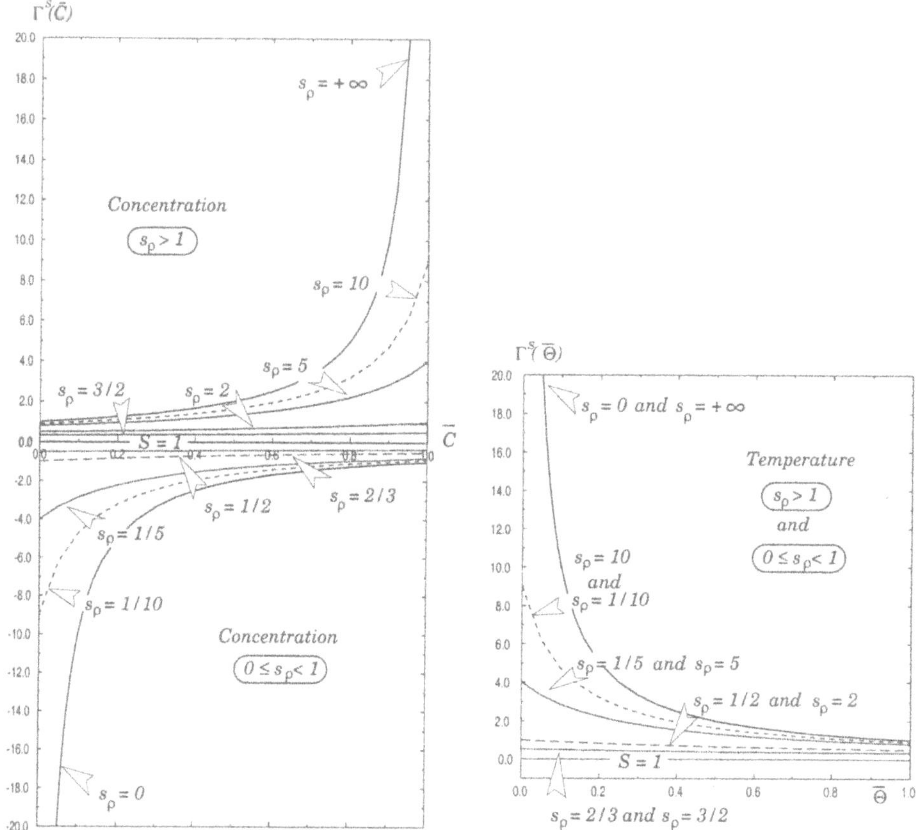


*Figure 4.1.* Indicator functions of density variance in variable concentration or temperature low-speed jets and mixing layers.

data of Blaisdell *et al.*[47], shows that the rms pressure level is a rapidly increasing function of the turbulence Mach number, since for $M_t = 1/2$ for instance $\sqrt{\overline{p'^2}}/\overline{P} \simeq 0.37$, and equals about 1.21 for $M_t = 1$. This last value is just given for sake of comparison since its validity should be questionable due to the emergence of non-isentropic processes (shocklets) that are to be expected for such a value of the turbulence Mach number.

***The strong Reynolds analogy*** We turn now to situations, namely wall-bounded flow configurations such as compressible turbulent boundary layers and channel flows, in which turbulence develops a very different behavior in the fluctuations of density and temperature than the previous one. As illustrated in the first chapter, compressibility effects are now mainly induced by mean temperature and density variations due to gradients normal to the wall. Contrasting with the compressible homogeneous shear flow, what is observed now is that pressure fluctuations are negligible compared to density and temperature fluctuations.

A useful clue in the diagnostic of wall-type compressible effects was introduced in 1961 by Morkovin [339] with the so-called strong Reynolds analogy (SRA). It will be briefly recalled here according to the original presentation of the author, based on the following mean momentum and total enthalpy balance equations in a turbulent boundary layer:

$$\overline{\rho \frac{D}{Dt}} \overline{U}_i = \frac{\partial}{\partial x_j}(\overline{\tau}_{ij} - \overline{\rho}\, \overline{u'_i u'_j}) + pressure\ terms\ ,$$

$$\overline{\rho \frac{D}{Dt}} \overline{H} = \frac{\partial}{\partial x_j}(\overline{q}_j - \overline{\rho}\, \overline{u'_j h'}) + viscous\ terms\ ,$$

where $H = C_p T + U_i U_i/2$ is the total enthalpy, $\tau_{ij}$ and $q_j$ the viscous stress tensor and the heat conduction flux respectively. For sake of clarity, all extra correlations due to density fluctuations, in addition to the main turbulent fluxes $\overline{\rho}\, \overline{u'_i u'_j}$ and $\overline{\rho}\, \overline{u'_j h'}$ respectively, have not been explicited in the right-hand-side of these equations.

When both pressure and viscous terms are discarded, the SRA assumption states[12] that the solutions in total enthalpy and velocity for these equations are linked, according to:

$$\overline{H} - \overline{H}_w = \frac{\overline{q}_w}{\overline{\tau}_w} \overline{U} \quad \text{and} \quad h' = \frac{\overline{q}_w}{\overline{\tau}_w} u'\ .$$

The most popular consequence of the SRA is obtained for adiabatic walls ($\overline{q}_w = 0$), since in this case the last condition reduces to $h' = 0$. Assuming that the fluctuation in total temperature is nearly zero and using the linearized expression:

$$C_p \theta' + \overline{U} u' \simeq 0\ ,$$

it results that:

$$\frac{\theta'}{\overline{T}} \simeq -(\gamma - 1) M^2 \frac{u'}{\overline{U}}\ ,$$

where $\gamma$ is the isentropic coefficient and $M$ the Mach number of the mean flow. Thus,

$$\frac{\sqrt{\overline{\theta'^2}}}{\overline{T}} \Big/ \frac{\sqrt{\overline{u'^2}}}{\overline{U}} \simeq (\gamma - 1) M^2 \quad \text{and} \quad \frac{\overline{u'\theta'}}{\sqrt{\overline{u'^2}}\sqrt{\overline{\theta'^2}}} = -1\ . \tag{4.31}$$

[12] Different formulations of the SRA assumptions can be adopted. For more details, see the discussion of this question in Spina et al. [445] for instance.

Differentiating the equation of state and neglecting pressure fluctuations yields $\theta'/\overline{T} = -\rho'/\overline{\rho}$, so that the previous relations can be extended to density fluctuations:

$$\frac{\rho'}{\overline{\rho}} \simeq (\gamma-1)M^2\frac{u'}{\overline{U}} \quad \text{and} \quad \frac{\sqrt{\overline{\rho'^2}}}{\overline{\rho}} / \frac{\sqrt{\overline{u'^2}}}{\overline{U}} \simeq (\gamma-1)M^2 \,,$$

which are often associated with the so-called Very strong Reynolds analogy (VSRA).

When the SRA was postulated by Morkovin in 1961, the experimental data available for comparison were rather scarce. Since then, the assumptions introduced to derive SRA and VSRA have been validated by experimental measurements in supersonic boundary layers over *adiabatic* walls, as reviewed for instance by Gaviglio [176] in 1976. In 1977, as quoted by Bradshaw [52], there was adequate evidence that in boundary layers at $M_\infty < 5$, the SRA relations are obeyed. In 1994, Lele [291] discussed the conditions necessary for the SRA relations to be used.

For boundary layers over *non-adiabatic* walls, Gaviglio [177] reported in 1987 that the SRA was not adequate. Several authors, Gaviglio [177], Elena & Gaviglio [145], Rubesin [398], Huang *et al.* [221] have independently derived a relationship between $u'$ and $\theta'$ as follows:

$$\frac{\theta'/\overline{T}}{(\gamma-1)M^2 u'/\overline{U}} \approx \frac{1}{c(\partial \overline{T}_T/\partial \overline{T} - 1)} \,,$$

where $\overline{T}_T$ denotes the mean total temperature.
Good agreement with measurements is reported by Gaviglio and Rubesin with $c = 1$ and $c = 1.34$ respectively. In Huang *et al.*, it is shown that $c$ is approximately equal to the turbulence Prandtl number.

It is generally considered that the limit of validity of Morkovin's hypothesis corresponds to $\sqrt{\overline{\rho'^2}}/\overline{\rho} \simeq 0.1$. Within this limit, the effects of density fluctuations are taken as practically negligible, and the turbulence structure is virtually the same as at low speed, incompressible flows. A roughly similar conclusion holds on the application of the SRA to free mixing layers, but the limit of validity is reached for a lower Mach number value ($M_\infty \simeq 1.5$), where $\sqrt{\overline{u'^2}}/\overline{U} \simeq 0.3$. The recent measurements of Barre *et al.* [30] in a mixing layer developing between an outer supersonic flow ($M_1 = 1.8$) and an inner subsonic flow ($M_2 = 0.3$) confirm the validity of the SRA in such a flow.

Sarkar [409] in 1995 and Simone *et al.* [431] in 1997 suggested that compressibility effects, distinct from the variation of mean density, should occur differently in boundary layers and mixing layers. As introduced by Sarkar [409], the key parameter is the gradient Mach number[13] $M_g$ defined with respect to the mean velocity gradient, a turbulent length scale and the local speed of sound, so that specific compressibility effects are expected when $M_g \simeq \mathcal{O}(1)$. Now, as opposed to the free mixing layer, the gradient Mach number is generally small in supersonic boundary layers. For the DNS data of a compressible boundary layer at $M_\infty = 2.25$ on an adiabatic wall reported by So *et al.* [437], the gradient Mach number is only $M_g = 0.17$, so that the stabilizing effect of compressibility on turbulence is not predominant and Morkovin's hypothesis can still be considered valid for such a flow.

### 4.4.3. THE DENSITY VARIANCE TRANSPORT EQUATION

To conclude this section, we shall discuss the transport equation for the variance of density fluctuations. A general transport equation for $\overline{\rho'^2}$ can be deduced from the continuity equation. From this equation, it results that the density *fluctuation* is governed by:

$$\frac{\partial \rho'}{\partial t} + \overline{U}_j \frac{\partial \rho'}{\partial x_j} + \rho' \frac{\partial \overline{U}_j}{\partial x_j} + \frac{\partial (\rho u'_j)}{\partial x_j} - \frac{\partial (\overline{\rho u'_j})}{\partial x_j} = 0 \, .$$

Multiplying by $\rho'$ and after some algebra, the following equation, as given for instance in Lele [292], is obtained:

$$\frac{\mathrm{d}\overline{\rho'^2}}{\mathrm{dt}} = -\underbrace{\frac{\partial (\overline{\rho'^2 u'_j})}{\partial x_j}}_{(a)} - \underbrace{2\overline{\rho'^2}\frac{\partial \overline{U}_j}{\partial x_j}}_{(b)} - \underbrace{\overline{\rho'^2 \frac{\partial u'_j}{\partial x_j}}}_{(c)} - \underbrace{2\overline{\rho' u'_j}\frac{\partial \overline{\rho}}{\partial x_j}}_{(d)} - \underbrace{2\overline{\rho}\,\overline{(\rho' \frac{\partial u'_j}{\partial x_j})}}_{(e)} \, . \quad (4.32)$$

According to eq.(4.32), the variance of the density fluctuations is:

- *diffused* by turbulent transport $(a)$;
- *intensified*[14] or *reduced* by the rate of change of the volume particle within the *mean*$(b)$ and *fluctuating* $(c)$ motions;
- *produced* or *destroyed* by the turbulent mass flux acting against the *mean* density gradient $(d)$, and by the density dilatation correlation term $(e)$.

Hence, some clues to the in-depth interpretation of the source terms of the density variance transport equation could emerge from a further analysis of the mechanisms governing (i) the mean and fluctuating dilatation, (ii) the

[13] When the flow is not of thin shear layer type, the deformation rate Mach number is to be considered instead of $M_g$, as proposed by Lele [292].

[14] Intensification results from compression

turbulent mass flux and (iii) the correlation between density and dilatation fluctuations. The first point is addressed in the next section.

## 4.5. Non-isovolume evolution

### 4.5.1. THE GENERAL EQUATION

As suggested by eq.(4.15), a rational way to identifying the various sources contributing to non-solenoidal fluid motions consists in detailing the origins of density variations due to changes in pressure, temperature and concentration. This approach has been already used independently by Chassaing [82] and Lele [292] and will be reformulated here.

Indeed, taking the material logarithmic derivative of eq.(4.12) and substituting in eq.(4.15) yields:

$$\mathrm{div}\vec{U} = -\frac{1}{P}\frac{\mathrm{d}P}{\mathrm{dt}} + \frac{1}{T}\frac{\mathrm{d}T}{\mathrm{dt}} + \frac{a^*}{a^*C + b^*}\frac{\mathrm{d}C}{\mathrm{dt}} \,. \tag{4.33}$$

The right-hand-side of this equation can be simplified by substitution from the mass-fraction equation eq.(4.8), the energy equation eq.(4.10) and the equation of state eq.(4.12).

By combining the equation of state and the mass-fraction balance equation it is obtained:

$$\frac{a^*}{a^*C + b^*}\frac{\mathrm{d}C}{\mathrm{dt}} \equiv \frac{a^*T}{P} \times \rho\frac{\mathrm{d}C}{\mathrm{dt}} = \frac{a^*T}{P}\frac{\partial}{\partial x_j}(\rho\mathcal{D}\frac{\partial C}{\partial x_j}) \,. \tag{4.34}$$

Turning now to the enthalpy equation (4.10) and developing the left-hand-side, it comes, after dividing by $\rho C_p T \equiv \gamma P/(\gamma - 1)$:

$$\frac{1}{C_p}\frac{\mathrm{d}C_p}{\mathrm{dt}} + \frac{1}{T}\frac{\mathrm{d}T}{\mathrm{dt}} = \frac{\gamma-1}{\gamma}\frac{1}{P}\frac{\mathrm{d}P}{\mathrm{dt}} + \frac{\gamma-1}{\gamma P}\Phi_\nu + \frac{\gamma-1}{\gamma P}\frac{\partial}{\partial x_j}(\lambda\frac{\partial T}{\partial x_j}) \,. \tag{4.35}$$

Finally, substitution of eqs.(4.34) and (4.35) into (4.33) yields:

$$\mathrm{div}\vec{U} = \underbrace{-\frac{1}{\gamma P}\frac{\mathrm{d}P}{\mathrm{dt}}}_{(i)} \underbrace{-\frac{1}{C_p}\frac{\mathrm{d}C_p}{\mathrm{dt}}}_{(ii)} + \underbrace{\frac{\gamma-1}{\gamma P}\Phi_\nu}_{(iii)} + \underbrace{\frac{\gamma-1}{\gamma P}\frac{\partial}{\partial x_j}(\lambda\frac{\partial T}{\partial x_j})}_{(iv)} + \underbrace{\frac{a^*T}{P}\frac{\partial}{\partial x_j}(\rho\mathcal{D}\frac{\partial C}{\partial x_j})}_{(v)} \,. \tag{4.36}$$

Under the assumptions of the present study (considering in particular, non reactive mixture of perfect gases), this equation gives the general expression for the rate of change in an elementary volume of a "fluid particle" during

its *instantaneous* motion along its trajectory.
Basically, eq.(4.36) shows that any change in such an elementary volume results from four distinct sources: (i) pressure, (ii) changes in the specific heat at constant pressure, (iii) viscous dissipation, and molecular diffusion, including both heat (iv) and mass (v) transport. Since all terms scale as the inverse of time, eq.(4.36) can be regarded as a general balance equation in frequencies. The analysis of such frequencies is the question to be discussed hereafter.

An alternative form to eq.(4.36) can be obtained by introducing the dissipations $\Phi_\theta$ and $\Phi_\gamma$ corresponding to heat and mass transfer:

$$\Phi_\theta = \frac{\lambda}{\rho C_v}\left(\frac{\partial T}{\partial x_j}\right)^2 \quad \text{and} \quad \Phi_\gamma = \mathcal{D}\left(\frac{\partial C}{\partial x_j}\right)^2 .$$

The introduction of $\Phi_\theta$ is straightforward, recalling that $(\gamma-1)/(\gamma P) = (\rho C_p T)^{-1}$, so that:

$$\frac{\gamma-1}{\gamma P}\frac{\partial}{\partial x_j}(\lambda\frac{\partial T}{\partial x_j}) = \frac{1}{\rho C_p}\frac{\partial}{\partial x_j}(\frac{\lambda}{T}\frac{\partial T}{\partial x_j}) + \frac{\Phi_\theta}{\gamma T^2} .$$

Similarly, for the mass fraction, it can be first verified that:

$$\frac{a^*T}{P}\frac{\partial}{\partial x_j}(\rho\mathcal{D}\frac{\partial C}{\partial x_j}) = \frac{a^*T\,C}{P}\frac{\partial}{\partial x_j}(\frac{\rho\mathcal{D}}{C}\frac{\partial C}{\partial x_j}) + \frac{a^*T\,C}{P}\frac{\rho\mathcal{D}}{C^2}\left(\frac{\partial C}{\partial x_j}\right)^2 .$$

Now, from eq.(4.12), it is clear that:

$$\frac{a^*T\,C}{P} = \frac{1}{\rho} - \frac{1}{\rho_\beta} ,$$

where $\rho_\beta$ is the density of the pure species $\beta$.
Hence:

$$\frac{a^*T}{P}\frac{\partial}{\partial x_j}(\rho\mathcal{D}\frac{\partial C}{\partial x_j}) = \frac{1}{\rho}(1-\frac{\rho}{\rho_\beta})\frac{\partial}{\partial x_j}(\frac{\rho\mathcal{D}}{C}\frac{\partial C}{\partial x_j}) + (1-\frac{\rho}{\rho_\beta})\frac{\Phi_\gamma}{C^2} .$$

After substitution, the equivalent expression of eq.(4.36) is:

$$\begin{aligned}\mathrm{div}\vec{U} = -\frac{1}{\gamma P}\frac{\mathrm{d}P}{\mathrm{dt}} - \frac{1}{C_p}\frac{\mathrm{d}C_p}{\mathrm{dt}} + \frac{1}{\rho C_p}\frac{\partial}{\partial x_j}(\frac{\lambda}{T}\frac{\partial T}{\partial x_j}) + \frac{1}{\rho}(1-\frac{\rho}{\rho_\beta})\frac{\partial}{\partial x_j}(\frac{\rho\mathcal{D}}{C}\frac{\partial C}{\partial x_j}) \\ \frac{\gamma-1}{\gamma P}\Phi_\nu + \frac{\Phi_\theta}{\gamma T^2} + (1-\frac{\rho}{\rho_\beta})\frac{\Phi_\gamma}{C^2} .\end{aligned} \quad (4.37)$$

where the mechanical, thermal and mass dissipation terms are explicitly present.

### 4.5.2. DISCUSSION

To detail the time scales of the main source terms which appear in the expression of the divergence of the *instantaneous* velocity field, eq.(4.36), the pressure, dissipation, heat and mass diffusion terms will be considered separately.

***Pressure effect:*** When scaling any velocity component as $U_i \sim \mathcal{O}(U_0)$ $(i=1,2,3)$ and the spatial gradient as $\partial/\partial x_i \sim \mathcal{O}(1/L_0)$, the characteristic time scale of the rate of change in an elementary volume is simply:

$$\mathcal{T}_0 = L_0/U_0 \ .$$

Based on the corresponding expression in eq.(4.36), the time scale of the pressure term can be obtained as:

$$\mathcal{T}_1 = \frac{\gamma P}{|\mathrm{d}P/\mathrm{dt}|} \sim \frac{\rho_0 a_0^2}{P_{ref}} \mathcal{T}_p \ .$$

In the previous expression, $\gamma P$ is scaled as $\gamma P \sim \mathcal{O}(\rho_0 a_0^2)$, where $a_0$ denotes a given speed of sound in a fluid which density is $\rho_0$, while the material derivative of the pressure is scaled as $P_{ref}/\mathcal{T}_p$.
Thus, the characteristic time scale associated with the pressure term in eq.(4.36) can be obtained from a characteristic time scale of the pressure variation *and* a characteristic pressure scaling. As we shall see later on, there exist different methods in the scaling of the pressure.
Let us just give here two examples:

- *advective* scaling, for which the *advective* time scale is adapted to the pressure variation, i.e., $\mathcal{T}_p \sim L_0/U_0 \sim \mathcal{T}_0$, along with a dynamical scaling of the pressure, $P_{ref} \sim \rho_0 U_0^2$. Then:

$$\mathcal{T}_1 \sim \frac{1}{M_0^2} \mathcal{T}_0 \ ,$$

  where $M_0 = U_0/a_0$ is a given Mach number of the flow. This expression shows that the compressible effects on the rate of change in an elementary volume are $\mathcal{O}(M_0^2)$ during the advection of the fluid particle, a result which is relevant to low-speed motions;
- *acoustic* scaling, where a suitable estimation for the time variation of the pressure over the distance $l_a$ is $\mathcal{T}_p \sim l_a/a_0$. Then, with the same estimation for the pressure reference as before ($P_{ref} \sim \rho_0 U_0^2$):

$$\mathcal{T}_1 = \frac{1}{M_0} \frac{l_a}{L_0} \mathcal{T}_0 \ .$$

In this case, pressure effects can be present, even at moderate Mach numbers, in specific zones of the flow field, depending upon the local value of the length scales ratio $l_a/L_0$. In other words, the Mach number is not the *only one* parameter to be considered in such compressible flows, a result which was anticipated in the previous chapter when discussing incompressible limits to the general compressible equations as derived by Zank & Matthaeus [494], and Bayly *et al.* [40] for instance.

***Dissipation effect:*** The dissipation term (*iii*) in eq.(4.36) can be treated in the same way, yielding for instance:

$$\mathcal{T}_3 = \frac{\gamma P}{\gamma - 1}\frac{1}{\Phi_\nu} \sim \frac{\rho_0 a_0^2}{\gamma - 1}\frac{\mathcal{T}_\nu^2}{\rho_0 \nu_0} = \frac{1}{\gamma - 1}\frac{a_0^2}{\nu_0}\mathcal{T}_\nu^2 ,$$

where $\mathcal{T}_\nu$ denotes a charateristic time scale of the viscous dissipation.
The direct estimate $\mathcal{T}_\nu = \mathcal{T}_0 = L_o/U_o$ is not adopted here since, in the turbulent regime, a different scaling is to be considered for "mean" and "turbulent" contributions to the total mean dissipation [459].
Thus, without detailing, a general expression of the dissipation time scale can be simply taken as $\mathcal{T}_\nu = l_\nu/u^*$ where $l_\nu$ and $u^*$ are two appropriate length and velocity scales. Hence:

$$\mathcal{T}_3 = \frac{1}{\gamma - 1}\frac{1}{M^{*2}}\mathcal{T}_m , \tag{4.38}$$

where $\mathcal{T}_m = l_\nu^2/\nu_0$ is the characteristic time scale of the *molecular* (viscous) diffusion of momentum over the distance $l_\nu$, and $M^* = u^*/a_0$ a specific *turbulence* Mach number which differs from the previous "global" Mach number $U_0/a_0$ considered so far.
When applied to the *mean* dissipation contribution, with obviously $u^* \sim U_0$ and $M^* \sim M_0$, eq.(4.38) gives:

$$\mathcal{T}_3 = \frac{1}{\gamma - 1}\frac{1}{M_0^2}\mathcal{T}_m \equiv \frac{1}{\gamma - 1}\frac{1}{M_0^2}\left(\frac{l_\nu}{L_0}\right)^2 \times R_{e0} \times \mathcal{T}_0 , \tag{4.39}$$

where $R_{e0} = U_0 L_0/\nu_0$ is the Reynolds number based on the advective length scale $L_0$.
Here again, *mean* volume variation of a fluid particle can be driven *locally* in the flow field by *mean* viscous dissipation. For instance, a time scale balance $\mathcal{T}_3 \sim \mathcal{T}_0$ can be expected in those regions of the flow field where

$$l_\nu/L_0 \sim M_0/\sqrt{R_{e0}} .$$

When applied to the actual values of the Reynolds and Mach encountered in most usual applications, this estimate shows that such regions are not practically significant, excepting near wall zones (with, $R_{e0} = 10^6$, $l_\nu/L_0 = 10^{-3}$ and $10^{-2}$ for $M_0 = 1$ and 10 respectively).

***Heat diffusion:*** The time scale $\mathcal{T}_4$ linked with the heat diffusion term $(iv)$ in eq.(4.36) is given by:

$$\frac{1}{\mathcal{T}_4} = \frac{\gamma - 1}{\gamma P} \frac{\partial}{\partial x_j}(\lambda \frac{\partial T}{\partial x_j}) \equiv \frac{1}{\rho C_p T} \frac{\partial}{\partial x_j}(\lambda \frac{\partial T}{\partial x_j}) \sim \frac{\lambda_0}{\rho_0 C_{p0}} \times \frac{1}{l_\theta^2} ,$$

where $l_\theta$ is the characteristic length scale of heat transfer by molecular diffusion. Adopting the same length scale for all, heat, mass and momentum molecular effects, (i.e., here $l_\theta \sim l_\nu$), the fourth time scale is obviously:

$$\mathcal{T}_4 = \frac{l_\nu^2}{k_0} \equiv P_{r0} \times \mathcal{T}_m = P_{r0} \times R_{e0} \times \left(\frac{l_\nu}{L_0}\right)^2 \times \mathcal{T}_0 , \qquad (4.40)$$

where $k_0 = \lambda_0/(\rho_0 C_{p0})$ is the thermal diffusivity and $P_{r0} = \nu_0/k_0$ the reference value of the Prandtl number.
This relation shows that molecular diffusion of heat does not contribute significantly to the volume variation of a fluid particle, except in those regions of the flow field where $l_\nu/L_0 \sim (R_{e0} \times P_{r0})^{-1/2}$.

***Mass diffusion:*** To find out the expression of the characteristic time scale linked with the mass diffusion term $(v)$ in eq.(4.36), the scaling of the coefficient $a^*$ is to be addressed first. As shown by its definition — see section (4.1.3) —, $a^*$ is a constant, the value of which only depends on the molecular weights of the pure species. Accordingly, it will be scaled as $b \sim \mathcal{O}(P_0/(T_0 \rho_1))$ where $\rho_1$ denotes a *constant* density, different from the density scaling ($\rho_0$) used so far to estimate density dependent quantities which are *variable* within the flow field. Hence:

$$\frac{1}{\mathcal{T}_5} = \frac{a^* T}{P} \frac{\partial}{\partial x_j}(\rho \mathcal{D} \frac{\partial C}{\partial x_j}) \sim \frac{P_0}{T_0 \rho_1} \frac{T_0}{P_0} \frac{1}{l_m^2} \rho_0 \mathcal{D} .$$

Again, when adopting a single characteristic length scale for molecular transfer ($l_m = l_\nu$), the fifth time scale is:

$$\mathcal{T}_5 = s_\rho \times S_{c0} \times \mathcal{T}_m \equiv s_\rho \times S_{c0} \times R_{e0} \times \left(\frac{l_\nu}{L_0}\right)^2 \times \mathcal{T}_0 , \qquad (4.41)$$

where $s_\rho = \rho_1/\rho_0$ denotes a characteristic density ratio and $S_{c0} = \nu_0/\mathcal{D}_0$ is the reference Schmidt number.
Thus, the same conclusion as before for the thermal diffusion can be driven, the only difference coming from the presence of an additional independent parameter: the density ratio $s_\rho$.

## 4.6. Solenoidal condition in variable density turbulent flows

As far as turbulent flows are concerned, the previous analysis can be considered as applying to the "instantaneous" motion. As usual in turbulence, a statistical approach is introduced which leads to define some kind of "mean motion". The departure between the instantaneous and mean values is then attributed to the "fluctuating motion". The specific consequences of the density variation on the general averaging procedure will be addressed in more details in Chapter 5, and we are merely concerned in this section with pointing out some consequences of such an instantaneous analysis of the solenoidal condition when transposed to both mean and fluctuating motions.

Thus, for the present discussion, there is no need to go beyond a formal decomposition in mean and fluctuating parts, as simply given by:

$$U_i = \overline{U}_i + u'_i\,,\ P = \overline{P} + p',\ T = \overline{T} + \theta',\ C = \overline{C} + \gamma' \text{ and } \rho = \overline{\rho} + \rho'\ , \quad (4.42)$$

where an overbar denotes a mean value referring to the *same* statistical or ensemble average as the one used in *constant density* fluid motions. Consequently, all fluctuations are centered, so that:

$$\overline{u'}_i = \overline{p'} = \overline{\theta'} = \overline{\gamma'} = \overline{\rho'} = 0\ .$$

Hence, the first point to be discussed is concerned with the linkage between the divergence of mean and fluctuating velocity fields.

### 4.6.1. MEAN AND FLUCTUATING DIVERGENCE LINKAGE

*Divergence of the MEAN motion:* by averaging the continuity equation, it is obtained:

$$\frac{\partial \overline{U}_i}{\partial x_i} = \underbrace{-\frac{1}{\overline{\rho}}\left(\frac{\partial \overline{\rho}}{\partial t} + \overline{U}_i \frac{\partial \overline{\rho}}{\partial x_i}\right)}_{(i)} \underbrace{-\frac{1}{\overline{\rho}}\frac{\partial (\overline{\rho' u'_i})}{\partial x_i}}_{(ii)}\ . \quad (4.43)$$

Thus, the source terms which can make the *mean* motion to depart from the solenoidal condition originate from :

- the *mean motion*, due to the material rate of variation of the mean density along the mean flow: $\overline{\rho}^{-1} D\overline{\rho}/Dt$, with $D/Dt \equiv \partial/\partial t + \overline{U}_i \partial/\partial x_i$;
- the turbulent *fluctuating motion*, due to the turbulent mass flux $\overline{\rho' u'_i} \equiv \overline{\rho u'_i}$. This term is specific to variable density fluctuating motions[15].

[15] It should be emphasized that the presence of the turbulent mass flux in the present discussion is not dependent on the formalism which is adopted. The turbulent mass flux

The ratio of any given component of the turbulent mass flux, to the corresponding one of the mean momentum, can be estimated as:

$$\frac{\overline{\rho' u'_i}}{\overline{\rho}\,\overline{U}_i} \sim C \times I_\rho \times I_u \quad (\textit{no summation with respect to } i) \; ,$$

where $I_\rho = \sqrt{\overline{\rho'^2}}/\overline{\rho}$ and $I_u = \sqrt{\overline{u'^2}}/\overline{U}$ are the turbulent density and velocity intensities respectively, and $C$, $|C| \leq 1$, a correlation coefficient.

As shown from the data gathered in Table 4.5, and since the turbulent velocity intensity does not generally exceed 20% to 25% in most common flows (often much lower), it is generally concluded that the turbulent mass flux is ten times lower than the mean momentum.[16]

Thus the analysis of the source terms driving a non-solenoidal condition for the *mean* velocity field can be developed as before, based on a simplified form of eq.(4.43) where the second term (ii) is discarded. Hence from the discussion of the instantaneous situation, it can be inferred that two different classes of *mean variable density* flows can be considered, corresponding to:

- *quasi-isovolume mean* evolution, when considering low speed flows with low mean pressure variations and high turbulence Reynolds/Peclet numbers. In this case, the departure from the mean solenoidal condition only results in molecular diffusion effects which can be taken as negligible as compared with turbulent diffusion;
- *non-isovolume mean* evolution, as a consequence of mean pressure effects or mean temperature effects driven by viscous dissipation. This latter situation is only concerned with high Mach number flows. The former can be present either in high speed flows (supersonic flow over a compression ramp or in compression corners for instance) or low speed flows as in combustion chambers of reciprocating engines, where the

can just be introduced explicitly or implicitly into the expressions. When using density-weighted averages — the so-called Favre's averages —, this term is actually present, but appears implicitly in the equations, as part of the mass-weighted mean value of the velocity. With the notations detailed in the following chapter, the density-weighted decomposition of the velocity simply reads $U_i = \tilde{U}_i + u''_i$ from which it results that:

$$\frac{\partial \overline{U}_i}{\partial x_i} = \frac{\partial \tilde{U}_i}{\partial x_i} + \frac{\partial \overline{u''_i}}{\partial x_i} \, ,$$

since $u''_i$ is no longer a *centered* fluctuation. Thus, the divergence of the "*usual*" mean velocity requires $\overline{u''}_i$ which is directly linked with the turbulent mass flux of the present analysis, as we shall see in Chapter 5.

[16] This "global" conclusion must be adapted when applied to *2-D*, thin shear layer flows due to the distorted scaling for mean velocity components, $\overline{U}$ along the longitudinal or streamwise direction of advection $\gg \overline{V}$ along the transversal or spanwise direction of diffusion.

mean Mach number, based on the maximum piston speed is about 0.02 at 5,000 rpm (Cambon & Simone [68]).

*Divergence of the FLUCTUATING motion:* by introducing decompositions (4.42) into the continuity equation, it becomes, after some simple rearrangement:

$$\overline{\rho}\frac{\partial u_i'}{\partial x_i} = -\overline{\rho}\frac{\partial \overline{U}_i}{\partial x_i} - \rho'\frac{\partial \overline{U}_i}{\partial x_i} - \frac{D\overline{\rho}}{Dt} - \frac{D\rho'}{Dt} - u_i'\frac{\partial \overline{\rho}}{\partial x_i} - \frac{\partial(\rho' u_i')}{\partial x_i} .$$

After subtracting the mean continuity equation, the following equation is obtained:

$$\overline{\rho}\frac{\partial u_i'}{\partial x_i} = \underbrace{-\rho'\frac{\partial \overline{U}_i}{\partial x_i} - u_i'\frac{\partial \overline{\rho}}{\partial x_i} - \frac{D\rho'}{Dt}}_{(i)} \underbrace{- \frac{\partial \rho' u_i'}{\partial x_i} + \frac{\partial \overline{\rho' u'_i}}{\partial x_i}}_{(ii)} . \qquad (4.44)$$

Thus the non-solenoidal condition for the fluctuating motion results from two types of contributions which group terms that are $(i)$ linear in density and velocity fluctuations and $(ii)$ quadratic, associated with the turbulent mass flux.

### 4.6.2. APPROXIMATION TO THE FLUCTUATING DIVERGENCE

The question is now the identification of the source terms in the expression of $\partial u_i'/\partial x_i$. Due to the complexity resulting from the non-linearities in eq.(4.36), only a linearized approximation to this equation will be used in the present discussion. It is based on the alternative expression (4.37), when discarding $C_p$ variations and molecular diffusion terms and assuming constant physical properties of the fluid.
It formally reads to first order fluctuations:

$$\frac{\partial u_i'}{\partial x_i} \simeq \pi' + D_\nu' + D_\theta' + D_\gamma' , \qquad (4.45)$$

where $\pi'$ stands for the pressure term and $D_\nu'$, $D_\theta'$, $D_\gamma'$ for the mechanical, thermal and mass dissipation terms in eq.(4.37) respectively.
Considering the pressure term $\pi = (\gamma P)^{-1}\mathrm{d}P/\mathrm{dt}$, the first order contribution in pressure fluctuations can be taken as:

$$\pi' \equiv \pi - \overline{\pi} \simeq \frac{1}{\gamma\overline{P}}\frac{\mathrm{d}p'}{\mathrm{dt}} - \frac{p'}{\gamma\overline{P}^2}\frac{\mathrm{d}\overline{P}}{\mathrm{dt}} .$$

It should be noticed that without further approximations, no linearization has been introduced in the advective operator which, thus, still refers to

the instantaneous velocity, $d/dt = \partial/\partial t + U_i \partial/\partial x_i$.
With the same kind of approximations, it comes out, for the contribution of the viscous dissipation fluctuation:

$$D'_\nu \equiv \frac{\gamma-1}{\gamma}\left(\frac{\Phi_\nu}{P}\right)' \simeq \frac{\gamma-1}{\gamma}\left(\frac{\phi'_\nu}{\overline{P}} - \frac{\overline{\rho}(\overline{\phi}_\nu+\overline{\epsilon})}{\overline{P}} \times \frac{p'}{\overline{P}}\right),$$

where $\Phi_\nu \simeq \overline{\rho}\nu(S_{ij}S_{ij} - \frac{2}{3}S\delta_{ij})$ so that: $\overline{\Phi}_\nu \simeq \overline{\rho}(\overline{\phi}_\nu + \overline{\epsilon})$,
with $\overline{\phi}_\nu = \nu(\overline{S}_{ij}\overline{S}_{ij} - \frac{2}{3}\overline{S}\delta_{ij})$ , $\overline{\epsilon} = \nu(\overline{s'_{ij}s'_{ij}} - \frac{2}{3}\overline{\vartheta'^2}\delta_{ij})$
and $\phi'_\nu = \Phi_\nu - \overline{\Phi}_\nu$ .

A similar treatment of the thermal and mass dissipation terms in eq.(4.37) yields:

$$D'_\theta \equiv \frac{1}{\gamma}\left(\frac{\Phi_\theta}{T^2}\right)' \simeq \frac{\phi'_\theta}{\gamma\overline{T}^2} - 2\frac{\theta'}{\gamma\overline{T}}\frac{\overline{\phi}_\theta + \overline{\epsilon}_\theta}{\overline{T}^2},$$

$$D'_\gamma \equiv \left[(1-\frac{\rho}{\rho_\beta})\frac{\Phi_\gamma}{C^2}\right]' \simeq -\frac{\overline{\rho}}{\rho_\beta}\frac{\phi'_\gamma}{\overline{C}^2} - \left(2\frac{\gamma'}{\overline{C}}\frac{\overline{\rho}}{\rho_\beta} + \frac{\rho'}{\rho_\beta}\right)\frac{\overline{\phi}_\gamma + \overline{\epsilon}_\gamma}{\overline{C}^2},$$

where the thermal and mass dissipations of the fluctuating motion are defined as:

$$\overline{\epsilon}_\theta = \frac{\lambda}{\overline{\rho}C_v}\overline{\left(\frac{\partial\theta'}{\partial x_i}\right)^2} \quad \text{and} \quad \overline{\epsilon}_\gamma = \mathcal{D}\overline{\left(\frac{\partial\gamma'}{\partial x_i}\right)^2}.$$

Hence, after substitution of these various expressions in eq.(4.45) one obtains:

$$\begin{aligned}\frac{\partial u'_i}{\partial x_i} \simeq & -\frac{1}{\gamma\overline{P}}\frac{dp'}{dt} - \frac{p'}{\gamma\overline{P}^2}\frac{d\overline{P}}{dt} \\ & + \frac{\gamma-1}{\gamma}\frac{\phi'_\nu}{\overline{P}} + \frac{\phi'_\theta}{\gamma\overline{T}^2} - \frac{\overline{\rho}}{\rho_\beta}\frac{\phi'_\gamma}{\overline{C}^2} \\ & -\frac{\gamma-1}{\gamma}\frac{\overline{\rho}}{\overline{P}}(\overline{\phi}_\nu+\overline{\epsilon})\frac{p'}{\overline{P}} - \frac{2}{\gamma\overline{T}^2}(\overline{\phi}_\theta+\overline{\epsilon}_\theta)\frac{\theta'}{\gamma\overline{T}} - \frac{\overline{\phi}_\gamma+\overline{\epsilon}_\gamma}{\overline{C}^2}\left(2\frac{\overline{\rho}}{\rho_\beta}\frac{\gamma'}{\overline{C}} + \frac{\rho'}{\rho_\beta}\right) . \end{aligned} \quad (4.46)$$

***Discussion*** A two-fold series of remarks can be drawn from this last equation:

- dilatation fluctuations come from both *pressure* and *dissipative* effects. It should be recalled that, from the beginning, diffusion effects have been discarded from the present analysis which only applies to homogeneous situations or high turbulence Reynolds number flows;

- when used to estimate dilatation fluctuation correlations, i.e., terms like $\overline{f'\vartheta'}$, where $\vartheta' = \partial u'_i/\partial x_i$, this last expression shows that three types of correlations are involved, including (i) material derivatives — pressure terms, first line in the RHS of eq.(4.46) —, (ii) spatial derivatives, as part of the fluctuating dissipative terms in the second line, and (iii) second order moments, such as $\overline{f'p'}$, $\overline{f'\rho'}$... Based on the arguments introduced by Tennekes & Lumley [459] to estimate orders of magnitude in the incompressible moments equations, the classical result for high turbulence Reynolds number flows $\overline{\epsilon} \ll \overline{\phi}_\nu$ can be easily extended to all dissipation terms (iii). Thus it can be inferred that the dominant contribution in the third type of terms in eq.(4.46) comes from the dissipation terms of the fluctuating motion, $\overline{\epsilon}$, $\overline{\epsilon}_\theta$ and $\overline{\epsilon}_\gamma$.

Simplifications to eq.(4.46) have been proposed by Simone *et al.* [431] and Ristorcelli [394] for instance. When density fluctuations are much smaller than the mean density, and omitting all viscous and dissipative terms, the RHS terms in eq.(4.46) reduce to the first one (Simone *et al.* [431]). Discarding viscous effects, and focusing on the effects of compressibility on the vortical mode of the flow, not on the acoustic propagation problem, Ristorcelli [394] succeeded in 1997, in producing the following *diagnostic* constitutive relationship for the fluctuating dilatation:

$$-\gamma\frac{\partial u'_i}{\partial x_i} = \frac{\partial p'^{(1)}}{\partial t} + v_i'^{(1)}\frac{\partial p'^{(1)}}{\partial x_i} \,. \tag{4.47}$$

This results in a small perturbation procedure based on expansions from the incompressible situation of the form:

$$\begin{aligned} p' &= \epsilon^2\left[p'^{(1)} + \epsilon^2 p'^{(2)} + \ldots\right. \\ \rho' &= \epsilon^2\left[\rho'^{(1)} + \epsilon^2 \rho'^{(2)} + \ldots\right. \\ u'_i &= v_i'^{(1)} + \epsilon^2\left[w_i'^{(1)} + \epsilon^2 w_i'^{(2)} + \ldots\right. , \end{aligned}$$

where $p'$ and $\rho'$ denotes the non-dimensional pressure and density perturbations about a reference state $(P_\infty, \rho_\infty)$, i.e., $P = P_\infty(1+p')$ and similarly for $\rho$. The small parameter $\epsilon = \gamma^{1/2}M_t$ is related to the turbulence Mach number $M_t = \overline{u'_i u'_i}/(3a_\infty)$ with $a_\infty^2 = \gamma P_\infty/\rho_\infty$.

Several assumptions are used to derive eq.(4.47). As sketched in Fig. 4.2, it is assumed that the characteristic length scale $\ell$ of a turbulent eddy, which produces the pressure and density fluctuations in the medium, is small with respect to the length scale $\lambda$ associated with the propagation of pressure and density fluctuations ('*compact-source*' assumption). This assumption is a direct consequence of the time-scale equilibrium between acoustic and turbulence frequencies ($\hat{u}/\ell \sim a/\lambda$) along with the low turbulence Mach number condition, since in this case $\ell \ll \lambda \sim \ell/M_t$, letting $\hat{u} \equiv \overline{u'_i u'_i}/3$.

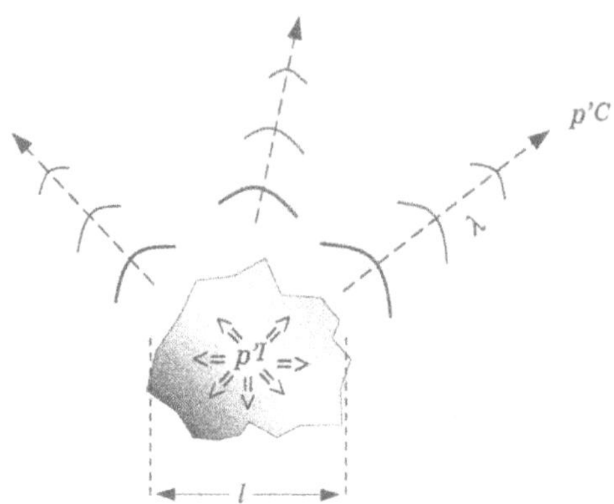


*Figure 4.2.* Characteristic scaling of the compact-source assumption according to Ristorcelli analysis.

At last, the underlying singular perturbation idea of the method is that, related to these two length scales, there exist two regions in the flow field:

- the *inner* region of size $\ell$ in which the major portion of the pressure field is associated with the vortical motions. In this region, the "incompressible" part of the pressure ($p'^I$ in the figure) dominates and the compressible part is felt instantaneously;
- the *outer* region, of size $\lambda$, in which the much smaller propagating acoustic pressure ($p'^C$) is the major component of the pressure field.

According to the pseudo-sound analysis of Ristorcelli, the zeroth-order equations show that the density and pressure fluctuations are linked by $p'^{(1)} = \gamma \rho^{(1)}$. The first-order inner expansion of the continuity equation yields eq.(4.47), which states that the dilatation fluctuation is diagnostically related to the rate of change of the incompressible fluctuating pressure following the fluctuating motion of a fluid particle (material derivative of the incompressible fluctuating pressure within the fluctuating motion).

## 4.7. Variable density situations in turbulent fluid motions

In order to identify different variable density effects in turbulent flows, it is worth recalling that two main issues are to be addressed. As shown just beneath the surface of the discussions throughout the previous sections, they concern:

- the pressure scaling,
- the time-scale estimation.

### 4.7.1. NORMALIZATION OF THE PRESSURE

From a general point of view, one can consider that there exist at least three normalizations of the pressure:

$$(a)\ P^* = \frac{P}{\rho_0 U_0^2}\,, \qquad (b)\ P^+ = \frac{L}{\mu U_0} P\,, \qquad (c)\ P^\dagger = \frac{P}{\rho v^2}\,. \qquad (4.48)$$

– The first one eq.(4.48a) refers to the *dynamical* estimation which is used in incompressible fluid motion, where the density $\rho_0$ is constant and $U_0$ stands for a characteristic velocity;
– The second expression correponds to a *viscous* estimation which is **globally** pertinent for creeping motions, or **locally** suitable near solid boundaries in high Reynolds number flows, since $P^+ = R_{e0} P^*$, where $R_{e0} = U_0 L/\nu_0$ is a given Reynolds number;
– The last one — eq.(4.48c)— can be considered as a *thermodynamic* estimation. Depending upon the definition of the characteristic velocity $v$, it can be viewed as either a "*thermal*" or a "*compressible*" normalization. The first definition corresponds for instance to $v^2 = C_p \Delta T$ where $\Delta T$ stands for a characteristic estimate of the global temperature variation. It is the choice adopted by Bayly *et al.* [40] when deriving models for weakly compressible flows (see previous Chapter).
When the characteristic velocity is taken as $\rho v^2 = \rho_0 a_0^2/\gamma$, where the speed of sound in a given reference state is $a_0^2 = \gamma P_0/\rho_0$, eq.(4.48c) yields a direct estimation of the compressible/incompressible normalizations, since $P^\dagger/P^* = \gamma M_0^2$, where $M_0 = U_0/a_0$ is a given Mach number.

In addition to this review of pressure normalizations, it should be added that:

- A given normalization can be adopted throughout the flow field, or restricted to local regions, as depicted in Ristorcelli [394], for instance;
- The previous normalizations are not necessarily suitable to estimating pressure *variations* or *fluctuations*. This led for instance Sarkar [409] to introduce an hybrid scaling, where the "incompressible" normalization $\rho_0 U_0^2$ is used for the instantaneous pressure scaling but not applied to its *time* variation, where the acoustic time-scale normalization $l/a_0$ is adopted.

### 4.7.2. TYPICAL TIME-SCALES RATIOS

As stressed in Chassaing [83] for instance, any transport equation in fluid mechanics eventually gives rise to a time-scale equilibrium analysis. Obviously, velocity gradients are the first candidate to the introduction of such characteristic time-scales or frequencies, but several other terms are to be considered to this extent in variable density fluid motions. Some of them, encountered in the previous sections which are relevant for the present discussion are listed in Table 4.6.

TABLE 4.6. Typical time-scales in variable density fluid turbulence.

| Appellation | Symbol | *Expression* |
|---|---|---|
| Mean deformation or distortion or shear time-scale | $\tau_D, \tau_S$ | $[\lvert\partial\overline{U}_i/\partial x_j\rvert_{max}]^{-1}$, $[\lvert S_{ij}\rvert_{max}]^{-1}$, $\lvert\frac{\partial\overline{U}}{\partial y}\rvert^{-1}$ |
| Energy containing eddy turn over time or turbulence time decay | $\tau_t$ | $\ell/\overline{k}^{1/2}$ |
| Energy containing eddy acoustic time-scale | $\tau_a$ | $\ell/a$ |

In order to find out which time-scales are dominant in any given variable density configuration, one classically introduces significant time-scales ratios. Such typical ratios are now reviewed, as an extension to the presentations nicely reported in Simone *et al.* [431] and Freund *et al.* [163].

***Dilatation/deformation ratio*** When the mean flow is neither kinematically homogeneous ($\partial\overline{U}_i/\partial x_j \neq 0$) nor isovolume, the mean velocity gradient tensor provides two different time-scalings, associated with (i) the mean dilatation ($|\partial\overline{U}_i/\partial x_i|$ and (ii) the most important mean deformation rate, $\max|\partial\overline{U}_i/\partial x_j|$ for $i=1,2,3$ and $j=1,2,3$. In shear dominated flows, the latter is associated with the dominant mean velocity gradient which reduces simply to $\partial\overline{U}_1/\partial x_2 \equiv \partial\overline{U}/\partial y$ in plane, thin, shear layers. The comparison of both time scales introduces the mean dilatation/deformation ratio:

$$N_{DD} = \frac{|\partial\overline{U}_i/\partial x_i|}{\max_{j=1,3}^{i=1,3} |\partial\overline{U}_i/\partial x_j|} .$$

***Mean flow Mach number*** A global role of compressibility in a fluid motion can be, as usual, accounted for with a given Mach number of the mean flow field:

$$M = <U> /a \, , \tag{4.49}$$

where $<U>$ stands for a reference velocity of the mean flow and $a$ is a the speed of sound in given state.
Such a Mach number simply compares $L/<U>$, a mean advective time-scale to the acoustic propagation time-scale $L/a$ over the *same* distance $L$.

It is directly introduced in any non-dimensional formulation of the general mean flows equations, and can be defined in several ways, depending upon the choices of the reference velocity and speed of sound: free-stream Mach number,$M_\infty$, in a boundary layer, convective Mach number, $M_c$, in mixing layers...
This parameter can be useful when describing and comparing compressibility effects in a *given* flow configuration, but poorly reflects behaviours which are specific to different flow configurations. For instance at the same value of the mean flow Mach number, a boundary layer does not exhibit the same large reduction in thickness growth rate as a mixing layer does. (see Chapter 2)

***Gradient Mach number*** A better comparison as above, of the characteristic time-scales associated with compressibility effects and mean flow motion, can be achieved when using the ratio $\tau_a/\tau_S$. When the characteristic time-scale $\tau_S$ can be taken from the mean strain or shear rate, viz. $S = |\partial\overline{U}/\partial y|$, this yields:

$$M_g = S\ell/a\ . \tag{4.50}$$

Different appellations to this parameter, which was first introduced in compressible, homogeneous shear flows, exist in the literature: *gradient* Mach number, as first introduced by Sarkar [409] in 1995, or *distortion* Mach number, as in Simone *et al.* [431] in 1997.
As shown by the former author, the gradient Mach number plays an important role in discriminating compressibility effects in boundary layers and mixing layers. This parameter is found many times larger in mixing layers than in boundary layers. In a turbulent compressible annular mixing layer (Freund *et al.* [163]) it is found to saturate at about $M_g \simeq 1.8$ to 2 within the range $0.8 \leq M_c \leq 1.8$ of the convective Mach number $M_c$. This value is to be compared with the corresponding one in a boundary layer flow. As estimated by Sarkar [409], it is found that $M_g \leq 0.2$ in a boundary layer at a free stream Mach number $M_\infty \leq 3$.

***Turbulence and rms Mach number*** To account for compressibility effects on the *fluctuating* motion, a turbulence Mach number is considered:

$$M_t = \hat{u}/a\ , \tag{4.51}$$

where $\hat{u}$ stands for a characteristic velocity of the turbulent fluctuations. This characteristic velocity can be taken as the rms value of a given fluctuating component (the longitudinal one in boundary layers for example), yielding the so-called rms Mach number $\sqrt{\overline{u'^2}}/a$. When the turbulence kinetic energy is used, viz. $\hat{u} = \overline{k}^{1/2}$, the turbulence Mach number is simply

$M_t = \tau_a/\tau_t$. It compares the characteristic time-scales of two distinct processes, turbulent diffusion and acoustic propagation, which take place over the *same* distance, which is not necessarily the characteristic length-scale of the energy containing eddies.
Since the turbulence Mach number reflects intrinsic compressibility in turbulence, it accounts for such effects which are associated with non solenoidal fluctuating velocity fields. It generally increases with the mean flow or gradient Mach number, as shown by the results reported in Table 4.7.

TABLE 4.7. Turbulence Mach number in various compressible turbulent flows.

| Flow configuration | Ref. | $M_\infty$, $M_c$ or $M_g$ | $M_t$ |
|---|---|---|---|
| Homogeneous shear flow (case A4) | Sarkar [409] 1995 | $M_g = 2.2$<br>$= 5$ | 0.44<br>0.64 |
| Turbulent channel flow | Huang *et al.* [221] 1995 | $M_\infty = 1.5$<br>$= 3$ | 0.17 (max)<br>0.26 (max) |
| Annular mixing layer | Freund *et al.* [163] 2000 | $M_c = 0.2$<br>$= 1.8$ | 0.1<br>0.54 |
| Boundary layer | Review by Spina *et al.* [445] 1994 | $M_\infty = 2.32$ - $2.87$<br>$= 7.2$<br>$= 9.4$ | ~0.21*[16] (max)<br>1*[16] (max)<br>1.16*[16] (max) |

### 4.7.3. LENGTH-SCALE INTERPRETATION

In incompressible turbulent flow, the energetic description can be obtained from the amount of turbulent kinetic energy per unit mass ($\overline{k} = \overline{u_i u_i}/2$) and its dissipation rate $\overline{\varepsilon}$ per unit mass, from which it directly results a turbulent time scale $\tau_t = \overline{k}/\overline{\varepsilon}$ and a turbulent length scale $L_t = \overline{k}^{3/2}/\overline{\varepsilon}$, the Batchelor dissipation length scale.
Such scales can be compared with those of the mean flow, and, considering in particular the time scale, with the significant mean deformation rate $T_{mean}$, i.e., $S^{-1}$ in thin shear layers for instance.

Excepting large scale motions, it is generally assumed that the significant time-scales of turbulence are much lower than those of the mean flow,

[16]The values reported in Spina *et al.* [445] are not those of the turbulence Mach number as defined here, but the rms value of the Mach number fluctuations $M' = M - \overline{M}$ where $M = U/a$. If the fluctuations in the speed of sound are neglected, $M' \approx \overline{u'}^{1/2}/\overline{a}$ According to the results reported in Smits & Dussauge [434] page 217, the maximum values correspond to a region in the boundary layer where $\overline{u'^2} \sim \overline{v'^2} \sim \overline{w'^2}$, so that in this case, it coincides approximately with $M_t$.

so that $\tau_t \ll T_{mean}$, since $\tau_t$ basically refers to the energy containing eddies. As noted by Lumley, this means that mean motion acts on a "turbulent material" which is always in equilibrium, i.e., turbulence adapts "instantaneously" to mean flow variations.

When the fluid can no longer be taken as incompressible, one has to be cautious when using such considerations. As shown in Chapter 1 for a turbulence passing through a shock, the Batchelor dissipation length scale $L_t$ exhibits a specific evolution, distinct from those of others characteristic length-scales, such as Taylor's micro and macro scales. In a compressible annular mixing layer, the DNS by Freund *et al.* [163] also show that $L_t$ behaves differently from $\ell$, a characteristic length scale associated with the energy wave number range of the spectrum. The fundamental reasons that could explain such discrepancies are not totally elucidated at the moment. However, it can be recalled that, in the incompressible regime, the scaling of the dissipation rate from the energy containing eddies is based on purely kinematic interactions, i.e., in which pressure fluctuations doesn't take part (the pressure propagates instantaneously). This situation could be changed in compressible fluid turbulence where pressure readjustment within turbulent eddies can takes place during a finite turn-over time, in addition to the action of vortex stretching, as suggested in [163].

### 4.7.4. TURBULENT FLOWS CLASSIFICATION OF VARIABLE DENSITY FLUID

In this final section, an attempt to organize the various situations depicted in Chapters 3 and 4 is presented. The classifications is based on the solenoidal condition for both mean and fluctuating velocity fields. It starts with such flows configurations in which the mean velocity field slightly departs from the mean solenoidal condition (quasi mean isovolume flows) and ends with non-isovolume fluctuating turbulence, i.e., turbulent motions where the divergence of the fluctuating velocity field cannot be neglected. The situation is summarized in Table 4.8, where $s_\rho$ and $s_\rho^*$ stands for a global and local density ratio respectively. $M^*$ is a local Mach number of the mean flow. $\Delta$ denotes the mean velocity divergence and $\vartheta'$ the fluctuating one.

It should be emphasized that the present classification is by no way correlated with an increasing evolution in the turbulent density intensity. From the comparison with data gathered in Table 4.5, it can be observed that large values of $I_\rho$ can be encountered in variable density fluid flows even when a quasi mean isovolume condition is fulfilled.

TABLE 4.8. Classification of some variable density turbulent flows.

| Velocity divergence | Flow configuration | Significant parameter |
|---|---|---|
| $\Delta \simeq 0$ <br> $\vartheta' \cong 0$ | * Low speed, isobaric mixing of different density fluids $[\Delta \sim \mathcal{O}(mol.diff.)]$ | $s_\rho$ |
| | * Low speed, adiabatic, perfect gas flow <br> * Shallow convection in the atmosphere (Boussinesq approx.) | |
| $\Delta = -\frac{1}{\rho^*}\frac{d\rho^*}{dt}$ <br> $\vartheta' \simeq 0$ | * Anelastic approximation for soundproof low speed flows. ($\rho^*$ is a suitable scaling for density variation, see Chap.3) | |
| $\Delta \neq 0$ <br> $\vartheta' \simeq 0$ | * Compressible turbulence for internal combustion engine applications | $M^*$, $s_\rho^*$ |
| | * Compressible boundary layers ($M_\infty \leq 3$, Morkovin's analogy) | |
| $\Delta \neq 0$ <br> $\vartheta' \neq 0$ | * Compressible high speed shear layers: <br> - Same density fluid <br> - Different density fluids | <br> $M_g$, $M_t$ <br> $M_g$, $M_t$, $s_\rho$ |

CHAPTER 5

# STATISTICAL AVERAGING IN VARIABLE DENSITY FLUID TURBULENT MOTION

*This chapter is devoted to deriving statistical transport equations governing averaged single point properties of the flow field. It is only concerned with first (mean) and second order moments. Two different formulations will be mainly considered, referring to mean mass-conservative and non conservative evolutions. It is not intended for extensive and detailed presentations of the different formalisms — using either the classical formulation of the equations ("standard", "mass-weighted" averaging,...) or the specific volume formulation —, but rather focuses on physical interpretation of density fluctuation correlations, according to mass conservative and non conservative mean flow analysis.*

## 5.1. Introduction

In studying turbulent motions of variable density fluids, the general statistical equations can be derived at least from two main formulations, according to whether *mass* or *volume* balance equations are considered. In both cases, the analysis is applied to a given body of fluid and is concerned with the expression of the amount of any transportable quantity. This introduces integrals over the mass $\mathcal{M}$ or the volume $\mathcal{V}$ of the body of fluid, i.e., formally:

$$(a) \quad A = \int_{\mathcal{M}} F(x_j,t)dm \quad = \quad \int_{\mathcal{V}} \rho(x_j,t)F(x_j,t)dv \; , \tag{5.1}$$

$$(b) \quad B = \int_{\mathcal{V}} G(x_j,t)dv \quad = \quad \int_{\mathcal{M}} v(x_j,t)G(x_j,t)dm \; , \tag{5.2}$$

where $F = F(x_i,t)$ — resp. $G = G(x_i,t)$ — refers to the same transportable quantity, taken as a function defined per unit mass and unit volume respectively.

The linkage between the elementary mass $dm$ and volume $dv$ can be equally expressed as $dm = \rho dv$ or $dv = v dm$, using the density ($\rho$) or the specific volume ($v \equiv 1/\rho$) as the flow field dependent function.

The specific volume analysis can be particularly convenient when dealing with buoyant flows or mass mixing with constant temperature and pressure conditions. In the latter situation, the equation of state, as recalled in the previous chapter for a binary mixing of perfect species, is simply $\rho = a\rho C + b$. It reduces exactly to a linear expression of the mass fraction as a function of the specific volume, viz.

$$C = \frac{1}{a} - \frac{b}{a}v \ ,$$

where it is recalled that $a$ and $b$ are two constants for a given binary mixture. More generally, using the specific volume as the flow-field dependent variable simplifies the form of the convective terms in the Reynolds-averaged balance equations: all terms involving turbulent mass fluxes $\overline{\rho' u'_i}$ are removed (see §5.3.3). On the other hand, pressure-volume correlations are introduced, e.g., $\overline{v'\partial p'/\partial x_i}$.
Specific volume formulation has been extensively developed in deriving averaged equations without and with discussion of the closure by Shih *et al.* [427] in (1987), Rey [384], Rey & Rosant [385], Aurier & Rey [26]. It has alson been adopted by Sandoval [407] in direct numerical simulation of homogeneous buoyant flows.
From now, only mass-integral formulations will be considered, using flow-field functions defined per unit mass.

## 5.2. Averaging of variable density transport equations

Based on a statistical or ensemble averaging operator, any instantaneous value of a given flow field function $F(x_j, t)$ can be decomposed as follows:

$$F(x_j, t) = \overline{F}(x_j, t) + f'(x_j, t) \ , \tag{5.3}$$

where $\overline{F}(x_j, t)$ is the mean value and $f'(x_j, t)$ denotes the random centered fluctuation:

$$\overline{f'}(x_j, t) = 0 \ .$$

This decomposition is usually called the "Reynolds average" or "Reynolds decomposition", an appellation which is a bit confusing[1], referring to the emended version [387] of Reynolds' original paper [386] published in 1895.

[1]In equation (4) of ref. [387], page 547, the mean axial velocity component is taken over a given *domain*: spatial averaging and not time averaging, as often reported. It is expressed as $\overline{u} = \Sigma(\rho u)/\Sigma(\rho)$, which is clearly a mass-weighted average. Despite this historical evidence, "Reynolds averaging" is used here with its modern and commonly adopted meaning.

Introducing decomposition (5.3) in the convective product $(\rho F U_j)$ of the transport term of $F$ (conservative form), we get:

$$A_j(F) \equiv \rho F U_j = \overline{\rho}\overline{F}\,\overline{U}_j + \overline{\rho}\overline{U}_j f' + \overline{\rho}\overline{F} u'_j + \overline{\rho} f' u'_j$$

$$+\rho'\overline{F}\,\overline{U}_j + \rho' f'\overline{U}_j + \rho' u'_j \overline{F} + \rho' f' u'_j \,. \tag{5.4}$$

Thus, when averaging any non linear, density-coupled term, correlations with density fluctuations (d.f.c.) are introduced as specific characteristics of variable density fluid turbulence, since:

$$\overline{A}_j(F) = \overline{\rho}\overline{F}\,\overline{U}_j + \overline{\rho' f'}\,\overline{U}_j + \overline{\rho' u'_j}\,\overline{F} + \overline{\rho}\,\overline{f' u'}_j + \overline{\rho' f' u'_j} \,. \tag{5.5}$$

As opposed to the constant density situation, the physical interpretation of the open set of mean transport equations is no longer unique[2] when the density changes. In this case, any definition of a "*mean evolution*", depends on a physical choice about the interpretation of the role of the additional correlations due to density variation. The mathematical definition of the *mean variation operator* directly results. Several proposals can be adopted to cope with such extra density correlations.

## 5.3. Mean motion in variable density fluid turbulence

When the density is constant, a physical interpretation can be easily given to the *averaged equations*, which instills sense to a so-called *mean motion.* In other words a *mean evolution* can be recovered from the averaged equations governing *mean statistical* values. This can be readily seen by looking at the momentum equation for instance. When averaging the incompressible Navier-Stokes equations, velocity correlations $\overline{u'_i u'_j}$ are introduced by the non linearity of the advection term. However, these **averaged** *Navier-Stokes* equations, can also be understood as the *Reynolds* equations, viz. balance equations governing the **mean** momentum of a given body of fluid moving with the **mean** velocity. Accordingly, the same correlations $\overline{u'_i u'_j}$ are now considered as external surface forces or stresses acting in the new *mean evolution.* This new physical interpretation completely differs from the mathematical advective origin of this term.

Due to additional density correlations, such a simple analysis does not directly apply to variable density fluid turbulence.

[2]Statistical averaging of the instantaneous equations is a mathematical operation. It is not aiming at providing a physical frame to the interpretation of the resulting *averaged equations*, according to some new *mean evolution*. This can only be done on physical groundings. In incompressible fluid flows, it is achieved by using spatial averaging, as introduced by O. Reynolds, assuming implicit ergodicity between ensemble and volume averages.

### 5.3.1. BINARY REGROUPING

A strong analogy with the constant density situation — removing in particular density fluctuation correlations (d.f.c.), say $\overline{\rho'\phi'}$ from mean transport expressions — can be obtained by using mass-weighted averages. Widely developed by Favre since 1958 [151], [152], [153], [154], [155], [156], this regrouping is now commonly called "Favre's averaging", a denomination which mainly refers to a recent historical context.

Mass-weighted[3] averaging basically consists in grouping any d.f.c into a *new macroscopic mean value* such as, for instance $\overline{\rho}\,\overline{U}_i$ and $\overline{\rho u'_i}$, in order to give

$$\overline{\rho}\,\widetilde{U}_i = \overline{\rho U_i} \equiv \overline{\rho}\overline{U}_i + \overline{\rho' u'_i}(\equiv \overline{\rho}\overline{U}_i + \overline{\rho u'_i}) \ . \tag{5.6}$$

Hence, by averaging any convective term $A_j\left(F\right) = \rho F U_j$, it is obtained that:

$$\overline{A}_j(F) = \overline{\rho}\,\widetilde{F}\widetilde{U}_j + \overline{\rho f'' u''_j} \ , \tag{5.7}$$

where the superscript $''$ denotes fluctuations with respect to the mass-weighted average,

$$f'' = F - \widetilde{F} \quad with \quad \widetilde{F} = \frac{\overline{\rho F}}{\overline{\rho}} \ .$$

Formally, eq.(5.6) includes no more nonlinear terms than for the constant density fluid motion. Hence, using mass-weighted averages leads to a *binary regrouping* of all convective non linearities. Such a formal analogy results from a strong physical argument, i.e., the *macroscopic* or mean *conservation of mass* in both constant and variable density fluid flows. Accordingly for the latter, the mean mass-weighted continuity equation reads:

$$\frac{\partial \overline{\rho}}{\partial t} + \frac{\partial\left(\overline{\rho}\widetilde{U}_j\right)}{\partial x_j} = 0 \ . \tag{5.8}$$

However, two points should be emphasized:

- *Mean* velocities $\overline{U}_i$ and $\widetilde{U}_i$ in constant and variable density fluid flows are not defined in the same way. Consequently, such kinematic properties as mean strain rate, mean vorticity etc., are not physically identical in both cases, a point that should be taken into consideration when using such quantities as arguments of closure schemes;
- The Favrian fluctuation is not centered: $\overline{f''} = -\overline{\rho f'}/\overline{\rho} \neq 0$ , (5.9)

  where $f' = F - \overline{F}$ denotes the fluctuation with respect to the Reynolds average. Thus, the formal incompressible analogy does not strictly

[3] In a mass balance of $F$ the volume-integral is concerned with $\rho F$. Hence, the mathematical formulation yields density-weighted expressions.

apply to open second-order-moment transport equations, as shown for instance by Chassaing [80]. Moreover, binary regrouping should also be concerned with physical interpretation and governing equations of such quantities as $\overline{u''_i}$, $\overline{u''_i u''_j}$...

### 5.3.2. TRANSPORT SELECTED REGROUPING

Introduced by Bauer *et al.* [39] in 1968, this second regrouping has been developed and used in second-order modeling by Ha Minh *et al.* [193] and [194] in 1981. The original idea is that, in any transport term like $\rho F U_j$, where $F$ is a scalar, vectorial or tensorial function of the velocity, — any momentum component or kinetic energy for instance —, the velocity component plays two roles: (i) convected function (inside $F$) and (ii) transporting agent (outside $F$).
Considering the momentum transport term $A_j(U_i) = \rho U_i U_j$, these two distinct roles can be brought into prominence by a mixed-weighted decomposition, selecting *Reynolds averaging* for the *advection* or transport velocity (say $U_j$) and *mass-weighted* averaging for the momentum (*convected*) component ($\rho U_i$).
This is achieved by introducing:

$$\overline{G}_i = \overline{\rho}\overline{U}_i + \overline{\rho' u'_i} \qquad \text{and} \qquad g'_i = \rho'\overline{U}_i + \overline{\rho} u'_i + \rho' u'_i - \overline{\rho' u'_i}\,, \tag{5.10}$$

where the fluctuation $g'_i$ is centered.
As demonstrated in [193] for instance, the Reynolds stress tensor associated with decomposition (5.10) is $-\overline{g'_i u'_j}$, is not symmetric. This is the major difference of the present regrouping with the two other ones and with the incompressible case.

### 5.3.3. TERNARY REGROUPING

The ternary regrouping was proposed by P. Chassaing [81] in 1985. It mainly aims at enlightening the thermodynamic incidence of density variation on d.f.c. terms, which cannot be inferred from a mechanical interpretation focusing only on the transport equations. Hence, after expliciting *all* d.f.c. terms, and primarily the turbulent mass flux in *all* equations, specific features of the correlations resulting from density fluctuations are obtained from the overall linkage due to the equation of state. A better interpretation of d.f.c. terms (see next chapter) within the framework of second-order modeling (see Chapter 11) is expected from such a thermodynamic basis.

According to the ternary formulation, and considering eq.(5.5) for example, all d.f.c.'s are made explicit and grouped into contributions disctinct from those originating from advective/convective non-linearities, which are

now involving the *instantaneous* value of the density, leading to '*mass-weighted*' second and higher order moments of *centered* (Reynolds) fluctuations.
Hence, the *ternary* regrouping suggested for any convective term takes the form:

$$\overline{A}_j\,(F) = \underbrace{\overline{\rho}\,\overline{F}\,\overline{U}_j}_{(a)} + \underbrace{\overline{\rho f' u'_j}}_{(b)} + \underbrace{\overline{\rho f'}\,\overline{U}_j + \overline{\rho u'_j}\,\overline{F}}_{(c)} \; . \tag{5.11}$$

In eq.(5.11), the first term is the contribution from the mean components, the second one $(b)$ is the *compressible* or *variable-density* equivalent to the classical second-order correlation. As shown in [81], open transport equations for such moments can be exactly derived. In the last term $(c)$, are grouped the first order d.f.c. contributions.

When applied to the continuity equation, this averaging procedure reflects the non linearity due to density variation, yielding:

$$\frac{\partial \overline{\rho}}{\partial t} + \frac{\partial \overline{\rho}\overline{U}_i}{\partial x_i} = -\frac{\partial (\overline{\rho' u'_i})}{\partial x_i} \; . \tag{5.12}$$

Hence, a turbulent mass flux $\overline{\rho' u'_i}$ is introduced in the same way as any d.f.c. contribution.

As compared with the binary regrouping, the following differences emerge:

- No direct *formal* analogy exists between transport equations of constant and variable density fluid motions due to d.f.c.;
- The *mean* continuity equation does not refer to a mean mass-conservative evolution: turbulent mass flux is introducing a non-zero *source term*;
- The *physical meaning* of all mean (macroscopic) quantities is the *same* for both constant and variable density situations.

### 5.3.4. KINETIC ENERGY AVERAGING

Some consequences of the previous formulations may appear more clearly when applied to the kinetic energy. Let us denote by $K = U_i U_i / 2$ the instantaneous value of kinetic energy per unit mass. The expressions of the *mean* values of $\overline{K}$ and $(\overline{\rho\,K})$ — the kinetic energy per unit volume — are given in Table 5.1, according to binary and ternary regroupings.
In incompressible fluid motion, the averaged kinetic energy per unit mass and volume *both* reduce to the sum of two contributions, referring to *mean* and *fluctuating* motions respectively. This is also the case when considering the mean value of (i) the kinetic energy per unit mass with a ternary

TABLE 5.1. Expressions of mean kinetic energy according to binary and ternary regrouping.

| Type of Regrouping | $\overline{K}$ | $\overline{\rho K}$ |
|---|---|---|
| Ternary | $\frac{1}{2}\overline{U}_i\overline{U}_i + \frac{1}{2}\overline{(u'_i u'_i)}$ | $\frac{1}{2}\overline{\rho}\overline{U}_i\overline{U}_i + \frac{1}{2}\overline{(\rho u'_i u'_i)} + \overline{U}_i\overline{\rho' u'_i}$ |
| Binary | $\frac{1}{2}\tilde{U}_i\tilde{U}_i + \frac{1}{2}\overline{(u''_i u''_i)} + \tilde{U}_i\overline{u''_i}$ | $\frac{1}{2}\tilde{U}_i\tilde{U}_i + \frac{1}{2}\overline{(u''_i u''_i)}$ |

regrouping *or* (ii) the kinetic energy per unit volume with a binary regrouping. In other words, turbulent mass flux contributions are present in the expressions of $\overline{\rho K}$ (ternary regrouping) and $\overline{K}$ (binary regrouping[4] ).

The formulas given in the previous table can be easily generalized to the mean value of the product of two scalar functions $F$ and $G$. The corresponding expressions are, for instance:

$$\overline{(\rho F G)} = \underbrace{\overline{\rho}\widetilde{F}\widetilde{G}}_{(a)} + \underbrace{\overline{\rho}\,\widetilde{f''g''}}_{(c)} , \tag{5.13}$$

$$\text{or} \quad \overline{(\rho F G)} = \underbrace{\overline{\rho}\,\overline{F}\,\overline{G}}_{(a')} + \underbrace{\overline{\rho' f'}\,\overline{G} + \overline{\rho' g'}\,\overline{F}}_{(b')} + \underbrace{\overline{\rho f' g'}}_{(c')} . \tag{5.14}$$

In eqs.(5.13) and (5.14), ($a$) and ($a$') account for the *mean* motion contribution, ($c$) and ($c$') for the *fluctuating* motion contribution. A d.f.c. term ($b$') is only present with the ternary regrouping formulation.

### 5.3.5. COMPARISON BETWEEN BINARY AND TERNARY REGROUPING

Although based on quite different mean motion analysis, all formulations are obviously algebraically equivalent. However, they differ with the way density fluctuations are physically accounted for, when considering statistical mean equations.
These are the two points to be addressed successively in the present section.

***Formal relationship*** The relationship between binary and ternary regrouping can be analytically detailed. Some expressions are given in Table 5.2.

[4]When Favre's averages are used (binary regrouping), a "third" mean quantity, analogous to a kinetic energy, is readily present viz. $\frac{1}{2}\overline{\rho}\,\overline{u''_i}\,\overline{u''_i}$. However, since it does not contribute to the average value of the instantaneous kinetic energy of the flow, it could be considered as an artifact of this type of formulation.

TABLE 5.2. Relationship between binary and ternary regrouping.

| Binary to Ternary | Ternary to Binary |
|---|---|
| $\overline{F} = \widetilde{F} + \overline{f''}$ | $\widetilde{F} = \overline{F} + \frac{\overline{\rho' f'}}{\overline{\rho}}$ |
| $f' = f'' - \overline{f''}$ | $f'' = f' - \frac{\overline{\rho' f'}}{\overline{\rho}}$ |
| $\overline{f'} = 0$ | $\overline{f''} = -\frac{\overline{\rho' f'}}{\overline{\rho}}$ |
| $\widetilde{f'} = \overline{\rho' f'}/\overline{\rho} = -\overline{f''}$ | $\widetilde{f''} (\equiv \overline{\rho f''}/\overline{\rho}) = 0$ |
| $\overline{\rho' f'} = -\overline{\rho}\, \overline{f''}$ | $\overline{\rho' f''} = \overline{\rho' f'}$ |
| $\overline{\rho' f' g'} = \overline{\rho' f'' g''} + 2\overline{\rho}\, \overline{f''}\, \overline{g''}$ | $\overline{\rho' f'' g''} = \overline{\rho' f' g'} - 2\frac{\overline{\rho' f'}\,\overline{\rho' g'}}{\overline{\rho}}$ |
| $\overline{\rho' f' g' h'} = \overline{\rho' f'' g'' h''}$<br>$-\overline{\rho' f'' g''}\, \overline{h''} - \overline{\rho' g'' h''}\, \overline{f''} - \overline{\rho' h'' f''}\, \overline{g''}$<br>$+\overline{\rho' f''}\, \overline{g''}\, \overline{h''} + \overline{\rho' g''}\, \overline{h''}\, \overline{f''} + \overline{\rho' h''}\, \overline{f''}\, \overline{g''}$ | $\overline{\rho' f'' g'' h''} = \overline{\rho' f' g' h'}$<br>$-\frac{\overline{\rho' f'}\,\overline{\rho' g' h'} + \overline{\rho' g'}\,\overline{\rho' h' f'} + \overline{\rho' h'}\,\overline{\rho' f' g'}}{\overline{\rho}}$<br>$+3\,\frac{\overline{\rho' f'}\,\overline{\rho' g'}\,\overline{\rho' h'}}{\overline{\rho}^2}$ |
| $\overline{f' g'} = \overline{f'' g''} - \overline{f''}\, \overline{g''}$ | $\overline{f'' g''} = \overline{f' g'} + \frac{\overline{\rho' f'}\,\overline{\rho' g'}}{\overline{\rho}^2}$ |
| $\widetilde{f' g'} = \widetilde{f'' g''} + \overline{f''}\, \overline{g''}$ | $\widetilde{f'' g''} = \overline{f' g'} + \frac{\overline{\rho' f' g'}}{\overline{\rho}} - \frac{\overline{\rho' f'}\,\overline{\rho' g'}}{\overline{\rho}^2}$ |
| $\overline{f' g' h'} = \overline{f'' g'' h''} + 2\,\overline{f''}\, \overline{g''}\, \overline{h''}$<br>$-\overline{f'' g''}\, \overline{h''} - \overline{g'' h''}\, \overline{f''} - \overline{h'' f''}\, \overline{g''}$ | $\overline{f'' g'' h''} = \overline{f' g' h'} - \frac{\overline{\rho' f'}\,\overline{\rho' g'}\,\overline{\rho' h'}}{\overline{\rho}^3}$<br>$-\frac{\overline{\rho' f'}\,\overline{g' h'} + \overline{\rho' g'}\,\overline{h' f'} + \overline{\rho' h'}\,\overline{f' g'}}{\overline{\rho}}$ |

Relations given in Table 5.2 can be easily obtained from the definitions of "Reynolds" and "Favre" decompositions of any statistical function of the flow field $F(x_j, t)$:

$$F = \overline{F} + f' = \widetilde{F} + f'' . \tag{5.15}$$

For instance, by multiplying eq.(5.15) by $\overline{\rho}$ and averaging, one obtains:

$$\overline{\rho}\,\overline{F} + \overline{\rho f'} = \overline{\rho}\,\widetilde{F} \qquad \text{or} \qquad \widetilde{F} = \overline{F} + \frac{\overline{\rho f'}}{\overline{\rho}} , \tag{5.16}$$

since, by definition of Favre averaging, $\overline{\rho f''} \equiv 0$. Then, introducing eq.(5.16) in eq.(5.15) it comes:

$$f'' = f' - \frac{\overline{\rho f'}}{\overline{\rho}} , \tag{5.17}$$

and, by averaging, since $\overline{f'} = 0$:

$$\overline{f''} = -\frac{\overline{\rho f'}}{\overline{\rho}} . \tag{5.18}$$

Thus, the correspondence between mean and fluctuating values according to Reynolds and Favre formulations is given by:

$$\widetilde{F} = \overline{F} + \frac{\overline{\rho f'}}{\overline{\rho}} \qquad \text{or} \qquad \overline{F} = \widetilde{F} + \overline{f''} , \tag{5.19}$$

and

$$f'' = f' - \frac{\overline{\rho f'}}{\overline{\rho}} \qquad \text{or} \qquad f' = f'' - \overline{f''} . \tag{5.20}$$

Now, multiplying eq.(5.17) by $g''$ one obtains:

$$f''g'' = f'g'' - \frac{\overline{\rho f'}}{\overline{\rho}} g'' = f'(g' - \frac{\overline{\rho g'}}{\overline{\rho}}) - \frac{\overline{\rho f'}}{\overline{\rho}} g'' ,$$

from which it is easy to deduce the correspondence between second order correlations:

$$\overline{f''g''} = \overline{f'g'} + \frac{\overline{\rho f'}}{\overline{\rho}} \frac{\overline{\rho g'}}{\overline{\rho}} \qquad \text{or} \qquad \overline{f'g'} = \overline{f''g''} + \overline{f''}\, \overline{g''} . \tag{5.21}$$

***Physical correspondence*** To clarify the physical comparison between binary and ternary regrouping, let us consider first the simple case of low speed, variable density flows at high turbulence Reynolds number. As shown in the previous chapter, discarding molecular transport terms leads to a mean isovolume evolution with respect to Reynolds averaging, that is:

$$\partial \overline{U}_i / \partial x_i = 0 .$$

In this case, it can be easily seen that the mean continuity equation reduces to:

$$(a) \quad \frac{\partial \overline{\rho}}{\partial t} + \widetilde{U}_k \frac{\partial \overline{\rho}}{\partial x_k} = -\overline{\rho} \frac{\partial \widetilde{U}_k}{\partial x_k} (\equiv +\overline{\rho} \frac{\partial \overline{u''}_k}{\partial x_k}) , \tag{5.22}$$

$$or \quad (b) \quad \frac{\partial \overline{\rho}}{\partial t} + \overline{U}_k \frac{\partial \overline{\rho}}{\partial x_k} = - \frac{\partial \overline{\rho u'}_k}{\partial x_k} , \tag{5.23}$$

according to a binary or ternary formulation respectively.

The physical interpretation of both equations, as sketched in Fig.5.1, is as follows:

a) By definition, the mean amount of mass of any given body of fluid enclosed in a material surface $S$ remains constant when following the local displacement of the surface moving at the local mass-weighted velocity $\widetilde{U}_k$. Thus, the mean density variation is purely due to mean "*volume*" variation, as clearly shown by the bulk source term in eq.(5.22), see Fig.5.1(a).

b) Since the mean Reynolds velocity field is solenoidal in the same situation,

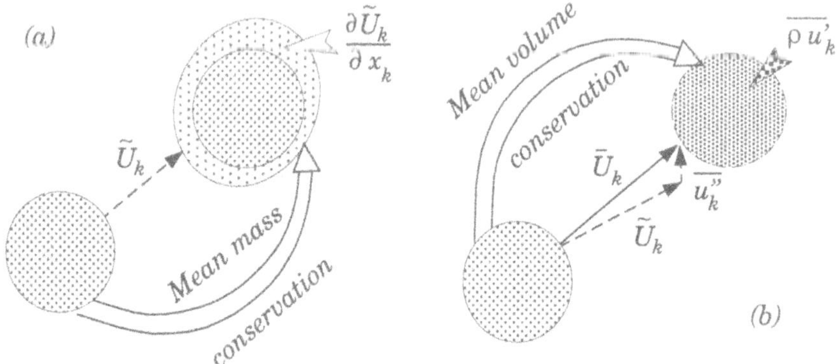


*Figure 5.1.* Physical analysis of mean density variation in a mean Reynolds solenoidal evolution: (a) Binary regrouping (b) Ternary regrouping.

it results that the mean lagrangian evolution of a given body of fluid is *volume conservative*, when referring to $\overline{U}_k$. Thus the mean density variation is totally due to *turbulent mass fluxes* across the surface of the domain, as a result of the source term in eq.(5.22), see Fig.5.1(b).

Recalling that, in any case $\overline{U}_k = \widetilde{U}_k + \overline{u''}_k$ it appears that, in the present situation, both formulations are equivalent, giving a unique physical interpretation of the origin of mean density variation. In other words, binary regrouping leads to a mean volume interpretation of the source term due to the mean velocity fluctuation $\overline{u''}_k$, which is viewed as a mass flux, with the ternary regrouping.

Now, for the general variable density situation where $\frac{\partial \overline{U}_i}{\partial x_i} \neq 0$, source terms in the mean continuity equations come from both mean bulk variations and turbulent mass flux, according to:

$$(b') \quad \frac{\partial \overline{\rho}}{\partial t} + \overline{U}_k \frac{\partial \overline{\rho}}{\partial x_k} = -\frac{\partial \overline{\rho u'}_k}{\partial x_k} - \overline{\rho} \frac{\partial \overline{U}_k}{\partial x_k} ,$$

using the ternary regrouping. Thus, this formulation allows to distinguishing between two separate origins of mean density variation, a possibility which is not available when adopting binary regrouping, since the mean continuity equation eq.(5.22) is unchanged.

## 5.4. Mean flow equations

In this section and the following ones, the derivation of the equations will not be generally detailed. Most of the demonstrations can be found in particular, in [157], [152], [153], [399], [299], [236].

The averaging procedure will be addressing the following "instantaneous" balance equations[5]:

$$\textit{Continuity}: \quad \frac{\partial \rho}{\partial t} + \frac{\partial(\rho U_k)}{\partial x_k} = 0 \, ,$$

$$\textit{Momentum}: \quad \frac{\partial \rho U_i}{\partial t} + \frac{\partial(\rho U_i U_j)}{\partial x_j} = \rho F_i - \frac{\partial P}{\partial x_i} + \frac{\partial \tau_{ij}}{\partial x_j} \, ,$$

$$\textit{Internal energy}: \quad \frac{\partial \rho e}{\partial t} + \frac{\partial(\rho e U_j)}{\partial x_j} = -P\frac{\partial U_i}{\partial x_i} + \tau_{ij}\frac{\partial U_i}{\partial x_j} + \frac{\partial q_i}{\partial x_i} \, ,$$

$$\textit{Enthalpy}: \quad \frac{\partial \rho h}{\partial t} + \frac{\partial(\rho h U_j)}{\partial x_j} = \frac{dP}{dt} + \tau_{ij}\frac{\partial U_i}{\partial x_j} + \frac{\partial q_i}{\partial x_i} \, .$$

The last two equations are non independent but using alternate expressions of the same first law of the thermodynamics with $e = C_v T$ or $h = C_p T$.
In the previous formulations, $d/dt \equiv \partial/\partial t + U_j \, \partial/\partial x_j$ denotes the material derivative, $\tau_{ij}$ stands for the viscous stress tensor of a Newton-Stokes fluid:

$$\tau_{ij} = 2\mu S_{ij} - \frac{2}{3}\mu\frac{\partial U_k}{\partial x_k}\delta_{ij} \equiv \mu(\frac{\partial U_i}{\partial x_j} + \frac{\partial U_j}{\partial x_i}) - \frac{2}{3}\mu\frac{\partial U_k}{\partial x_k}\delta_{ij} \, ,$$

and the heat conduction flux is $q_i = -\lambda \dfrac{\partial T}{\partial x_i}$,

where $\lambda$ is the thermal conductivity of the fluid according to Fourier's law.

### 5.4.1. BINARY REGROUPING

$$\textit{Mean continuity}: \quad \frac{\partial \overline{\rho}}{\partial t} + \frac{\partial(\overline{\rho}\widetilde{U}_k)}{\partial x_k} = 0 \, . \tag{5.24}$$

By definition the *mean* motion referring to a density-weighted macroscopic velocity is mass conservative.

$$\textit{Mean momentum}: \quad \frac{\partial(\overline{\rho}\widetilde{U}_i)}{\partial t} + \frac{\partial(\overline{\rho}\widetilde{U}_i \widetilde{U}_j)}{\partial x_j} = \overline{\rho}\widetilde{F}_i - \frac{\partial \overline{P}}{\partial x_i} - \frac{\partial(\overline{\rho}\, \widetilde{u''_i u''_j})}{\partial x_j} + \frac{\partial \overline{\tau}_{ij}}{\partial x_j} \, . \tag{5.25}$$

This equation is formally similar to the "incompressible" one. The non linear *advection* only results in one additional term in the mean density-weighted momentum balance equation. Its effect can be accounted for by *external* stresses: the turbulent or "Reynolds" stresses $-\overline{\rho}\,\widetilde{u''_i u''_j} \equiv -\overline{\rho u''_i u''_j}$.

[5]These equations can be found in any Textbook on basic Fluid Mechanics, such as Chassaing [83], page 121, for instance.

By taking the scalar product of the mean momentum equation by the density-weighted velocity $\widetilde{U}_i$, one obtains the equation governing the kinetic energy of the mean motion:

*Mean kinetic energy:*

$$\frac{\partial(\frac{1}{2}\overline{\rho}\widetilde{U}_i\widetilde{U}_i)}{\partial t} + \frac{\partial[(\frac{1}{2}\overline{\rho}\widetilde{U}_i\widetilde{U}_i)\,\widetilde{U}_j]}{\partial x_j} = \underbrace{\overline{\rho}\widetilde{U}_i\widetilde{F}_i}_{(i)} - \underbrace{\frac{\partial(\overline{P}\widetilde{U}_i)}{\partial x_i}}_{(ii)} + \underbrace{\overline{P}\frac{\partial\widetilde{U}_i}{\partial x_i}}_{(a)}$$

$$- \underbrace{\frac{\partial(\overline{\rho}\,\widetilde{u_i''u_j''}\,\widetilde{U}_i)}{\partial x_j}}_{(iii)} + \underbrace{\overline{\rho}\,\widetilde{u_i''u_j''}\,\frac{\partial\widetilde{U}_i}{\partial x_j}}_{(c)} + \underbrace{\frac{\partial(\overline{\tau}_{ij}\,\widetilde{U}_i)}{\partial x_j}}_{(iv)} - \underbrace{\overline{\tau}_{ij}\,\frac{\partial\widetilde{U}_i}{\partial x_j}}_{(b)} \tag{5.26}$$

Referring to the corresponding equation in incompressible fluid motion, the physical meaning of the various right-hand-side terms in eq.(5.26) is as follows:

(i) power of the mean external body forces in the mean motion;
(ii) power of the mean external pressure forces in the mean motion;
(iii) power of the Reynolds stresses acting through the mean motion;
(iv) power of the mean external viscous stresses in the mean motion;
(*a*) mean pressure-dilatation energy transfer. This term is specific to non mean solenoidal evolution;
(*b*) mean motion viscous dissipation;
(*c*) energy exchange with turbulence kinetic energy ("shear production").

The roles of (*a*), (*b*) and (*c*) terms will be explicited in a next section.

*Mean internal energy:*

$$\frac{\partial(\overline{\rho}\,\widetilde{e})}{\partial t} + \frac{\partial(\overline{\rho}\,\widetilde{e}\,\widetilde{U}_j)}{\partial x_j} = - \underbrace{\overline{P}\frac{\partial\widetilde{U}_i}{\partial x_i}}_{(a)} - \underbrace{\overline{P\frac{\partial u_i''}{\partial x_i}}}_{(e)}$$

$$+ \underbrace{\overline{\tau}_{ij}\,\frac{\partial\widetilde{U}_i}{\partial x_j}}_{(b)} + \underbrace{\overline{\tau_{ij}\frac{\partial u_i''}{\partial x_i}}}_{(d)} + \underbrace{\frac{\partial\overline{q_i}}{\partial x_i}}_{(i)} - \underbrace{\frac{\partial(\overline{\rho}\,\widetilde{e''u_j''})}{\partial x_j}}_{(ii)} \tag{5.27}$$

The physical meaning of the right-hand-side terms in eq.(5.27) is as follows:

($a$) mean pressure-dilatation term. Since it is found with an opposite sign in the kinetic energy equation of the mean motion, this term corresponds to an energy transfer which is specific to non mean solenoidal evolution;

($e$) pressure-fluctuation dilatation correlation. This term is specific to non solenoidal *fluctuation velocity* fields;

($b$) viscous heat production due to mechanical dissipation by the *mean* motion. This term appears with an opposite sign in the kinetic energy equation of the mean motion;

($d$) viscous heat production due to mechanical dissipation by the *fluctuating* motion. This term, with an opposite sign, accounts for the turbulence dissipation in the turbulence kinetic energy balance;

(i) mean external heat source by conduction;

(ii) turbulent diffusion due to the turbulent flux $-\overline{\rho}\widetilde{e''u''_j} \equiv -\overline{\rho e''u''_j}$ where $e''$ stands for the internal energy fluctuation with respect to the mean density-weighted value ($e''=e-\widetilde{e}$).

### 5.4.2. TERNARY REGROUPING

$$Mean\,continuity: \quad \frac{\partial\overline{\rho}}{\partial t}+\frac{\partial(\overline{\rho}\overline{U}_k)}{\partial x_k}=-\frac{\partial(\overline{\rho' u'_k})}{\partial x_k}\,. \tag{5.28}$$

As expected, the turbulent mass flux $\overline{\rho' u'_k} \equiv \overline{\rho u'_k}$ appears as a source term in this equation, when using conventional averages.

$$Mean\,momentum: \quad \frac{\partial(\overline{\rho}\overline{U}_i)}{\partial t}+\frac{\partial(\overline{\rho}\overline{U}_i\overline{U}_j)}{\partial x_j}=\overline{\rho}\overline{F}_i-\frac{\partial\overline{P}}{\partial x_i}-\frac{\partial(\overline{\rho u'_i u'_j})}{\partial x_j}+\frac{\overline{\tau}_{ij}}{\partial x_j}-A_i\,, \tag{5.29}$$

where it is assumed that there is no turbulent fluctuation of the external body forces.

By identifying the turbulence stresses to $-\overline{\rho u'_i u'_j}$, all contributions arising from the d.f.c. are grouped in the additional term $A_i$:

$$A_i=\frac{\partial\overline{\rho u'_i}}{\partial t}+\frac{\partial}{\partial x_j}(\overline{\rho u'_i}\,\overline{U}_j+\overline{\rho u'_j}\,\overline{U}_i)\,.$$

*Mean kinetic energy*:

The equation governing the kinetic energy of the mean motion is obtained by taking the scalar product of the mean momentum equation (5.29) by the mean velocity $\overline{U}_i$.

$$\frac{\partial(\frac{1}{2}\overline{\rho}\overline{U}_i\overline{U}_i)}{\partial t}+\frac{\partial[(\frac{1}{2}\overline{\rho}\overline{U}_i\overline{U}_i)\overline{U}_j]}{\partial x_j}=\underbrace{\overline{\rho}\overline{U}_i\overline{F}_i}_{(i')}-\underbrace{\frac{\partial(\overline{P}\overline{U}_i)}{\partial x_i}}_{(ii')}+\underbrace{\overline{P}\frac{\partial\overline{U}_i}{\partial x_i}}_{(a')}-A \tag{5.30}$$

$$-\underbrace{\frac{\partial(\overline{\rho u'_i u'_j}\,\overline{U}_i)}{\partial x_j}}_{(iii')}+\underbrace{\overline{\rho u'_i u'_j}\frac{\partial\overline{U}_i}{\partial x_j}}_{(c')}+\underbrace{\frac{\partial(\overline{\tau}_{ij}\overline{U}_i)}{\partial x_j}}_{(iv')}-\underbrace{\overline{\tau}_{ij}\frac{\partial\overline{U}_i}{\partial x_j}}_{(b')},$$

$$\text{with}\quad A=A_i\overline{U}_i=\overline{U}_i(\frac{\partial\overline{\rho u'_i}}{\partial t}+\frac{\partial\overline{\rho u'_i}\,\overline{U}_j}{\partial x_j})+\frac{\partial}{\partial x_j}(\frac{1}{2}\overline{U}_i\overline{U}_i\overline{\rho u'_j})\ .$$

Excepting the additional contribution $A$, the physical meaning of all other terms in the right-hand-side of eq.(5.30) is the same as in eq.(5.26) and is not explicited here again.

*Mean internal energy*:

$$\frac{\partial(\overline{\rho}\overline{e})}{\partial t}+\frac{\partial(\overline{\rho}\overline{e}\,\overline{U}_j)}{\partial x_j}=-\underbrace{\overline{P}\frac{\partial\overline{U}_i}{\partial x_i}}_{(a')}-\underbrace{\overline{p'\frac{\partial u'_i}{\partial x_i}}}_{(e')}$$

$$+\underbrace{\overline{\tau}_{ij}\frac{\partial\overline{U}_i}{\partial x_j}}_{(b')}+\underbrace{\overline{\tau'_{ij}\frac{\partial u'_i}{\partial x_i}}}_{(d')}+\underbrace{\frac{\partial\overline{q}_i}{\partial x_i}}_{(i')}-\underbrace{\frac{\partial(\overline{\rho e'u'_j})}{\partial x_j}}_{(ii')}-B\ , \tag{5.31}$$

$$\text{with}\quad B=\frac{\partial\overline{\rho e'}}{\partial t}+\frac{\partial}{\partial x_j}(\overline{\rho e'}\,\overline{U}_j+\overline{\rho u'_j}\,\overline{e})\ .$$

The physical meaning of the right-hand-side terms in eq.(5.31) is the same as the one for the corresponding terms in eq.(5.27). Indeed, the two equations only differ from the presence of:

- explicit d.f.c. terms $\overline{\rho e'}$ and $\overline{\rho u'_j}$ in $B$;
- the pressure *fluctuation* instead of the *instantaneous* value of the pressure[6] in (*e'*);
- the *fluctuation* of the viscous stress tensor in (*d'*) instead of its *instantaneous* value.

[6]This 'formally minor' change could have more serious consequences in the modeling issue of such terms, due to the presence of $\overline{P\partial u''_i/\partial x_i}$ in the binary formulation, which has no equivalent in the ternary regrouping.

### 5.5. Turbulent mass flux equation

Turbulent mass fluxes $\overline{\rho u'_i}$ are explicitly present in the mean motion equations based on conventional averaging. When density-weighted averaging is adopted, such fluxes also appear in the transport equations for second and higher order velocity moments through $\overline{u''}_i \equiv -\overline{\rho u'_i}/\overline{\rho}$.
At least for sake of modeling, it could be worthwhile having to one's disposal an exact transport equation for $\overline{u''}_i$ (binary regrouping) or $\overline{\rho u'_i}$ (ternary regrouping). Since this equation is not as usual as the previous ones, its derivation is briefly recalled here, using ternary formalism.
Since $\overline{\rho u'_i} = \overline{(\rho U_i)} - \overline{\rho}\overline{U}_i$ the wanted transport equation is obtained from the difference between the averaged transport equation of momentum $\overline{(\rho U_i)}$ and the transport equation of the mean flow momentum $(\overline{\rho}\overline{U}_i)$.
The first one is simply:

$$\frac{\partial\overline{(\rho U_i)}}{\partial t} + \frac{\partial\overline{(\rho U_i U_j)}}{\partial x_j} = \overline{(\rho F_i)} - \frac{\partial\overline{P}}{\partial x_i} + \frac{\partial\overline{\tau}_{ij}}{\partial x_j} .$$

The second one is deduced from the mean continuity equation, introducing the following expression of the mean momentum equation:

$$\frac{\partial\overline{U}_i}{\partial t} + \overline{U}_j\frac{\partial\overline{U}_i}{\partial x_j} = \overline{F}_i - \overline{\left(\frac{1}{\rho}\frac{\partial P}{\partial x_i}\right)} + \overline{\left(\frac{1}{\rho}\frac{\partial\tau_{ij}}{\partial x_i}\right)} - \overline{u'_j\frac{\partial u'_i}{\partial x_j}} . \tag{5.32}$$

Hence, by combining $\overline{U}_i\times$eq.(5.28) $+\,\overline{\rho}\times$eq.(5.32), and after some algebraic manipulations, one obtains see e.g. (Purwanto [377]):

$$\frac{\partial(\overline{\rho u'_i})}{\partial t} + \frac{\partial(\overline{\rho u'_i}\,\overline{U}_j)}{\partial x_j} = -\frac{\partial(\overline{\rho u'_i u'_j})}{\partial x_j} - \overline{\rho u'_j}\frac{\partial\overline{U}_i}{\partial x_j} + \overline{\rho}\,\overline{(u'_j\frac{\partial u'_i}{\partial x_j})}$$

$$-\overline{\left(\frac{\rho'}{\rho}\frac{\partial P}{\partial x_i}\right)} + \overline{\left(\frac{\rho'}{\rho}\frac{\partial\tau_{ij}}{\partial x_i}\right)} . \tag{5.33}$$

When using a binary regrouping, the corresponding equation is:

$$\frac{\partial(\overline{\rho}\,\overline{u''}_i)}{\partial t} + \frac{\partial(\overline{\rho}\,\overline{u''}_i\,\widetilde{U}_j)}{\partial x_j} = -\frac{\partial(\overline{\rho}\,\widetilde{u''_i u''_j})}{\partial x_j} - \overline{\rho}\,\overline{u''}_j\frac{\partial\widetilde{U}_i}{\partial x_j} - \overline{\rho}\,\overline{(u''_j\frac{\partial u''_i}{\partial x_j})}$$

$$+\overline{\left(\frac{\rho'}{\rho}\frac{\partial P}{\partial x_i}\right)} - \overline{\left(\frac{\rho'}{\rho}\frac{\partial\tau_{ij}}{\partial x_i}\right)} . \tag{5.34}$$

## 5.6. Reynolds stress transport equation

### 5.6.1. BINARY REGROUPING

Different formulations of the equation governing the Reynolds stress tensor $\bar{\rho}\widetilde{u''_i u''_j}$ can be considered, according to various rearrangement of its terms. The expression given here is chosen as to enlightening energy transfer between mean kinetic and internal energies (see section 5.8).
It reads:

$$\frac{\partial}{\partial t}(\bar{\rho}\widetilde{u''_i u''_j}) + \frac{\partial}{\partial x_k}(\bar{\rho}\widetilde{u''_i u''_j}\widetilde{U}_k) = \underbrace{-\frac{\partial}{\partial x_k}(\bar{\rho}\widetilde{u''_i u''_j u''_k})}_{(i)} \underbrace{- \bar{\rho}(\widetilde{u''_i u''_k}\frac{\partial \widetilde{U}_j}{\partial x_k} + \widetilde{u''_j u''_k}\frac{\partial \widetilde{U}_i}{\partial x_k})}_{(ii)}$$

$$\underbrace{-\frac{\partial(\overline{P}\,\overline{u''_j})}{\partial x_i} - \frac{\partial(\overline{P}\,\overline{u''_i})}{\partial x_j}}_{(iii)} \underbrace{-\frac{\partial(\overline{p' u''_j})}{\partial x_i} - \frac{\partial(\overline{p' u''_i})}{\partial x_j}}_{(iv)} + \underbrace{\overline{P(\frac{\partial u''_i}{\partial x_j} + \frac{\partial u''_j}{\partial x_i})}}_{(v)}$$

$$+\underbrace{[\frac{\partial(\overline{\tau_{ik} u''_j})}{\partial x_k} + \frac{\partial(\overline{\tau_{jk} u''_i})}{\partial x_k}]}_{(vi)} - \underbrace{[\overline{\tau_{ik}\frac{\partial u''_j}{\partial x_k}} + \overline{\tau_{jk}\frac{\partial u''_i}{\partial x_k}}]}_{(vii)} . \qquad (5.35)$$

The physical description of the right-hand-side terms in eq.(5.35) is as follows:

($i$) turbulent diffusion;
($ii$) mean motion "shear" coupling;
($iii$) mean motion pressure coupling;
($iv$) pressure action in the fluctuating motion;
($v$) pressure strain correlation, referring to the *instantaneous* value of the pressure;
($vi$) viscous action in the fluctuating motion;
($vii$) "dissipation".

For sake of modeling (Chapter 11), a different formulation of eq.(5.35) can be adopted. It is based on the following rearrangement:

$$(iii) + (v) = -\overline{u''_i}\frac{\partial \overline{P}}{\partial x_j} - \overline{u''_j}\frac{\partial \overline{P}}{\partial x_i} + \overline{p'\left(\frac{\partial u''_i}{\partial x_j} + \frac{\partial u''_j}{\partial x_i}\right)} ,$$

$$(iv) = -\frac{\partial}{\partial x_k}(\overline{p' u''_j}\,\delta_{ik}) - \frac{\partial}{\partial x_k}(\overline{p' u''_i}\,\delta_{jk}) ,$$

$$(vi) + (vii) = \overline{u''_i}\frac{\partial \overline{\tau}_{jk}}{\partial x_k} + \overline{u''_j}\frac{\partial \overline{\tau}_{ik}}{\partial x_k} + \frac{\partial}{\partial x_k}(\overline{\tau'_{ik} u''_j} + \overline{\tau'_{jk} u''_i}) - (\overline{\tau'_{ik}\frac{\partial u''_j}{\partial x_k}} + \overline{\tau'_{jk}\frac{\partial u''_i}{\partial x_k}}) .$$

Using the previous expressions in eq.(5.35) and introducing the kinematic turbulent stress tensor $R_{ij} = \widetilde{u''_i u''_j} \equiv \overline{\rho u''_i u''_j}/\overline{\rho}$, the corresponding equation is:

$$\frac{\partial(\overline{\rho}R_{ij})}{\partial t} + \frac{\partial(\overline{\rho}R_{ij}\widetilde{U}_k)}{\partial x_k} = -\overline{\rho}(R_{ik}\frac{\partial\widetilde{U}_j}{\partial x_k} + R_{jk}\frac{\partial\widetilde{U}_i}{\partial x_k}) + \overline{p'\left(\frac{\partial u''_i}{\partial x_j} + \frac{\partial u''_j}{\partial x_i}\right)}$$

$$-\frac{\partial}{\partial x_k}\left(\overline{\rho u''_i u''_j u''_k} + \overline{p' u''_i}\,\delta_{jk} + \overline{p' u''_j}\,\delta_{ik} - (\overline{\tau'_{ik} u''_j} + \overline{\tau'_{jk} u''_i})\right)$$

$$+\overline{u''_i}\left(\frac{\partial\overline{\tau}_{jk}}{\partial x_k} - \frac{\partial\overline{P}}{\partial x_j}\right) + \overline{u''_j}\left(\frac{\partial\overline{\tau}_{ik}}{\partial x_k} - \frac{\partial\overline{P}}{\partial x_i}\right) - (\overline{\tau'_{ik}\frac{\partial u''_j}{\partial x_k}} + \overline{\tau'_{jk}\frac{\partial u''_i}{\partial x_k}}) \quad (5.36)$$

### 5.6.2. TERNARY REGROUPING

According to the ternary regrouping analysis, the variable density equivalent expression to the incompressible Reynolds stress tensor is $-\overline{\rho u'_i u'_j}$. The corresponding transport equation reads (Chassaing [81]):

$$\frac{\partial}{\partial t}(\overline{\rho u'_i u'_j}) + \frac{\partial}{\partial x_k}(\overline{\rho u'_i u'_j}\,\overline{U}_k) = \underbrace{-\frac{\partial}{\partial x_k}(\overline{\rho u'_i u'_j u'_k})}_{(i')} \underbrace{- \overline{\rho u'_i u'_k}\frac{\partial\overline{U}_j}{\partial x_k} + \overline{\rho u'_j u'_k}\frac{\partial\overline{U}_i}{\partial x_k}}_{(ii')}$$

$$+ \underbrace{(\overline{\rho u'_j}\,\overline{F}_i + \overline{\rho u'_i}\,\overline{F}_j}_{(iii')} \underbrace{- \frac{\partial(\overline{p' u'_j})}{\partial x_i} - \frac{\partial(\overline{p' u'_i})}{\partial x_j}}_{(iv')} + \underbrace{\overline{p'(\frac{\partial u'_i}{\partial x_j} + \frac{\partial u'_j}{\partial x_i})}}_{(v')}$$

$$+\underbrace{[\frac{\partial(\overline{\tau'_{ik} u'_j})}{\partial x_k} + \frac{\partial(\overline{\tau'_{jk} u'_i})}{\partial x_k}]}_{(vi')} - \underbrace{[\overline{\tau'_{ik}\frac{\partial u'_j}{\partial x_k}} + \overline{\tau'_{jk}\frac{\partial u'_i}{\partial x_k}}]}_{(vii')} - A_{ij}\ , \quad (5.37)$$

$$\text{with} \quad A_{ij} = \overline{\rho u'_i}\frac{\mathrm{d}\overline{U}_j}{\mathrm{dt}} + \overline{\rho u'_j}\frac{\mathrm{d}\overline{U}_i}{\mathrm{dt}}\ ,$$

where d/dt stands for the "mean" material derivative $\partial/\partial t + \overline{U}_j\,\partial/\partial x_j$.

The physical description of the right-hand-side terms in eq.(5.37) is the same as in eq.(5.35). The two equations only differ from the presence of:

- the explicit contribution of the external body forces coupled with the turbulent mass flux ($iii'$), and the absence of mean pressure terms ($iii$);
- the pressure and viscous stresses *fluctuations* instead of their *instantaneous* values in ($v'$) - ($v$), ($vi'$) - ($vi$) and ($vii'$) - ($vii$);

– the turbulent mass fluxes $\overline{\rho u'_j}$ in the additional contribution $A_{ij}$.

## 5.7. Turbulence kinetic energy

Transport equations governing the turbulence kinetic energy according to binary and ternary formulations are deduced by contraction ($i = j$ and summation) of eqs.(5.35) and (5.37).

### 5.7.1. BINARY REGROUPING

Denoting the *instantaneous* value of the turbulence kinetic energy of density-weighted velocity fluctuations by $k = u''_i u''_i/2$, the mean value, according to Favre's averaging, is simply $\widetilde{k}(\equiv \frac{1}{2}\widetilde{u''_i u''_i}) = \frac{1}{2}\overline{\rho u''_i u''_i}/\overline{\rho}$. It is governed by the following balance equation:

$$\frac{\partial}{\partial t}(\overline{\rho}\,\widetilde{k}) + \frac{\partial}{\partial x_j}(\overline{\rho}\,\widetilde{k}\widetilde{U}_j) = -\underbrace{\frac{\partial}{\partial x_j}(\overline{\rho}\,\widetilde{k\,u''_j})}_{(i)} - \underbrace{\overline{\rho}\,\widetilde{u''_i u''_j}\frac{\partial \widetilde{U}_i}{\partial x_j}}_{(c)} - \underbrace{\frac{\partial(\overline{P}\,\overline{u''}_i)}{\partial x_i}}_{(ii)} \tag{5.38}$$

$$-\underbrace{\frac{\partial(\overline{p' u''}_i)}{\partial x_i}}_{(iii)} + \underbrace{\overline{P\frac{\partial u''_i}{\partial x_i}}}_{(e)} + \underbrace{\frac{\partial(\overline{\tau_{ij} u''_i})}{\partial x_j}}_{(iv)} - \underbrace{\overline{\tau_{ij}\frac{\partial u''_i}{\partial x_j}}}_{(d)}.$$

The physical meaning of the right-hand-side terms in eq.(5.38) is as follows:

($i$) turbulent diffusion;

($c$) "shear production", this term is present with an opposite sign in the kinetic energy equation of the mean motion eq.(5.26);

($ii$) external power of mean pressure forces acting through the fluctuating motion;

($iii$) external power of pressure fluctuation in the fluctuating motion;

($e$) pressure correlation with dilatation fluctuation. The opposite term is present in the mean internal energy equation correlation eq.(5.27);

($iv$) external power of fluctuating viscous forces in the fluctuating motion;

($d$) turbulent dissipation.

### 5.7.2. TERNARY REGROUPING

Similar contraction of eq.(5.37) yields:

$$\frac{\partial(\frac{1}{2}\overline{\rho u'_i u'_i})}{\partial t} + \frac{\partial(\frac{1}{2}\overline{\rho u'_i u'_i}\,\overline{U}_j)}{\partial x_j} = -\underbrace{\frac{\partial(\frac{1}{2}\overline{\rho u'_i u'_i u'_j})}{\partial x_j}}_{} - \underbrace{\overline{\rho u'_i u'_j}\frac{\partial \overline{U}_i}{\partial x_j}}_{} + \underbrace{\overline{\rho u'_i}\,\overline{F}_i}_{} \tag{5.39}$$

$$
-\underbrace{\frac{\partial(\overline{p'u'_i})}{\partial x_i}}_{(iii')}\overset{(i')}{} + \underbrace{\overline{p'\frac{\partial u'_i}{\partial x_i}}}_{(e')} + \underbrace{\frac{\partial(\overline{\tau'_{ij}u'_i})}{\partial x_j}}_{(iv')}\overset{(c')}{} - \underbrace{\overline{\tau'_{ij}\frac{\partial u'_i}{\partial x_j}}}_{(d')}\overset{(ii')}{} - C\,,
$$

$$
\text{with} \quad C = \frac{\partial(\overline{\rho u'_i})}{\partial t} + \overline{U}_j\frac{\partial(\overline{\rho u'_i})}{\partial x_j} \equiv \overline{\rho u'_i}\,\frac{\mathrm{d}\overline{U}_i}{\mathrm{dt}}\,.
$$

Apart from the explicit contribution of the turbulent mass flux in $C$, the analogies and differences in the physical meaning of the right-hand-side terms in eq.(5.39), as compared with those in eq.(5.38) are:

($i$') turbulent diffusion, analogous to ($i$);
($c$') "shear production", analogous to ($c$);
($ii$') external power of mean body forces acting through the fluctuating motion. This term disappears from the kinetic balance equation in a binary analysis. Under Boussinesq's approximation, it can be shown (Chassaing [80]) that the effect of body forces (Ternary regrouping) are accounted for by the mean pressure term ($ii$) in the binary formulation;
($iii$') external power of pressure fluctuation in the fluctuating motion, similar to ($iii$);
($e$') pressure *fluctuation* correlation with dilatation fluctuation. It should be noticed that the *instantaneous* value of the pressure appears in ($e$);
($iv$') external power of fluctuating viscous forces in the fluctuating motion, analogous to ($iv$) with a similar remark as above;
($d$') turbulent dissipation, analogous to ($d$) with a similar remark as above.

## 5.8. Energy transfer in compressible turbulent flows

### 5.8.1. BINARY REGROUPING

We are now able to identify the different components of the energy transfer that takes place in the turbulent motion of a compressible fluid. As opposed to the incompressible situation, where the analysis is only involving the kinetic energy balance equations, energy coupling in variable density fluid motions is involving three disctinct equations for mean internal energy, mean kinetic energy and turbulence kinetic energy.
According to a Favrian analysis, the corresponding equations (5.26), (5.27) and (5.38) can be rewritten as follows:

$$
\frac{\partial(\overline{\rho}\,\widetilde{e})}{\partial t} + \frac{\partial(\overline{\rho}\,\widetilde{e}\,\widetilde{U}_j)}{\partial x_j} \qquad = -(a) + (b) \qquad + (d) - (e) + A.T(1).
$$

$$\frac{\partial(\frac{1}{2}\overline{\rho}\widetilde{U}_i\widetilde{U}_i)}{\partial t}+\frac{\partial[(\frac{1}{2}\overline{\rho}\widetilde{U}_i\widetilde{U}_i)\,\widetilde{U}_j]}{\partial x_j} = +(a)\;-\;(b)-(c) \qquad +A.T(2).$$

$$\frac{\partial}{\partial t}(\overline{\rho}\,\widetilde{k})+\frac{\partial}{\partial x_j}(\overline{\rho}\,\widetilde{k}\widetilde{U}_j) = +(c)\;-\;(d)+(e)\;+A.T(3).$$

with

$$(a)=\overline{P}\frac{\partial\widetilde{U}_i}{\partial x_i} \quad (b)=\overline{\tau}_{ij}\frac{\partial\widetilde{U}_i}{\partial x_i} \quad (c)=-\overline{\rho}\,\widetilde{u''_i u''_j}\frac{\partial\widetilde{U}_i}{\partial x_j} \quad (d)=\overline{\tau_{ij}\frac{\partial u''_i}{\partial x_j}} \quad (e)=\overline{P\frac{\partial u''_i}{\partial x_i}} \tag{5.40}$$

$A.T(\alpha)$, $\alpha$=1 to 3 denote three additional terms which are specific to each equation, as denoted by roman figures in the underbrace labels of eqs.(5.26), (5.27) and (5.38).

From the previous equations, two main conclusions can be drawn about:

- the "external" supply of *total* mean energy;
- the "internal" energy transfer between three *separate* contributions.

***Total mean energy*** The equation governing the mean value of the total amount of internal *plus* kinetic energy, viz.

$$\overline{\mathcal{E}}=\overline{\rho(e+\frac{1}{2}U_iU_i)}\equiv\overline{\rho}\,\widetilde{e}+\frac{1}{2}\overline{\rho}\,\widetilde{U}_i\widetilde{U}_i+\frac{1}{2}\overline{\rho}\,\widetilde{u''_i u''}_i\,,$$

is

$$\begin{aligned}\frac{\partial\overline{\mathcal{E}}}{\partial t}+\frac{\partial(\overline{\mathcal{E}}\widetilde{U}_j)}{\partial x_j} = & -\frac{\partial(\overline{\rho}\,\widetilde{e''u''_j})}{\partial x_j}-\frac{\partial(\overline{\rho}\,\widetilde{ku''_j})}{\partial x_j}\;\ldots\ldots\; Diffusion\\ & +\frac{\partial\overline{q_i}}{\partial x_i}\;\ldots\ldots\; Thermal\,power\\ & -\frac{\partial(\overline{\rho}\,\widetilde{u''_i u''_j}\,\widetilde{U}_i)}{\partial x_j}\;\ldots\ldots\; Turbulence\,stress\,power\\ & -\frac{\partial(\overline{P}\,\widetilde{U}_i)}{\partial x_i}-\frac{\partial(\overline{P}\,\overline{u''}_i)}{\partial x_i}-\frac{\partial(\overline{p'u''}_i)}{\partial x_i}\;\ldots\ldots\; Pressure\,power\\ & +\frac{\partial(\overline{\tau}_{ij}\,\widetilde{U}_i)}{\partial x_j}+\frac{\partial(\overline{\tau_{ij}u''}_i)}{\partial x_j}\;\ldots\ldots\; Viscous\,stress\,power\end{aligned} \tag{5.41}$$

Eq.(5.41) is simply deduced by summation of eqs.(5.26), (5.27) and (5.38), where it is clear that $\sum_{\alpha=1}^{\alpha=3} A.T(\alpha)$ is the only term which is contributing to a net rate of change of $\overline{\mathcal{E}}$. According to the first law of thermodynamics, eq.(5.41) states that, on average, the rate of variation of the mean total energy of a fluid particle is balanced by *diffusion* and *external* power *exchanges* of thermal energy (restricted here to heat conduction transfer) and mechanical energy due to turbulence stresses, viscous stresses and pressure forces.

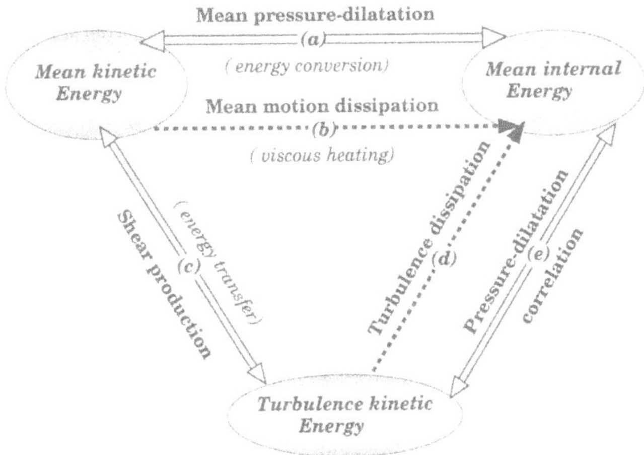


*Figure 5.2.* Energy transfer in compressible turbulent flows (binary regrouping analysis).

***Internal transfer of energy*** The total external supply of energy is converted into (i) mean internal energy (ii) kinetic energy of the mean motion and (iii) turbulence kinetic energy. Hence, an "internal" exchange of energy takes place between these three energy contributions, as depicted and quantified by the redistribution terms $(a)$ to $(e)$ in eqs.(5.40).
The situation can be sketched as illustrated in Fig.5.2, adapted from Huang *et al.* [221] and Ristorcelli [394] inter alia. It appears to be *formally*[7] similar to the one encountered in incompressible turbulence, as far as viscous dissipation by mean and turbulent motions $(b)$ - $(d)$ and shear production $(c)$ and the concerned.
The main significant differences, specific to compressible fluid motion, occur from internal energy conversion into mean and turbulence kinetic energy by pressure forces, acting through $(a)$ mean and $(e)$ fluctuating dilatation respectively.

## 5.8.2. TERNARY REGROUPING

When using density-weighted averages, a *single* turbulent mass flux contribution appears *explicitly* in the formulation of the general energy balance (see, for instance, the mean pressure term in eq.(5.41) ). Since all other contributions of d.f.c. are embedded in the macroscopic mean values, binary regrouping makes difficult to identify *all* d.f.c. contributions. To this regard and to some extent it could be worthwhile using the ternary regrouping, despite a bit more cumbersome formulation.

[7]This does not presume that the corresponding mechanisms are the same for both incompressible and compressible situations.

In this case, the averaged amount of the total energy is:

$$\overline{\mathcal{E}} = \overline{\rho(e + \frac{1}{2}U_iU_i)} \equiv \overline{\rho}(\overline{e} + \frac{1}{2}\overline{\rho}\,\overline{U}_i\overline{U}_i) + \frac{1}{2}\overline{\rho u'_i u'_i} + \overline{\rho e'} + \overline{\rho u'_i}\,\overline{U}_i\,. \quad (5.42)$$

As before, the equations governing the mean internal energy ($\overline{e}$), the mean kinetic energy ($\overline{\rho}\overline{U}_i\overline{U}_i/2$) and the turbulence kinetic energy ($\overline{\rho u'_i u'_i}/2$) — (5.30), (5.31) and (5.39) respectively — can be rewritten as follows:

$$\frac{\partial(\overline{\rho}\,\overline{e})}{\partial t} + \frac{\partial(\overline{\rho}\,\overline{e}\,\overline{U}_j)}{\partial x_j} = -(a') + (b') \quad +(d') - (e') + A.T(1) - C1$$

$$\frac{\partial(\frac{1}{2}\overline{\rho}\overline{U}_i\overline{U}_i)}{\partial t} + \frac{\partial[(\frac{1}{2}\overline{\rho}\overline{U}_i\overline{U}_i)\,\overline{U}_j]}{\partial x_j} = +(a') - (b') - (c') \quad + A.T(2) - C2$$

$$\frac{\partial}{\partial t}(\frac{1}{2}\overline{\rho u'_i u'_i}) + \frac{\partial}{\partial x_j}(\frac{1}{2}\overline{\rho u'_i u'_i U_j}) = \quad +(c') - (d') + (e) + A.T(3) - C3$$

with

$$(a') = \overline{P}\frac{\partial\overline{U}_i}{\partial x_i} \quad (b') = \overline{\tau}_{ij}\frac{\partial\overline{U}_i}{\partial x_i} \quad (c') = -\overline{\rho u'_i u'_j}\frac{\partial\overline{U}_i}{\partial x_j} \quad (d') = \overline{\tau'_{ij}\frac{\partial u'_i}{\partial x_j}} \quad (e') = \overline{p'\frac{\partial u'_i}{\partial x_i}} \quad (5.43)$$

and denoting by $C(\alpha)$, $\alpha$=1 to 3, the following expressions:

$$C1 = \frac{\partial(\overline{\rho e'})}{\partial t} + \frac{\partial(\overline{\rho e'}\,\overline{U}_j)}{\partial x_j}, \quad C2 = \overline{U}_i(\frac{\partial\overline{\rho u'_i}}{\partial t} + \frac{\partial(\overline{\rho u'_i}\,\overline{U}_j)}{\partial x_j}), \quad C3 = \overline{\rho u'_i}(\frac{\partial\overline{U}_i}{\partial t} + \overline{U}_j\frac{\partial\overline{U}_i}{\partial x_j})$$

Now, it can be easily verified that:

$$C1 + C2 + C3 = \frac{\partial(\overline{\rho e'})}{\partial t} + \frac{\partial(\overline{\rho e'}\,\overline{U}_j)}{\partial x_j} + \frac{\partial(\overline{\rho u'_i}\overline{U}_i)}{\partial t} + \frac{\partial(\overline{\rho u'_i}\overline{U}_i\overline{U}_j)}{\partial x_j},$$

which is identically equal to the mean material variation of the d.f.c. contributions to $\overline{\mathcal{E}}$ in eq.(5.42) (last two terms in the right-hand-side).

Hence, by summation of (5.30), (5.31) and (5.39), one obtains:

$$\frac{\partial \overline{\mathcal{E}}}{\partial t}+\frac{\partial(\overline{\mathcal{E}}\,\overline{U}_j)}{\partial x_j}=-\frac{\partial(\overline{\rho e' u'_j})}{\partial x_j}-\frac{\partial(\frac{1}{2}\overline{\rho u'_i u'_i u'_j})}{\partial x_j}\text{.........................} \mathit{Diffusion}$$

$$-\frac{\partial}{\partial x_j}[(\overline{e}+\frac{1}{2}\overline{U}_i\overline{U}_i)\overline{\rho u'_j}]\text{..........................} \mathit{D.f.c.\,diffusion}$$

$$+\frac{\partial \overline{q_i}}{\partial x_i}\text{.......................................................} \mathit{Thermal\,power}$$

$$+\overline{\rho}\,\overline{U}_i F_i+\overline{\rho u'_i}F_i\text{..................................} \mathit{Body\,force\,power}$$

$$-\frac{\partial(\overline{\rho u'_i u'_j}\,\overline{U}_i)}{\partial x_j}\text{............................} \mathit{Turbulence\,stress\,power}$$

$$-\frac{\partial(\overline{P}\,\overline{U}_i)}{\partial x_i}-\frac{\partial(\overline{p' u'_i})}{\partial x_i}\text{................................} \mathit{Pressure\,power}$$

$$+\frac{\partial(\overline{\tau}_{ij}\overline{U}_i)}{\partial x_j}+\frac{\partial(\overline{\tau'_{ij}u'_i})}{\partial x_j}\text{...........} \mathit{Viscous\,stress\,power} \quad (5.44)$$

Quite surprisingly, the result looks formally rather simple, as compared with the similar equation (5.41) obtained with the binary derivation. Now, as far as one is discussing energy transfer on the basis of eqs.(5.30), (5.31) and (5.39), the physical interpretation of the rolc of d.f.c. is as follows:

- *energy coupling (i):* on average, the five terms ($a'$) to ($e'$) do not contribute to the variation of the net amount of the mean total energy. Even though $\overline{\mathcal{E}}$ is no longer equal to the sum of the three mean contributions $\overline{\rho}\,\overline{e}$, $\overline{\rho}\,\overline{U}_i\overline{U}_i/2$ and $\overline{\rho u'_i u'_i}/2$, these terms still correspond to a *first type* of energy coupling in which density fluctuation correlations play no part;
- *energy coupling (ii):* all d.f.c. effects on energy coupling result in the additional advective contributions $C1$ to $C3$;
- *energy diffusion:* D.f.c. are also directly contributing to the diffusion of the kinetic energy of the mean motion (see additional diffusion term in the second line of eq.(5.44);
- *external exchange (i):* As opposed to eq.(5.41), no d.f.c. contribution to the mechanical energy supply from pressure forces is present;
- *external exchange (ii):* Turbulent mass flux is introducing an explicit contribution to the external mechanical power of body forces.

To summarize, when using a ternary regrouping, most of the departure from the incompressible situation in the energy exchange and coupling comes from that part of d.f.c. which is associated with additional transport and/or diffusion terms, in agreement with the advective origin of such density fluctuation correlations. On the other hand, with a mean mass-conservative formulation (binary regrouping), these advective d.f.c. contributions are removed from the mean balance equations. D.f.c. effects mostly result in the presence of a specific contribution to the mechanical power due to pressure forces.

## 5.9. Averaging the constitutive schemes

We restrict the analysis to the classical schemes of molecular diffusion, i.e., the Newton-Stokes and Fourier laws for momentum and heat transfer. Averaging the corresponding expressions, the mean viscous stresses ($\overline{\tau}_{ij}$) and heat conduction flux ($\overline{q}_i$) read:

$$(a) \quad \overline{\tau}_{ij} = 2(\overline{\mu S_{ij}}) + \frac{2}{3}(\overline{\mu \frac{\partial U_k}{\partial x_k}} \delta_{ij}) \,, \qquad (b) \quad \overline{q_i} = -\overline{\lambda \frac{\partial T}{\partial x_i}} \,, \tag{5.45}$$

where it should be noticed that a conventional averaging has been used. Now, in non constant density fluid motions, both the dynamic viscosity ($\mu$) and thermal conductivity ($\lambda$) are to be considered as *variable* physical properties in the flow field. Developing the previous relations yields, for instance:

$$\overline{\tau}_{ij} = 2\overline{\mu}\overline{S}_{ij} + \frac{2}{3}\overline{\mu}\frac{\partial \overline{U}_k}{\partial x_k}\delta_{ij} + 2(\overline{\mu' s'_{ij}}) + \frac{2}{3}(\overline{\mu' \frac{\partial u'_k}{\partial x_k}}) \,. \tag{5.46}$$

This equation can be simplified by assuming that correlations with fluctuations of the molecular transfer coefficients are negligible, viz.

$$\frac{\overline{\mu' f'}}{\overline{\mu}\overline{F}} \ll 1 \,, \qquad \frac{\overline{\lambda' f'}}{\overline{\lambda}\overline{F}} \ll 1 \,. \tag{5.47}$$

Consequently the mean value of the molecular fluxes are:

$$(a) \quad \overline{\tau}^R_{ij} = 2\overline{\mu}\overline{S}_{ij} + \frac{2}{3}\overline{\mu}\frac{\partial \overline{U}_k}{\partial x_k}\delta_{ij} \,, \qquad (b) \quad \overline{q}^R_i = -\overline{\lambda}\frac{\partial \overline{T}}{\partial x_j} \,. \tag{5.48}$$

Expression (5.48a) was used, for instance, by Sarkar *et al.* [411] in 1990.

When using Favre averaging, Vandromme and Ha Minh [469] assumed in 1983, that the viscous stresses can be expressed as:

$$\overline{\tau}^F_{ij} = 2\overline{\mu}\widetilde{S}_{ij} + \frac{2}{3}\overline{\mu}\frac{\partial \widetilde{U}_k}{\partial x_k}\delta_{ij} \,. \tag{5.49}$$

Here again, Favrian correlations with fluctuations of the molecular transfer coefficients are taken negligible ($\overline{\mu' f''} \ll \overline{\mu}\widetilde{F}$). Since $\widetilde{F} = \overline{F} + \overline{\rho' f'}/\overline{\rho}$, eq.(5.49) can be directly compared with eq.(5.48a), that is:

$$\overline{\tau}_{ij}^{F} = \overline{\tau}_{ij}^{R} + 2\overline{\mu}(\frac{\overline{\rho' s'_{ij}}}{\overline{\rho}} + \frac{1}{3}\frac{\partial(\overline{\rho' u'_k}/\overline{\rho})}{\partial x_k}\delta_{ij}) \ . \tag{5.50}$$

Thus with $\overline{\tau}_{ij}^{F}$, the d.f.c. terms are driving mean viscous stresses, which is not the case with $\overline{\tau}_{ij}^{R}$.

*CHAPTER 6*

# SOME BASIC VARIABLE DENSITY MECHANISMS IN TURBULENT FLOWS

*The aim of this chapter is to point out the existence and emphasize the understanding of some of those properties which make variable density turbulent flows different from the incompressible ones.*
*The equation governing the instantaneous vorticity is first derived in the general case: fluid with variable density* and *nonconstant physical properties. Then, new vorticity generation mechanisms, as compared with the constant density situation, are discussed.*
*The second section deals with correlations with density fluctuations (d.f.c.) which are necessarily introduced in any statistical treatment of the instantaneous Navier-Stokes equations. The "diffusive" role of such d.f.c. is analyzed and discussed in low speed flows.*
*The last part of the chapter is devoted to specific mechanisms associated with dilutation fluctuations in pressure-correlation and dissipation terms. They are analyzed in relation with their contributions in various energy balance equations.*

## 6.1. Introduction

To get a tractable picture of the dynamic and energetic statistical properties of a turbulent field, one has to combine one point and two or multi-points statistical analysis. One point analysis, based on single-point averaged equations, is basically local: when restricted to second order moments, it is only concerned with mechanisms which are present in momentum and energy balance equations, such as convection, turbulent transport or diffusion, production, dissipation, redistribution...
The second type of approach, which can be based on space-correlations, structure functions or spectra, aims at investigating interactions, energy transfer, dissipation... among the "structures" of the turbulent agitation.

Widely developed in studying incompressible fluid turbulence, both approaches can also be applied to turbulent motions of variable density fluid. Accordingly, when adopting the first approach, one is mainly interested in identifying variable density — and/or compressibility — effects that

*explicitly* emerge from terms which are present in the considered variable-density, local equations.
The second approach [431], [68], is more specifically devoted to investigating the nature of the same type of effects which is *implicitly* reflected by *structural* modifications of turbulence through the pressure field, the anisotropy of the fluctuating velocity field...

The present chapter is only concerned with the first approach. Some peculiarities of scalar fields structural properties will be presented in Chapter 7, assuming a passive contaminant behavior.

## 6.2. Vorticity in variable density fluid motion

Vorticity generation, and consequently enstrophy production are major features of turbulent flows. In constant density, incompressible fluid motion, enstrophy increases only by vortex stretching or shrinking. Since the rates of both mechanisms are proportional to the enstrophy itself, they are not acting in irrotational flows. For variable density fluid motions, additional mechanisms are present in the vorticity dynamics, which contribute to the generation and growth of enstrophy. Provided that pressure and density gradients are not parallel, a baroclinic torque, for instance, is generated, which creates or damps vorticity even in an irrotational fluid motion. To find out the origin, and get a better understanding of such mechanisms, we shall first derive the general transport equation of the vorticity vector in variable density fluid motions.

### 6.2.1. THE INSTANTANEOUS VORTICITY EQUATION

The instantaneous vorticity vector $\vec{\Omega} = \vec{\text{curl}}\vec{V}$ is governed by a transport equation which can be derived from the Navier-Stokes equation, rewritten here in classical vector notations:

$$\rho\left(\frac{\partial\vec{V}}{\partial t} + \vec{\text{grad}}\left(\frac{V^2}{2}\right) + \vec{\text{curl}}\vec{V}\wedge\vec{V}\right) = \rho\,\vec{\text{grad}}\,\mathcal{U} - \vec{\text{grad}}\,P + \mu\vec{\Delta}\vec{V} + \frac{\mu}{3}\vec{\text{grad}}(\text{div}\vec{V}) + E.V.V. \quad (6.1)$$

Here, E.V.V. stands for extra viscous terms due to spatial viscosity variations:

$$E.V.V. = \left(\frac{\partial U_i}{\partial x_j} + \frac{\partial U_j}{\partial x_i}\right)\frac{\partial\mu}{\partial x_j} - \frac{2}{3}\left(\frac{\partial U_j}{\partial x_j}\right)\frac{\partial\mu}{\partial x_i} \equiv 2\overline{\overline{S}} \odot \vec{\text{grad}}\mu - \frac{2}{3}\text{div}\vec{V}\,\vec{\text{grad}}\mu\ .$$

In eq.(6.1), the external body forces are supposed to derive from a potential scalar field $\mathcal{U}$, and $\vec{\Delta}$ denotes the laplacian operator:

$$\vec{\Delta}. = \vec{\text{grad}}(\text{div}.) - \vec{\text{curl}}(\vec{\text{curl}}.) \quad (6.2)$$

By introducing $\vec{\Omega}$ and using the previous identity, eq.(6.1) becomes:

$$\frac{\partial\vec{V}}{\partial t}+\vec{\text{grad}}(\frac{V^2}{2})+\vec{\Omega}\wedge\vec{V}=\vec{\text{grad}}\,\mathcal{U}-\frac{1}{\rho}\vec{\text{grad}}(P-\frac{4}{3}\mu\,\text{div}\vec{V})-\frac{\mu}{\rho}\vec{\text{curl}}\,\vec{\Omega}\,.$$
$$+\,E.V.V./\rho \tag{6.3}$$

Taking the curl of eq.(6.3) and recalling that $\vec{\text{curl}}(\vec{\text{grad}}.)\equiv\vec{0}$ and $\vec{\text{curl}}(f\vec{A})\equiv f\vec{\text{curl}}\,\vec{A}+\vec{\text{grad}}\,f\wedge\vec{A}$, the following relation is obtained:

$$\frac{\partial\vec{\Omega}}{\partial t}+\vec{\text{curl}}(\vec{\Omega}\wedge\vec{V})=-\vec{\text{grad}}(\frac{1}{\rho})\wedge\vec{\text{grad}}(P-\frac{4}{3}\mu\,\text{div}\vec{V})-\vec{\text{curl}}(\nu\vec{\text{curl}}\,\vec{\Omega})$$
$$+\,\vec{\text{curl}}(E.V.V./\rho)\,. \tag{6.4}$$

Now, it can be observed that:

$$\begin{aligned}\vec{\text{curl}}(\nu\,\vec{\text{curl}}\,\vec{\Omega})&\equiv\vec{\text{curl}}[\frac{1}{\rho}(\mu\,\vec{\text{curl}}\,\vec{\Omega})]\\&=\frac{1}{\rho}\vec{\text{curl}}(\mu\,\vec{\text{curl}}\,\vec{\Omega})+\vec{\text{grad}}(\frac{1}{\rho})\wedge(\mu\vec{\text{curl}}\,\vec{\Omega})\\&=\frac{\mu}{\rho}\vec{\text{curl}}(\vec{\text{curl}}\,\vec{\Omega})+\frac{1}{\rho}(\vec{\text{grad}}\mu\wedge\vec{\text{curl}}\,\vec{\Omega})+\vec{\text{grad}}(\frac{1}{\rho})\wedge(\mu\vec{\text{curl}}\,\vec{\Omega})\\&=-\nu\vec{\Delta}\vec{\Omega}+\vec{\text{grad}}(\frac{1}{\rho})\wedge(\mu\vec{\text{curl}}\,\vec{\Omega})+\frac{1}{\rho}(\vec{\text{grad}}\mu\wedge\vec{\text{curl}}\,\vec{\Omega})\,.\end{aligned}$$

Upon substituting the previous expression in eq.(6.4), one obtains:

$$\frac{\partial\vec{\Omega}}{\partial t}+\vec{\text{curl}}(\vec{\Omega}\wedge\vec{V})=\nu\vec{\Delta}\vec{\Omega}-\vec{\text{grad}}(\frac{1}{\rho})\wedge[\vec{\text{grad}}(P-\frac{4}{3}\mu\,\text{div}\vec{V})+\mu\vec{\text{curl}}\,\vec{\Omega}]$$
$$+\,E.V.V.*\,, \tag{6.5}$$

where $\qquad E.V.V.*=\vec{\text{curl}}(E.V.V./\rho)-(\vec{\text{grad}}\mu\wedge\vec{\text{curl}}\,\vec{\Omega})/\rho\,.$

The usual transport form of the left hand side terms in eq.(6.5) can be easily deduced by using the following identity:

$$\vec{\text{curl}}(\vec{A}\wedge\vec{B})=\overline{\overline{grad}}\vec{A}\odot\vec{B}-\vec{B}\text{div}\vec{A}-\overline{\overline{grad}}\vec{B}\odot\vec{A}+\vec{A}\text{div}\vec{B}\,.$$

Here $\odot$ is the product obtained by contraction upon the derivative coordinate. Hence:

$$\frac{\partial\vec{\Omega}}{\partial t}+\overline{\overline{grad}}\vec{\Omega}\odot\vec{V}=\overline{\overline{grad}}\vec{V}\odot\vec{\Omega}+\nu\vec{\Delta}\vec{\Omega}-\vec{\Omega}\text{div}\vec{V}$$
$$-\vec{\text{grad}}(\frac{1}{\rho})\wedge[\vec{\text{grad}}(P-\frac{4}{3}\mu\,\text{div}\vec{V})+\mu\vec{\text{curl}}\,\vec{\Omega}]\,.+E.V.V.* \tag{6.6}$$

From the definition of $E.V.V.*$, it is clear that:

$$E.V.V.* = \frac{1}{\rho}\overrightarrow{\mathrm{curl}}(E.V.V.) + \overrightarrow{\mathrm{grad}}(\frac{1}{\rho}) \wedge E.V.V. - \frac{1}{\rho}\overrightarrow{\mathrm{grad}}\,\mu \wedge \overrightarrow{\mathrm{curl}}\,\vec{\Omega}\ .$$

Upon substituting in eq.(6.6), one obtains:

$$\begin{aligned}\frac{\partial\vec{\Omega}}{\partial t} + \overline{\overline{grad}}\vec{\Omega} \odot \vec{V} &= \overline{\overline{grad}}\vec{V} \odot \vec{\Omega} + \nu\vec{\Delta}\vec{\Omega} - \vec{\Omega}\,\mathrm{div}\vec{V} \\ -\overrightarrow{\mathrm{grad}}(\frac{1}{\rho}) \wedge &\underbrace{[\overrightarrow{\mathrm{grad}}(P - \frac{4}{3}\mu\,\mathrm{div}\vec{V}) + \mu\overrightarrow{\mathrm{curl}}\vec{\Omega} - E.V.V.]}_{(i)} + E.V.V.**\ ,\end{aligned}$$

where

$$E.V.V.** = \frac{1}{\rho}(\overrightarrow{\mathrm{curl}}(E.V.V.) - \overrightarrow{\mathrm{grad}}\,\mu \wedge \overrightarrow{\mathrm{curl}}\vec{\Omega}\ .$$

Now, from eq.(6.3), it is clear that the underbrace terms in the previous equation are simply:

$$(i) = \rho(\frac{\mathrm{d}\vec{V}}{\mathrm{dt}} - \overrightarrow{\mathrm{grad}}\,\mathcal{U})\ ,$$

Thus, the final form of the vorticity transport equation[1] is:

$$\begin{aligned}\frac{\mathrm{d}\vec{\Omega}}{\mathrm{dt}} \equiv \frac{\partial\vec{\Omega}}{\partial t} + \overline{\overline{grad}}\vec{\Omega} \odot \vec{V} &= \underbrace{\overline{\overline{grad}}\vec{V} \odot \vec{\Omega}}_{(a)} + \underbrace{\nu\vec{\Delta}\vec{\Omega}}_{(b)} - \underbrace{\vec{\Omega}\,\mathrm{div}\vec{V}}_{(c)} \\ &+ \underbrace{\frac{\overrightarrow{\mathrm{grad}}\,\rho}{\rho} \wedge (\frac{\mathrm{d}\vec{V}}{\mathrm{dt}} - \overrightarrow{\mathrm{grad}}\,\mathcal{U})}_{(d)} + E.V.V.**\ .\end{aligned} \qquad (6.7)$$

Several comments can be added to this equation.

- For solenoidal, constant density flows, the right hand side of this equation reduces to the first two terms $(a)$ and $(b)$. The second one, $(b)$ is the classical diffusion term by molecular viscosity. The first one, which is identically zero in two-dimensional flows, accounts for vorticity *stretching* (or shrinking) and *tilting*. The first effect results in a change of the *modulus* of the vorticity vector, the second in a change of its *orientation*;

[1]Without detailing the viscous terms and using tensor notation, an alternative, more compact form of this equation is (see [3], for instance):

$$\frac{d\Omega_i}{dt} = S_{ij}\Omega_j - \Omega_i S_{jj} + \epsilon_{iql}\frac{1}{\rho^2}\frac{\partial\rho}{\partial x_q}\frac{\partial P}{\partial x_l} + \epsilon_{iql}\frac{\partial}{\partial x_q}(\frac{1}{\rho}\frac{\partial\tau_{lj}}{\partial x_j})\ ,$$

where $S_{ij}$ is the strain rate tensor, $\tau_{lj}$ the viscous stress tensor and $\epsilon_{iql}$ the alternate third rank pseudo-tensor.

- The third term $(c)$ is specific to non-solenoidal velocity field motion. Depending on the bulk volume variation of the fluid particle, it acts so as to increase or reduce the *modulus* of the vorticity vector. As opposed to the isovolume case, in non-solenoidal situations the vortex stretching is no longer the only mechanism which can change the intensity of the vorticity;
- For *inviscid* fluid motions, only one additional contribution, specific to the density variation, is introduced by $(d)$ in the vorticity transport equation. It is called the *baroclinic* source $(B.S.)$:

$$B.S. = -\overrightarrow{\mathrm{grad}}(\frac{1}{\rho}) \wedge \overrightarrow{\mathrm{grad}}\, P \equiv \frac{1}{\rho^2}\overrightarrow{\mathrm{grad}}\,\rho \wedge \overrightarrow{\mathrm{grad}}\, P\,. \tag{6.8}$$

  Thus, provided that density and pressure gradients are not aligned, vorticity can be generated from baroclinic torques in compressible or variable density fluid motions, even in initially irrotational flows. This source term vanishes in a *barotropic* evolution, where the pressure is a function of the density, say $P = f(\rho)$. Indeed:

$$\overrightarrow{\mathrm{grad}}\, P = \overrightarrow{\mathrm{grad}}[f(\rho)] \equiv f'(\rho)\,\overrightarrow{\mathrm{grad}}\,\rho\,,$$

  so that density and pressure gradients are necessarily aligned in such a situation.
- In general (*viscous* fluids), variable density effects are clearly associated with three terms $(b)$, $(d)$ and E.V.V.** in eq.(6.7). Accordingly, viscosity effects can also introduce changes in vorticity: (i) in addition to the baroclinic source $(d)$, and (ii) through the extra terms E.V.V.**, due to spatial variations of the viscosity.

A simpler alternative form to the previous vorticity transport equation (6.7) can be obtained by introducing the *specific vorticity* $\vec{\Omega}/\rho$ (see, for instance, Smits & Dussauge [434]). Indeed, taking the material derivative of the specific vorticity, one obtains:

$$\frac{d}{dt}(\frac{\vec{\Omega}}{\rho}) = \frac{1}{\rho}\frac{d\vec{\Omega}}{dt} - \frac{1}{\rho^2}\vec{\Omega}\frac{d\rho}{dt} \equiv \frac{1}{\rho}\frac{d\vec{\Omega}}{dt} + \frac{1}{\rho}\vec{\Omega}\,\mathrm{div}\vec{V}\,,$$

where the last equality directly results from the continuity equation. Thus, by dividing both sides of eq.(6.7) by $\rho$, it comes:

$$\begin{aligned}\frac{\mathrm{d}}{\mathrm{dt}}(\frac{\vec{\Omega}}{\rho}) &= \overline{\overline{grad}}\vec{V} \odot (\frac{\vec{\Omega}}{\rho}) + \frac{\nu}{\rho}\vec{\Delta}\vec{\Omega}\\ &+\frac{\overrightarrow{\mathrm{grad}}\,\rho^2}{\rho} \wedge (\frac{\mathrm{d}\vec{V}}{\mathrm{dt}} - \overrightarrow{\mathrm{grad}}\,\mathcal{U}) + \frac{E.V.V.**}{\rho}\,.\end{aligned} \tag{6.9}$$

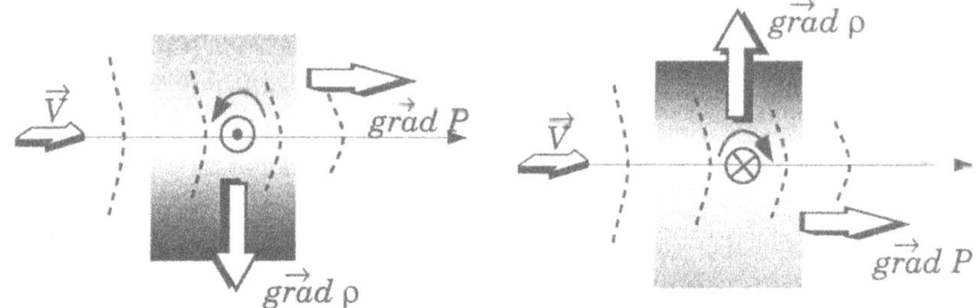


*Figure 6.1.* Sign of the vorticity generation by baroclinic torque in a horizontal, pressure driven, stratified, duct flow.

The previous formulation — eq.(6.9)— makes clear the analogy between the constant and variable density transport equations, when substituting the *specific* vorticity $\vec{\Omega}/\rho$ — variable density situation — by the usual kinematic vorticity $\vec{\Omega}$ — constant density situation. Thus in variable density flows, stretching a vortex tube does not necessarily increases its vorticity as lon as the specific vorticity is concerned. As already mentioned, vorticity generation/destruction by vortex stretching/shrinking cannot be deduced independently, i.e., without introducing density variations due to changes in the volume of the fluid element.

### 6.2.2. BAROCLINIC GENERATION IN A STRATIFIED DUCT FLOW

When density variation results from a vertical stratification in the gravity field ($\vec{g}$), the density gradient can be taken as $\overrightarrow{\text{grad}}\,\rho = \delta\rho\vec{g}/U_0^2$ where $U_0$ denotes a velocity scale, and $\delta$ a pure constant. The baroclinic source term ($B.S.$) is then given by:

$$B.S. = \frac{\delta}{\rho U_0^2}\vec{g} \wedge \overrightarrow{\text{grad}}\,P\,.$$

Thus in a horizontal duct flow, a baroclinic torque exists which creates positive or negative vorticity, depending upon the sign of the density stratification, as sketched in figure 6.1.

### 6.2.3. BAROCLINIC GENERATION THROUGH A CURVED SHOCK

It is well known that vorticity is generated behind a curved shock when a uniform inviscid fluid flow passes through it. This a direct consequence of the baroclinic generation in the vorticity equation eq.(6.7). According to Gibb's relation $T\delta s = \delta h - \delta P/\rho$, the pressure gradient can be expressed

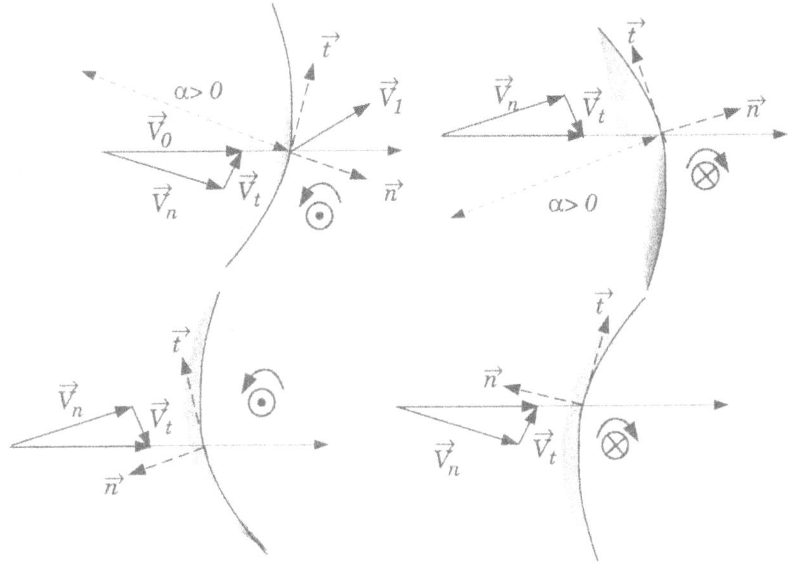


*Figure 6.2.* Sign of the vorticity generation by baroclinic torque in a flow passing through a curved shock: concave, top and convex, bottom.

with respect to entropy ($s$) and enthalpy ($h$) gradients as follows:

$$\frac{1}{\rho}\overrightarrow{\text{grad}}\, P = \overrightarrow{\text{grad}}\, h - T\overrightarrow{\text{grad}}\, s \ .$$

Thus, the baroclinic source term — eq.(6.8) — becomes:

$$B.S.(\equiv \overrightarrow{\text{curl}}(\frac{1}{\rho}\overrightarrow{\text{grad}}\, P) = \overrightarrow{\text{curl}}(T\,\overrightarrow{\text{grad}}\, s) = \overrightarrow{\text{grad}}\, T \wedge \overrightarrow{\text{grad}}\, s \ .$$

Now, as the flow passes through the shock, the entropy increases, so that $\overrightarrow{\text{grad}}\, s$ is positive in the flow direction. Since the total enthalpy remains constant through the shock, considered as an adiabatic process, the temperature gradient is related to the velocity field by $C_p\,\overrightarrow{\text{grad}}\, T = -\overrightarrow{\text{grad}}(V^2/2)$. From these remarks, it can be inferred that the sign of the baroclinic source term depends on the geometry of the flow field near the shock wave.

As shown in Kida and Orszag [246], it can been concluded that, for a given inviscid, steady, irrotational flow in front of a shock of a given radius of curvature ($\alpha$), the sign of the vorticity generated behind the shock depends both on the angle of attack of the flow and the curvature of the shock, i.e., the sign of $\alpha$. As sketched in Fig.6.2, for a concave shock ($\alpha > 0$), a positive vorticity is created if the tangential component of the incident velocity is of the same sign of the unit vector tangent to the shock.

## 6.3. D.f.c. and turbulent mass flux

As shown in the previous chapter, density fluctuations introduce specific correlations in the mean motion transport equations, as compared with the constant density situation. These are the so-called density fluctuation correlations (d.f.c.): $\overline{\rho' f'} \equiv -\overline{\rho}\overline{f''}$, for any scalar fluctuation $f'$ or $f''$. These d.f.c.'s include, in particular, the turbulent mass flux $\overline{\rho' u'_i} \equiv -\overline{\rho}\overline{u''_i}$. The formal equivalence between the previous expressions based on centered ($f'$, $u'_i$) or density weighted ($f''$, $u''_i$) fluctuations makes clear that the presence of such quantities in the averaged equations is *intrinsic* to variable density turbulent flow, i.e., independent of the averaging procedure (density-weighted or not) which is adopted.
A proper physical interpretation of the general role of d.f.c. terms is not easy. With a binary regrouping (see, for instance, Favre [153]), the turbulent mass flux is embodied in the "new", macroscopic, mass-weighted mean velocity. With such a mechanical interpretation, the contribution of the turbulent mass flux to the mean momentum balance equation is entirely implicit. However, such a result does not apply to all statistical equations, since an explicit contribution from $\overline{u''_i}$ is present in the transport equations for second order moments, for instance.
On the other hand, when averaging the continuity equation using centered statistical decompositions for both density and velocity, $\overline{\rho' u'_i} \equiv \overline{\rho u'_i}$ acts as a turbulent transport (by velocity fluctuations) of the instantaneous mass per unit volume. This interpretation may suggest to adopt a gradient type diffusion modeling, a scheme which may yield inconsistent predictions[2].

A deeper insight of the role of d.f.c. can be obtained when enlarging the frame of the analysis from a mechanical to a thermodynamic point of view, as suggested by the ternary formulation (Chapter 5) and illustrated by the following two examples. In these examples, exact expressions of density fluctuation correlations can be derived from the equation of state.

### 6.3.1. EXACT EXPRESSIONS OF D.F.C. IN LOW SPEED FLOWS

***Isothermal and isobaric mass mixing.*** Let us first consider the correlation of mass-fraction and density fluctuations $\overline{\rho' \gamma'}$ in an ideal mixing situation of two pure species, at constant pressure and temperature. As recalled in Chapter 4, the equation of state can be written as:

$$\overline{\rho} + \rho' = a\left(\overline{\rho}\overline{C} + \rho'\overline{C} + \rho\gamma'\right) + b\ , \tag{6.10}$$

[2]Driscoll *et al.* [134] concluded, from direct measurements of the axial and radial components of the turbulent mass flux in a turbulent non-premixed flame, that a first-gradient diffusion submodel for the axial correlation was in contradiction with the experimental data (see Chapter 10).

making use of Reynolds decomposition for both density and mass fraction, with $\overline{C} + \gamma' = C$. Here, and for a given binary mixture of pure species, $a$ and $b$ are two constant coefficients . Averaging eq.(6.10) directly yields:

$$\overline{\rho} = a\left(\overline{\rho}\overline{C} + \overline{\rho\gamma'}\right) + b \,, \tag{6.11}$$

which, by subtraction from eq.(6.10) gives:

$$\rho' = \frac{a}{1 - a\overline{C}}\left(\rho\gamma' - \overline{\rho\gamma'}\right) \,. \tag{6.12}$$

Then, multiplying this relation by the product of $n$ scalar fluctuations $f_1' f_2' ... f_n'$ and averaging, one can obtain:

$$\overline{\rho' f_1' f_2' ... f_n'} = \frac{a}{1 - a\overline{C}}\left(\overline{\rho\gamma' f_1' f_2' ... f_n'} - \overline{\rho\gamma'} \times \overline{f_1' f_2' ... f_n'}\right) \,. \tag{6.13}$$

This general expression of the $n^{th}$ order correlation with density fluctuation reduces to a simpler form for any first order moment. In this case, the second term in the right hand side of eq.(6.13) is zero, due to the centered fluctuation condition ($\overline{f_\alpha'} = 0$, $\forall\, \alpha = 1, ... n$). Hence:

$$\overline{\rho' f'} = \frac{a}{1 - a\overline{C}}\overline{\rho\gamma' f'} \,. \tag{6.14}$$

This expression clearly states that first order d.f.c. with any scalar fluctuation ($\overline{\rho' f'}$) is directly proportional to the second order density weighted correlation of the considered scalar fluctuation with the mass-fraction fluctuation ($\overline{\rho\gamma' f'}$). This results similarly applies to any component of the turbulent mass flux, viz.

$$\overline{\rho' u_i'} = \frac{a}{1 - a\overline{C}}\overline{\rho\gamma' u_i'} \,. \tag{6.15}$$

***Isobaric thermal mixing.*** Similar expressions can be derived for the temperature-d.f.c. in low speed motion when considering temperature mixing with an isobaric fluctuation condition ($p' \cong 0$). Hence, the instantaneous equation of state can be rewritten as:

$$\overline{P} = r(\overline{\rho}\overline{T} + \rho'\overline{T} + \rho\theta') \,,$$

making use of Reynolds decomposition $\overline{T} + \theta' = T$ for the temperature. By subtracting the mean value, the following expression[3] for the density fluctuation is obtained:

$$\rho' = -\frac{1}{\overline{T}}(\rho\theta' - \overline{\rho\theta'}) \,. \tag{6.16}$$

[3]Simplified expressions, under nearly incompressible or weakly compressible assumptions have been discussed in Chapters 3 and 4 (see, for instance, Chapter 4-§4.4.1). They correspond to linearized approximations to eq.(6.16).

Hence, making the appropriate products and averaging, any temperature-d.f.c. can be deduced, such as, for instance:

$$\overline{\rho' f'} = -\frac{1}{\overline{\overline{T}}}\,\overline{\rho\theta' f'} \qquad \text{and} \qquad \overline{\rho' u_i'} = -\frac{1}{\overline{\overline{T}}}\,\overline{\rho\theta' u_i'}\,. \tag{6.17}$$

To summarize, the following exact expressions apply to d.f.c. in the two previously considered situations:

$$\frac{\overline{\rho\gamma'}}{\overline{\overline{\rho\gamma'^2}}} = \frac{\overline{\rho u_i'}}{\overline{\overline{\rho\gamma' u_i'}}} = \frac{a}{1 - a\overline{\overline{C}}}\ (a), \qquad \frac{\overline{\rho\theta'}}{\overline{\overline{\rho\theta'^2}}} = \frac{\overline{\rho u_i'}}{\overline{\overline{\rho\theta' u_i'}}} = -\frac{1}{\overline{\overline{T}}}\ (b). \tag{6.18}$$

## 6.3.2. ANALYSIS OF D.F.C. IN LOW SPEED FLOWS

Direct measurements of d.f.c. terms are rather scarce. Most of the available data concern simple turbulent shear flows, such as boundary layers and jets, and mainly focus on the streamwise component of the turbulent mass flux and density correlations with temperature or mass fraction. They suggest that the $\overline{\rho' u'}$ correlation between the density fluctuation and the streamwise velocity fluctuation is generally low, as compared to the product of the mean values $\overline{\rho}\overline{U}$. In a helium/air jet, Zhu *et al.* [503] found that the maximum of the absolute centerline value of $|\overline{\rho' u'}/\overline{\rho}\overline{U}|_{\mathcal{C}}$ was about $0.3 \times 10^{-2}$. This result was confirmed later by So *et al.* [436]. Similarly, the measurements of Larue & Libby ( [269] 1977, [270] 1980) in a turbulent boundary layer with slot injection of helium, gave a maximum value of $\overline{\rho' u'}/\overline{\rho}\overline{U} \approx 1.4 \times 10^{-2}$. However, such low values are not representative of any d.f.c. Concerning the mass-fraction/density correlation, for instance, an interesting result emerges from the investigation of Sautet (1992). For pure H2/air jet mixing, *negative* values of density-concentration fluctuation correlation $\overline{\rho'\gamma'}$ are measured. The absolute value along the axis is as much as 40% of $(\bar{\rho}\bar{C})_{\mathcal{C}}$ at the downstream location $x/D_0 = 18$. On the contrary, when the density ratio of the exhausting jet is greater than unity, as for the pure CO2/air jet mixing, this correlation is found to be positive.

Thus, from the reported data, there is no definitive experimental evidence to support or contradict the idea that quantities like $\bar{\rho}\bar{F}$ and $\overline{\rho f'}$ have the same type of variation throughout the flow field, nor to assess whether or not any d.f.c. is always negligible.

Previous relations (6.18) can be used to get information on d.f.c.'s in simple thin shear layer flows, and in jet flows more particularly. This case has been studied in Chapter 4, §-4.4.1, where it has been shown that exact expressions of d.f.c.'s can be derived from second-order moments, by using

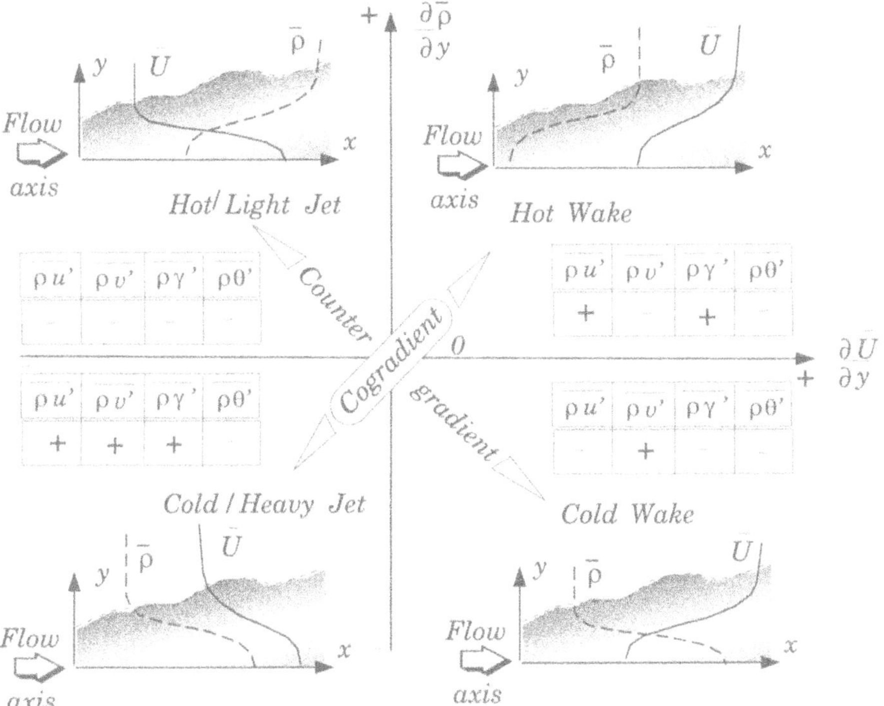


*Figure 6.3.* Signs of d.f.c. in co- and counter-gradient low speed, variable density jets and wakes.

the so-called 'indicator functions', defined in eqs (4.25) and (4.29) for mass-fraction and temperature, respectively. Now, it can be easily observed that the right hand side expression in eq.(6.18a) is identical to the mass-fraction indicator function $\Gamma^S(\overline{C})$ and to minus the temperature indicator function $\Gamma^S(\overline{\Theta})$ in eq.(6.18b). Hence, a first conclusion immediately emerges:
The signs of d.f.c. with mass-fraction and temperature, $\overline{\rho'\gamma'}$ and $\overline{\rho'\theta'}$, respectively depend on those of the corresponding indicator functions, and not of the mean mass-fraction and temperature gradients.
From the results given in Chapter 4, a second conclusion, illustrated in Fig. 6.3, can be drawn from the variations of $\Gamma^S(\overline{C})$ and $\Gamma^S(\overline{\Theta})$ plotted in Fig. 4.1:

- temperature-density fluctuations are *always negatively* correlated;
- mass-fraction-density fluctuations are *negatively* correlated in *light* jets ($s_\rho < 1$), but *positively* correlated in *heavy* ones ($s_\rho > 1$).

Thus, as far as mass-fraction and temperature d.f.c.'s are concerned, no *general* analogy exists between the signs of such moments in hot/*light* and cold/*heavy* situations.

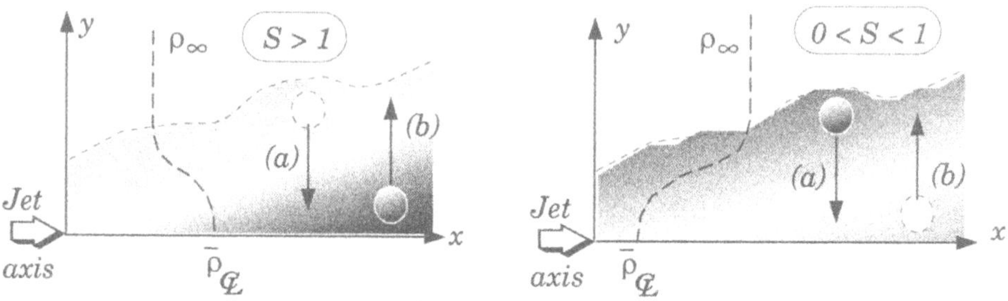


*Figure 6.4.* Sketch of co- and counter-gradient mixing in variable density jets.

Relations (6.18) can also be used to get the signs of second order cross-correlations with velocity fluctuations $\overline{\rho\gamma' u_i'}$ and $\overline{\rho\theta' u_i'}$, provided that the signs of the turbulent mass fluxes are known. This can be achieved from simple phenomenological considerations, referring to the two jet mixing situations sketched in figure 6.4.

In such flows, spanwise mixing at any given downstream location, results from turbulent transport of fluid lumps moving towards the axis Fig.6.4(a) or towards the outer edge of the jet (b). When $s_\rho > 1$, instantaneous events occurring in case (a) correspond to statistically dominant situations of inwards motions ($v' < 0$) of relatively lighter lumps of fluid ($\rho' < 0$). On the average, this case yields a positive correlation between density and radial velocity fluctuations. To complete the analysis, four different cases must be considered in the same way, leading to the results summarized in Table 6.1.

TABLE 6.1. Signs of d.f.c. in variable density jets; cases (a) and (b) as depicted in Fig.6.4.

| | | $\rho'$ | $u'$ | $v'$ | $\overline{\rho' u'}$ | $\overline{\rho' v'}$ |
|---|---|---|---|---|---|---|
| $s_\rho > 1$ | (a) | - | - | - | + | + |
| | (b) | + | + | + | + | + |
| $0 < s_\rho < 1$ | (a) | + | - | - | - | - |
| | (b) | - | + | + | - | - |

Owing to eq.(6.18), it can be concluded from the results given in Table 6.1 that second order cross-correlations $\overline{\rho\gamma' u'}$ and $\overline{\rho\gamma' v'}$ are always positive in such free jets.

The previous conclusions for jet flows can be generalized to other types of free thin shear layer flows, such as wakes and mixing layers (see Joly [235]). As depicted in figure 6.3, it appears that, if density and temperature fluctuations are obviously always negatively correlated, the signs of the other three d.f.c. are different, according to the flow configuration. It can be observed that:

- the axial mass flux $\overline{\rho' u'}$ and the mass-fraction d.f.c. $\overline{\rho'\gamma'}$ have the same signs;
- the signs of $\overline{\rho' u'}$ and $\overline{\rho'\gamma'}$ are both changed when any *one* mean gradient (velocity *or* density) is inverted;
- the sign of radial mass flux $\overline{\rho' v'}$ is only sensitive to mean density gradient inversion.

Although qualitative, these results clearly enhance the specific influence of d.f.c. in variable density turbulent flows, contrasting with a simple gradient type diffusion process. In addition, some quantitative consequences can be derived, based on suitable turbulence modeling, as discussed in [82] and [86].

## 6.4. Energetic peculiarities in variable density turbulence

This section is concerned with variable density and/or compressibility effects that are introduced in the averaged balance equations. As compared with the incompressible situation, it will be shown that energetic peculiarities in variable density fluid motions are partly due to additional terms which are *explicitly* introduced in the kinetic and internal energy equations, as a consequence of the density variation. A second part of the differences between both situations occurs from *implicit* density effects which are associated with terms that are formally common to the corresponding equations. These points will be now detailed. The equations referring to the instantaneous motion are briefly recalled. Mean motion equations will be considered using a density-weighted formulation for sake of clarity. The analysis will refer to kinetic energy and internal energy balances. Alternative presentations, based on the enthalpy balance equation, can be found in Lele [291] and Favre *et al.*[157] inter alia.

### 6.4.1. INSTANTANEOUS SITUATION

The equations governing the instantaneous values of the kinetic energy $\rho K \equiv \rho U_i U_i/2$ and the internal energy $\rho e \equiv \rho C_v T$ of a fluid particle (see previous chapter), can be rewritten as follows:

$$\frac{\partial}{\partial x_j}(-PU_i\delta_{ij} + \tau_{ij}U_i) = \frac{\mathrm{d}}{\mathrm{dt}}(\rho K) - \underbrace{P\frac{\partial U_i}{\partial xi}} + \underbrace{\tau_{ij}\frac{\partial U_i}{\partial x_j}} , \tag{6.19}$$

$$\frac{\partial q_j}{\partial x_j} = \frac{\mathrm{d}}{\mathrm{dt}}(\rho e) + \overbrace{P\frac{\partial U_i}{\partial xi}} - \overbrace{\tau_{ij}\frac{\partial U_i}{\partial x_j}} , \tag{6.20}$$

where d/dt stands for the instantaneous material derivative.
According to such a formulation (Morkovin [339]), the left hand sides of these equations represent, per unit time, the instantaneous energy supply to a moving fluid particle due to (i) the total mechanical work done on the particle by external forces[4] eq.(6.19), and (ii) the total heat transfer resulting from thermal conduction[5] with the outer fluid, eq.(6.19) .

By addition of eqs. (6.19) and (6.20), the total amount of these external mechanical and thermal powers are entirely converted into substantial variations of kinetic and internal energy of the fluid particle. Now, when the two balance equations are considered separately, two coupling terms are found.
Considering the kinetic energy balance, for instance, it appears that part of the external mechanical supply is used to change the instantaneous amount of kinetic energy. Another part ($\tau_{ij}\partial U_i/\partial x_j$) is irreversibly lost from the mechanical system through viscosity (viscous dissipation or viscous heating). A third one $P\partial U_i/\partial x_i$ is changed into a mechanically recoverable internal energy of the fluid, as generated by compression-expansion of the fluid element along its path. This last term only occurs when the volume of the fluid element changes. Moreover, in isovolume motions of constant density fluid, viscous heating is necessarily weak, so that kinetic and internal energy balances can be decoupled. This is no longer possible with non solenoidal velocity fields.

These considerations on the instantaneous fluid motion will be now extended to the mean flow analysis of the turbulent motion.

### 6.4.2. MEAN ENERGY LINKAGE

The equations governing (i) the kinetic energy of the mean motion ($2\widetilde{K}=\widetilde{U}_i\widetilde{U}_i$), (ii) the mean kinetic energy of the fluctuating motion ($2\widetilde{k}=\overline{\rho u_i'' u_i''}\equiv\overline{\rho}\,\widetilde{u_i'' u_i''}$) and (iii) the mean internal energy ($\widetilde{e}$) are[6] (see Chapter 5, eqs. 5.26, 5.27 and 5.38):

$$\frac{D(\overline{\rho}\widetilde{K})}{Dt}=-\frac{\partial(\overline{P}\widetilde{U}_i)}{\partial x_i}-\frac{\partial(\overline{\rho}\widetilde{u_i''u_j''}\widetilde{U}_i)}{\partial x_j}+\frac{\partial(\overline{\tau}_{ij}\widetilde{U}_i)}{\partial x_j}+(a)-(b)-(c)\,,$$

$$\frac{D(\overline{\rho}\widetilde{k})}{Dt}=-\frac{\partial(\overline{Pu_i''})}{\partial x_i}-\frac{\partial(\overline{\rho}\widetilde{k\,u_j''})}{\partial x_j}+\frac{\partial(\overline{\tau_{ij}u_i''})}{\partial x_j}+(c)-(d)+(e)\,,$$

$$\frac{D(\overline{\rho}\,\widetilde{e})}{Dt}=\frac{\partial\overline{q}_j}{\partial x_j}-\frac{\partial}{\partial x_j}(\overline{\rho}\,\widetilde{e''u_j''})-(a)+(b)+(d)-(e)\,,$$

[4]External body forces are not considered; such forces add a volume source term.
[5]Radiation, which adds a volume source term is not considered here.
[6]The mechanical power of the external body forces is not considered.

where $D(\overline{\rho}\widetilde{F})/Dt \equiv \partial(\overline{\rho}\widetilde{F})/\partial t + \partial(\overline{\rho}\widetilde{F}\widetilde{U}_j)/\partial x_j$.
Hence, in variable density fluid motions, the energy linkage results in the five coupling terms:

$$(a) = \overline{P}\frac{\partial \widetilde{U}_i}{\partial x_i} \quad (b) = \overline{\tau}_{ij}\frac{\partial \widetilde{U}_i}{\partial x_j} \quad (c) = -\overline{\rho}\,\widetilde{u''_i u''_j}\frac{\partial \widetilde{U}_i}{\partial x_j} \quad (d) = \overline{\tau_{ij}\frac{\partial u''_i}{\partial x_j}} \quad (e) = \overline{P\frac{\partial u''_i}{\partial x_i}}\,.$$

In incompressible turbulence, only three coupling terms are present, i.e., production $(c)$ and dissipation terms in the mean $(b)$ and fluctuating $(d)$ motions.
Thus, it can be concluded that the differences between constant and variable density situations are due to (i) changes introduced by density variations in terms that are common to both situations and (ii) the presence of specific additional terms in the variable density situation.

### 6.4.3. PRESSURE-COUPLING

The sum of the pressure terms $(a)$ and $(e)$, which are specific to non solenoidal motions, can be written as:

$$(a) + (e) = \overline{P}\frac{\partial \widetilde{U}_i}{\partial x_i} + \overline{P\frac{\partial u''_i}{\partial x_i}} \equiv \underbrace{\overline{P}\frac{\partial \widetilde{U}_i}{\partial x_i} + \overline{P}\frac{\partial \overline{u''}_i}{\partial x_i}}_{\equiv \overline{P}\frac{\partial \overline{U}_i}{\partial x_i}} + \overline{p'\frac{\partial u''_i}{\partial x_i}}\,.$$

Thus the whole pressure coupling consists in mean and fluctuating contributions:

- The mean pressure part represents the rate of change of the mechanically recoverable internal energy of the fluid, generated by compression-expansion ing the mean motion[7]. When positive — mean flow undergoing volume expansion —, it yields a *direct* supply of (mean and turbulent) kinetic energy, in addition to the one present in the corresponding isovolume situation. Part of this additional contribution can be *indirectly* gained by the fluctuating motion, from the production term, for instance;
- The fluctuating pressure coupling term is the *pressure-dilatation* correlation. Its expression is formally independent on the type of velocity averaging which is adopted (centered fluctuation $u'_i$ or Favrian fluctuation $u''_i$) since $u''_i \equiv u'_i - \overline{\rho' u'_i}/\overline{\rho}$ :

$$\Pi_d \equiv \overline{p'\frac{\partial u''_i}{\partial x_i}} \equiv \overline{p'\frac{\partial u'_i}{\partial x_i}}\,.$$

[7]The presence of $\overline{P}\partial\overline{u''}_i/\partial x_i$ in the mean contribution is sometimes considered as an artifact of the density-weighted formulation.

$\Pi_d$ is the only term which is *explicitly* introduced in the turbulence kinetic energy transport equation by compressible turbulence effects. When negative, it represents a loss of turbulence kinetic energy and a gain of mean internal energy.

### 6.4.4. DILATATION DISSIPATION

We turn now to the dissipation coupling term ($d$) in the kinetic equation of the fluctuating motion, as defined in §6.4.2. Assuming a Newton-Stokes fluid behavior[8], this viscous dissipation rate in a Favre-averaged formulation reads

$$\overline{\varepsilon}^F \equiv \overline{\rho}\,\overline{\epsilon}^F = \overline{\tau_{ij}\frac{\partial u_i''}{\partial x_j}} = 2\overline{(\mu S_{ij} s_{ij}'')} - \frac{2}{3}\overline{(\mu \frac{\partial U_l}{\partial x_l}\vartheta'')}\ , \qquad (6.21)$$

where $\vartheta'' = \partial u_l''/\partial x_l$.
A widely used simplification to the previous expression — Liou & Shih, [299], Wilcox, [484] *inter alia* — consists in neglecting correlations with viscosity fluctuations. The assessment of the effects of such additional correlations from DNS, as quoted by Lele [291], shows that they are actually small.
Hence, with a shortened symbolic notation, it becomes

$$\overline{\mu S s''} = \overline{\rho\nu S s''} \simeq \overline{\nu}\widetilde{S}\,\overline{\rho s''} + \overline{\nu}\overline{\rho s'' s''} \equiv \overline{\nu}\overline{\rho s'' s''}\ ,$$

since $\overline{\rho s''} \equiv 0$, due to Favre decomposition.
Thus, the turbulent dissipation rate reduces to

$$\overline{\rho}\,\overline{\epsilon}^F = 2\overline{\nu}(\overline{\rho s_{ij}'' s_{ij}''} - \frac{1}{3}\overline{\rho\vartheta''\vartheta''}) = 2\overline{\mu}(\frac{\overline{\rho s_{ij}'' s_{ij}''}}{\overline{\rho}} - \frac{1}{3}\frac{\overline{\rho\vartheta''\vartheta''}}{\overline{\rho}}) \equiv 2\overline{\mu}(\widetilde{s_{ij}'' s_{ij}''} - \frac{1}{3}\widetilde{\vartheta''^2})\ . \qquad (6.22)$$

It is worth recalling that, when Reynolds decomposition is used, the equivalent expression of the turbulent dissipation rate, neglecting again viscosity fluctuation correlations, is (see, for instance, Sarkar *et al.* [416]):

$$\overline{\rho}\,\overline{\epsilon}^R = \overline{\tau_{ij}'\frac{\partial u_i'}{\partial x_j}} \equiv \overline{\tau_{ij}' s_{ij}'} \simeq 2\overline{\mu}\overline{(s_{ij}' - \vartheta'/3\delta_{ij})s_{ij}'} = 2\overline{\mu}(\overline{s_{ij}' s_{ij}'} - \frac{1}{3}\overline{\vartheta'^2})\ , \qquad (6.23)$$

where $\vartheta' = \partial u_l'/\partial x_l$.
Following, for instance, in Lele [291] and Sarkar *et al.* [416], let us introduce the fluctuating vorticity tensor $w_{ij}' = \frac{1}{2}(\frac{\partial u_i'}{\partial x_j} - \frac{\partial u_j'}{\partial x_i})$, so that $s_{ij}' = \frac{\partial u_i'}{\partial x_j} - w_{ij}'$.
It can be easily deduced that:

[8] The bulk viscosity is zero. Otherwise, an additional dissipation rate is introduced, by any volume change, due to that bulk viscosity.

$$\overline{s'_{ij}s'_{ij}} = \overline{w'_{ij}w'_{ij}} + \overline{\frac{\partial u'_i}{\partial x_j}\frac{\partial u'_j}{\partial x_i}} \, , \tag{6.24}$$

along with:
$$\overline{\frac{\partial u'_i}{\partial x_j}\frac{\partial u'_j}{\partial x_i}} = \frac{\partial^2}{\partial x_i \partial x_j}(\overline{u'_i u'_j}) - 2\frac{\partial}{\partial x_j}(\overline{\vartheta' u'_j}) + \overline{\vartheta'^2} \, , \tag{6.25}$$

or:
$$\overline{\frac{\partial u'_i}{\partial x_j}\frac{\partial u'_j}{\partial x_i}} = \frac{\partial}{\partial x_i}(\overline{u'_j \frac{\partial u'_i}{\partial x_j}}) - \frac{\partial}{\partial x_j}(\overline{\vartheta' u'_j}) + \overline{\vartheta'^2} \, .$$

Adopting eq.(6.25) as in ref.[416] and after substituting into eqs. (6.24) and (6.23), the following expression of the dissipation rate is obtained:

$$\overline{\rho}\,\overline{\epsilon}^R = \underbrace{\frac{4}{3}\overline{\mu}\overline{\vartheta'^2}}_{(i)} + \underbrace{2\overline{\mu}(\overline{w'_{ij}w'_{ij}})}_{(ii)} + \underbrace{2\overline{\mu}(\frac{\partial^2 \overline{u'_i u'_j}}{\partial x_i \partial x_j} - 2\frac{\partial \overline{\vartheta' u'_j}}{\partial x_j})}_{(iii)} \, . \tag{6.26}$$

The last contribution (*iii*) vanishes under spatial homogeneity conditions. The second term in eq.(6.26) can be rewritten as $2\overline{\mu}\overline{w'_{ij}w'_{ij}} = \overline{\mu}\overline{w'_k w'_k}$, where $w'_k$ is the fluctuating vorticity vector. It is the only one present in homogeneous isovolume turbulent flows. It is thus regarded as the *incompressible* dissipation, or the *solenoidal* dissipation, according to Sarkar *et al.* [416] terminology.
The first term (*i*) in eq.(6.26) is specific to non-isovolume turbulence. It is called the *dilatational* or *compressible* dissipation (Zeman [495], Sarkar *et al.* [416]).

Hence, according to the formal analogy between eqs.(6.22) and (6.23), it follows from eq.(6.26) that, under spatial homogeneity conditions, the dissipation reduces to:

$$\overline{\rho}\,\overline{\epsilon}^R = \overline{\rho}\,\overline{\epsilon}^R_s + \overline{\rho}\,\overline{\epsilon}^R_d \, , \qquad \overline{\rho}\,\overline{\epsilon}^F = \overline{\rho}\,\overline{\epsilon}^F_s + \overline{\rho}\,\overline{\epsilon}^F_d \, , \tag{6.27}$$

with:
$$\overline{\rho}\,\overline{\epsilon}^R_s = \overline{\mu}\,\overline{\omega'_k \omega'_k} \, , \qquad \overline{\rho}\,\overline{\epsilon}^F_s = \overline{\mu}\,\widetilde{\omega''_k \omega''_k} \, ,$$

$$\overline{\rho}\,\overline{\epsilon}^R_d = \frac{4}{3}\overline{\mu}\,\overline{\vartheta'^2} \, , \qquad \overline{\rho}\,\overline{\epsilon}^F_d = \frac{4}{3}\overline{\mu}\,\widetilde{\vartheta''^2} \, ,$$

where subscripts $_d$ and $_s$ refer to dilatational (compressible) and solenoidal (isovolume) dissipations, respectively.

According to Sarkar *et al.* [416], it can be shown from standard order of magnitude estimates that the approximate expression in eq.(6.27) is also asymptotically exact for high turbulence Reynolds number flows.
From a physical point of view, the underlying ideas associated with decompositions in eq.(6.27) are that (i) the concept of Kolmogorov's cascade applies to the solenoidal dissipation, while (ii) the dilatational part can

contribute to the augmentation of the solenoidal dissipation by direct (additional) compressible effects, bypassing the solenoidal energy cascade. Such effects were first associated with the presence of shocklets, by Zeman [495] in 1990 for compressible mixing layers (see next section).
As mentioned in Chapter 2, the dilatational dissipation was taken for a short while as responsible for the stabilizing effect of compressibility in free shear layers. It is clear now that the amount of $\overline{\epsilon}_d$ is too weak to provide a quantitatively valuable explanation to such a result. In an annular mixing layer for example, the recent DNS data of Freund *et al.* [163] show that $\overline{\epsilon}_d/\overline{\epsilon}_s$ is less than 2% at a convective Mach number $M_c=1.5$.

## 6.5. Shocklets

A rapid change in velocity from supersonic to subsonic regime usually takes place through a shock wave. This is a well known feature of compressibility in high speed flows past solid bodies, either in the supersonic regime, where any part of the flow field can be concerned, or in the subsonic one, where the phenomenon can take place locally. Such shock waves are very thin surfaces[9] where pressure, density and temperature of the fluid vary at extremely high rates. When the flow is passing through a shock, entropy is produced along with a strong local heating, so that the shock is basically an important dissipating region. The significant parameter is the Mach number $M = U/a$ i.e., the ratio of the mean flow velocity to the local speed of sound.

As far as compressible turbulence is concerned, and since turbulence is a flow property, one can question whether such a phenomenon could be experienced by the fluctuating motion itself. In any turbulent motion, due to the presence of various eddies, velocity fluctuations actually exist in the flow field. Hence, strong instantaneous interactions between regions of locally high speed and low speed fluctuations are generated, which can be modified by compressibility effects. In other words, and introducing a turbulence Mach number $M_t = \hat{u}/a$, where $\hat{u}$ stands for a characteristic scale of the velocity fluctuations, one can question about a possible influence of that parameter on eddies behavior, and more generally on the modification of the instantaneous patterns of the fluctuating flow-field.

As a matter of fact, small shocks associated with turbulent eddies were first observed by Passot & Pouquet [359] in 1987. From direct numerical simulations of *2-D* compressible homogeneous flows, these authors pointed out the distinct possibility of shock-like structures, appearing randomly

[9]The characteristic length scale of the variations of the flow-field functions through the shock (the "thickness" of the shock) is of the same order as the mean free path of the molecular agitation, Candel [70]. If $\Delta u$ is the difference in the normal velocities across the shock and $\lambda$ is the thickness of the density interface, it can be estimated for a 1-D normal shock that $\rho\lambda\Delta u/\mu \sim \mathcal{O}(1)$, [495].

in the flow domain, even when the turbulence (rms) Mach number $M_t$ is subsonic. This is because the instantaneous Mach number can be more than twice the average value $M_t$, making instantaneous supersonic events possible. Such local shock-like structures are commonly known as "*eddy shocklets*" and briefly denoted now as "*shocklets*". However, it is still difficult to define what a shocklet is and how it appears.

The existence of eddy shocklets in *3-D* isotropic turbulence was confirmed by Lee *et al.* [284] in 1991. In free shear layers, shocklets were observed at different convective Mach numbers, depending whether the flow configuration was two or three dimensional. In *2-D* flows, the formation of shocklets was observed at about $M_c = 0.6$ to 0.7 (Lele [289] 1989, Fouillet [161] 1991, Lumpp [310] 1994). In *3-D* flows, the value of the convective Mach number where shocklets were observed is much higher: $M_c = 1.2$ in plane mixing layers (Vreman *et al.* [477] 1996, Sauvage [419] 2000) and $M_c = 1.54$ in an annular mixing layer (Freund *et al.* [163] 2000).
Finally, experimental evidence of the existence of shocklets was first provided by Papamoschou [355] in 1995, in a counterflowing supersonic mixing layer.

In addition to their definition which is not easy to give, it is not always easy to ascertain the presence of eddy shocklets from another type of shock, such as weak oblique shock (Freund et al [163]). Of course, such regions are associated with strong negative dilatation and steep variations of the flow field functions. As sketched in Zeman [495] — see figure §6.3 —, shock-like fronts of typical size $\lambda$ can occur within a turbulent eddy of characteristic length-scale $L$. The ratio $\lambda/L$ can be assessed from the 1-D estimate of the thickness of the density interface, $\rho\lambda\Delta u/\mu \sim \mathcal{O}(1)$, and the Prandtl-Meyer relation for the velocities through the shock, $u_1 u_2 \equiv u_1^2(1 - \Delta u/u_1) = a_*^2$, where $a_*$ is the speed of sound at the 'sonic' temperature $T_*$ ($T_*/T_0 = 2/(\gamma+1)$, where $T_0$ is the stagnation temperature). Hence:

$$\frac{\lambda}{L} \sim \frac{1}{R_t} \frac{M_*^2}{M_*^2 - 1},$$

where $R_t = u_1 L/\nu$ is a turbulence Reynolds number based on the upstream velocity, and $M_* = u_1/a_*$ the corresponding Mach number referring to the sonic condition. This relation shows that, in numerical simulations, the shocklet could be difficult to obtain at high turbulence Reynolds numbers without a high resolution.

Even though this simple physical model of an eddy shocklet could be very crude, such shock-like fronts have three markedly specific characteristics: viscous flow, unsteadiness and finite normal expansion (finite width), so that the classical tabulated normal shock jump conditions do not necessarily apply strictly.

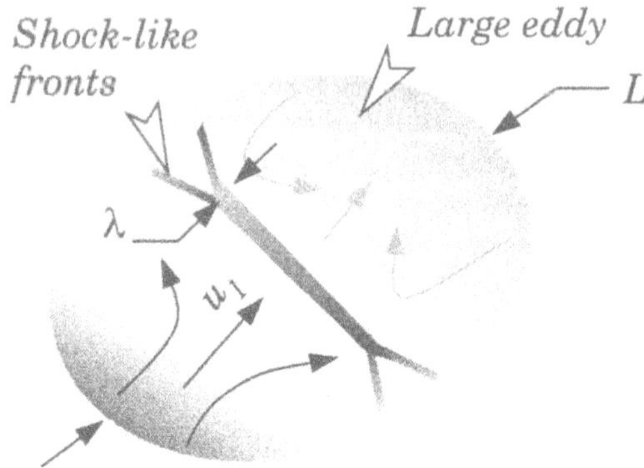


*Figure 6.5.* Sketch of shock-like structure in a turbulent eddy, adapted from Zeman [495].

The following example, taken from the *3-D* direct numerical simulations of a temporally evolving compressible mixing layer by Sauvage [419], is used to illustrate some aspects of such a flow feature. Four values of the convective Mach number, ranging from 0.4 to 1.6, have been considered in this study. The iso-contours of the local Mach number in a cross section of the mixing layer at $M_c = 1.2$ are plotted in figure 6.6. The local Mach number is defined here as $(U^2 + V^2 + W^2)^{1/2}/(\gamma P/\rho)^{1/2}$. Centered at a non dimensional location of about (7.5 ; −3), a steep variation of the local Mach number can be observed. It was checked that there exists a strong non-zero velocity component normal to this front, corresponding to an upstream local normal Mach number $M_{n1} = 1.26$.

An enlargement of the part of the flow field, capturing the shocklet detected in the previous figure (see arrows), is shown in Fig. 6.7(a), taking now the non dimensional local pressure as the significant parameter. When moving from right to left (decreasing $x$) the flow undergoes strong compression, as confirmed by the normal variations plotted in Fig. 6.7(b).

In this figure, the variations, normal to the shock, of the pressure and normal Mach number $M_n$ are plotted at the location $x = 7.68$, $y = -2.66$ (see arrows). The computed pressure jump $P_2/P_1$ is lower than the value expected from the classical normal jump conditions (1.685 for $M_{n1} = 1.26$). On the other hand, for such an upstream normal Mach number, the expected normal Mach number past the shock is 0.807, which is in rather good agreement with the computed value.

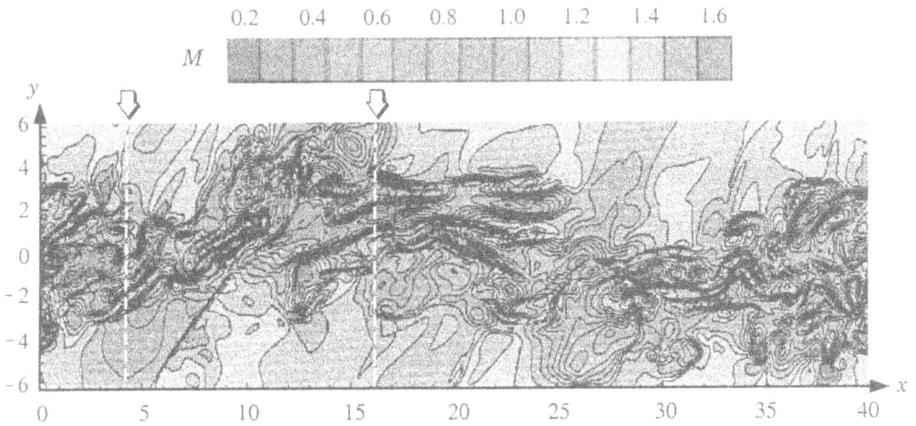


*Figure 6.6.* Local Mach number contours in a cross section of a temporally evolving mixing layer at $M_c = 1.2$Γ adapted from [419].

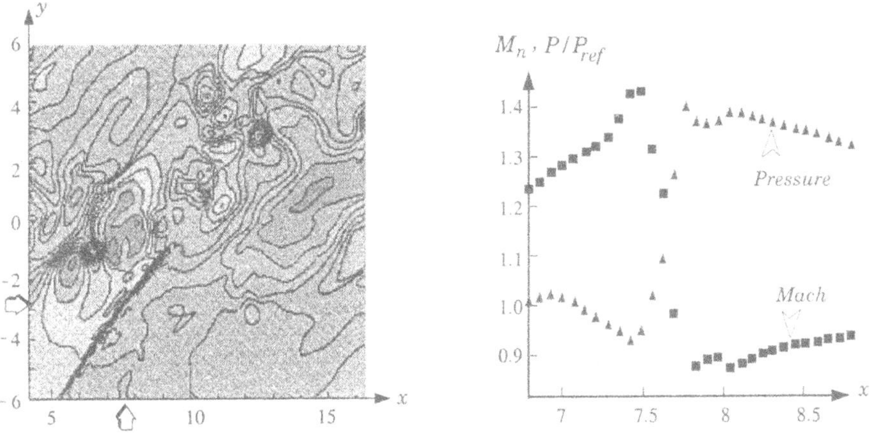


*Figure 6.7.* Local pressure contours (a) and normal variations across the shocklet (b)Γ adapted from [419].

CHAPTER 7

# RELATIVE BEHAVIOUR OF VELOCITY AND SCALAR STRUCTURE FUNCTIONS IN TURBULENT FLOWS

*This chapter reviews in a critical manner the existing analytical framework for describing the behaviour of velocity and scalar structure functions in turbulent flows. The assumptions which underpin this framework are only likely to be validated at very large Reynolds numbers and for relatively homogeneous and isotropic flows. These conditions are unlikely to apply in the laboratory. The major emphasis is on the likely dependence of second-order structure functions (or equivalently spectra) on both the Taylor microscale Reynolds number* $R_\lambda$ *and other parameters, such as the large scale anisotropy or the dissipation timescale ratio or, more generally, the initial conditions of the flow. Measurements strongly indicate that the influence of* $R_\lambda$ *and of the other parameters cannot be ignored. The retention of the non-homogeneity of the flow in the Navier-Stokes and heat transport equations provides a better idea of how large the magnitude of* $R_\lambda$ *should be before the "asymptotic" results of Kolmogorov and Yaglom may be attained. Special attention is given to a suitable framework which allows velocity and scalar fluctuations to be compared meaningfully. The analogy between scalar and energy structure functions (or spectra) appears to work well for flows with a continuous injection of turbulent energy and scalar variance.*

## 7.1. Introduction

Considerable attention has been given by physicists and engineers to the study of the small-scale turbulent motion due, in no small measure, to the impact of the similarity hypotheses of Kolmogorov [255] in 1941 or K41 and their subsequent modifications in 1962, (Kolmogorov [257] or K62 and Obhukov [350], or O62), to account for the spatio-temporal intermittency of the turbulent energy dissipation rate $\epsilon$. As noted by Sreenivasan [446], K41 presupposes that the small scales are isotropic and universal and can therefore be understood independently of the large scales which are specific to a given flow and are usually strongly anisotropic. Unfortunately, the reality is in marked contrast with this supposition. The main assumptions (the Reynolds number should be very large and the small scales are isotropic) which underpin K41 and K62 are not satisfied at the Reynolds numbers normally encountered in the laboratory. In order to achieve a better understanding of the small-scale motion, its dependence on the Reynolds number, the nature of the flow, including the effect of initial conditions, and, more particularly the anisotropy of the large-scale motion needs to be properly accounted for. It is also important, if not essential, that, in the longer term, K41 and K62 are replaced by a theoretical framework, based on the Navier-Stokes equations, against which laboratory data can be appraised with less ambiguity than is currently possible. In the shorter term, since it is almost inevitable that several aspects of K41 and K62 will continue to be used by turbulence practitioners, it seems worthwhile to have a realistic appreciation of the limitations of K41 and K62, especially when the interest extends to or includes the behaviour of passive scalars[1] as they are advected by turbulent flows. This is the main objective of the present chapter; whereas the major part of this volume is dedicated to variable density flows, it is somewhat premature to consider the behaviour of the small-scale motion in such flows, in view of the paucity of any hard experimental or numerical data currently available. We limit our focus to constant density flows, with the expectation that the present considerations should provide the platform for a more realistic assessment of small-scale velocity and scalar fluctuations in turbulent flows where changes in density may be dynamically important (see for example Chapter 1).

The phenomenology of small-scale turbulence was reviewed by Sreenivasan & Antonia [447], where a few important characteristics of the passive scalar were presented, albeit briefly. More recently, the behaviour of passive scalars in turbulent flows has been addressed explicitly by Warhaft [481] and Shraiman & Siggia [430]. Warhaft correctly stresses the need to interpret more cautiously the Kolmogorovian framework, in view of the way local isotropy is violated, both at inertial and dissipation scales, especially by the passive scalar. Apart from further consolidating this view, the present chapter provides a realistic basis for describing and assessing the behaviour of the passive scalar when information on the velocity fluctuations is already available.

No attempt is made here to consider the characteristics of the large-scale scalar motion, as this would almost merit a separate dedicated treatment. Nor do we examine the implications of relatively recent results for the scalar field, especially those obtained from direct numerical simulations (DNS) of canonical boundary layer and fully developed channel and pipe flows, on the modelling of the scalar transport. The reader is referred to, inter alia, [247, 248, 229, 311, 42, 241, 242, 243, 57]. Some of the previous papers also

[1] Warhaft [481] defines a passive scalar as a diffusive contaminant that is present in such low concentration that it has no effect on the dynamics of the flow.

consider the effect of the Prandtl number (or Schmidt number) $Pr$ ($\equiv \nu/\kappa$, $\nu$ and $\kappa$ are the momentum and thermal diffusivities respectively of the fluid). As well as providing insight into the physical mechanisms involved in the relationship between momentum and scalar transports, the DNS should help with the testing and development of scalar transport models, such as proposed for example by Launder [273].

## 7.2. Analytical Framework

An appropriate starting point for investigating the small-scale characteristics of a passive scalar in a turbulent flow is the analytical framework (K41) established by Kolmogorov [255]. The first two similarity hypotheses proposed by Kolmogorov describe the behaviour of turbulence associated with scales which lie within both the dissipative range (DR) and the inertial range (IR). The first of these relates to the DR and states that the pdf of $\delta u$ (or indeed $\delta v$ or $\delta w$) is uniquely determined by $\langle \epsilon \rangle$, the mean energy dissipation rate, and $\nu$. One consequence of this hypothesis is that

$$\langle (\delta u^*)^n \rangle = f_{un}(r^*) , \tag{7.1}$$

where $\delta u \equiv u(x+r) - u(x)$ is the increment between longitudinal velocity fluctuations at two spatial locations separated by a distance $r$. Angular brackets denote averaging with respect to time. The asterisk denotes normalization by the Kolmogorov velocity scale $U_K \equiv (\nu \langle \epsilon \rangle)^{1/4}$ and/or the Kolmogorov length scale $\eta \equiv \nu^{3/4} \langle \epsilon \rangle^{-1/4}$. The second hypothesis relates to the IR which Kolmogorov identified with $\eta \ll r \ll L$, where $L$ is an integral length scale of the turbulence. In this range, the effect of $\nu$ can be ignored and as a consequence

$$\langle (\delta \alpha^*)^n \rangle = C_{\alpha n} r^{*n/3} , \tag{7.2}$$

where $\alpha \equiv u$, $v$ or $w$ and $C_{\alpha n}$ can be identified with the Kolmogorov constants for $\alpha$. According to K41, $C_{un}$ may depend on the flow macrostructure. It is important to recall that the two hypotheses require that

(i) the Reynolds number is "very" large

(ii) the small scales are isotropic.

Evidently, the two conditions are not unrelated since the expectation of local isotropy becomes more plausible as the Reynolds number increases. It is common to use $R_\lambda$ [$= u'\lambda_u/\nu$, where $\lambda_u \equiv u'/(\partial u/\partial x)'$ is the longitudinal Taylor microscale; a prime denotes the rms value] or longitudinal Taylor microscale Reynolds number as the appropriate Reynolds number, although, ideally, $u'$ and $\lambda_u$ should be replaced by less directionally sensitive quantities (see Section 6). It is perhaps also worth noting that, implicit in K41, is the assumption that $\langle \epsilon \rangle$ is independent of $\nu$. Such an assumption is reasonable when $R_\lambda$ is large enough. The turbulent kinetic energy, which is cascaded down from the larger to smaller eddies, is removed essentially at the smallest scales through viscous dissipation. The energy flux across the large scales can then be interpreted as a measure of $\langle\epsilon\rangle$ and the time scale for $\langle\epsilon\rangle$ can be taken to be of the same order of magnitude as the characteristic time scale $(L/u')$ of the energy containing eddies. The dimensionless parameter $C_\epsilon \equiv \langle\epsilon\rangle L/u'^3$, whose magnitude is of order 1, has been found to become independent of $R_\lambda$ only when the latter is sufficiently large (e.g. [446]); the asymptotic value of $C_\epsilon$ depends on the nature of the flow and the initial conditions (e.g. [8]).

Hypotheses analogous to the first two similarity hypotheses of K41 were proposed by Obukhov [349] or O49 for a passive temperature field which is advected and mixed by locally isotropic turbulence (a detailed discussion of the O49 hypotheses is given by Monin and Yaglom [333]). The first of these essentially states that for sufficiently large $R_\lambda$ and $Pe$, where $Pe$ is the turbulent Péclet number $u'\lambda_\theta/\kappa$ [$\lambda_\theta \equiv \theta'/(\partial\theta/\partial x)'$ is the longitudinal thermal microscale], the joint pdf of $\delta u$ and $\delta\theta$ depends uniquely on $\langle\epsilon\rangle$, the mean temperature dissipation rate $\langle\chi\rangle$ ($\equiv \kappa\langle(\partial\theta/\partial x_i)^2\rangle$), $\nu$ and $\kappa$. One consequence of this is a relation analogous to eq.(7.1), i.e.

$$\langle(\delta\theta^*)^n\rangle = f_{\theta n}(r^*, Pr) \tag{7.3}$$

where the asterisk on $\theta$ denotes normalization by the temperature scale $\theta_K \equiv \langle\chi\rangle^{1/2}\eta^{1/2}U_K^{-1/2}$. In the second hypothesis, the dependence on $\nu$ and $\kappa$ is ignored for scales within the IR and a relation analogous to eq.(7.2) follows, viz.

$$\langle(\delta\theta^*)^n\rangle = C_{\theta n} r^{*n/3} \ . \tag{7.4}$$

Like $C_{un}$, the premultipliers $C_{\theta n}$ depend on the flow macrostructure. Expressions analogous to eq.(7.1)–(7.4) can be written for the spectral domain since structure functions and spectra are related by a Fourier transformation, e.g.

$$\langle(\delta\beta)^2\rangle = 2\int_0^\infty \phi_\beta(k_1)[1 - \cos(k_1 r)]dk_1 \ . \tag{7.5}$$

In eq.(7.5), $\beta$ stands for either $u$, $v$, $w$ or $\theta$ and the spectral density $\phi_\beta(k_1)$ is normalized so that $\int_0^\infty \phi_\beta(k_1)dk_1 = \langle\beta^2\rangle$, where $k_1$ is the one-dimensional wavenumber. The spectral premultiplier which corresponds to $C_{\theta 2}$ is usually referred to as the Obukhov–Corrsin constant. Corrsin [101] first identified the behaviour of $\phi_\theta(k_1)$ in the IR (or inertial-convective range), viz. $\phi_\theta(k_1) \sim k_1^{-5/3}$, which is the spectral equivalent of relation (4) [for $n = 2$]. When $Pr \simeq 1$, the case with which we are concerned here, the smallest length scale $\eta_\theta$ for the scalar is indistinguishable from $\eta$. It is also worth underlining that, in addition to the requirements of large $R_\lambda$ and local isotropy, the Péclet number should also be large. The dimensionless parameter $C_\chi \equiv \langle\chi\rangle L/\langle\theta^2\rangle u'$ is expected to become constant at sufficiently large $R_\lambda$ and $Pe$ but, like $C_\epsilon$, will depend on the nature of the flow and initial conditions (e.g. [490]).

To account for the spatial fluctuations (or intermittency) in the energy dissipation rate [268], Kolmogorov [257] (or K62) and Obukhov [350] modified dimensional arguments of K41 by replacing $\langle\epsilon\rangle$ by the variable $\epsilon_r$ where the subscript $r$ denotes averaging over a scale $r$. In this new framework (K62), eq.(7.2) is replaced by

$$\langle(\delta\alpha^*)^n\rangle = C^+_{\alpha n}\langle\epsilon_r^{*n/3}\rangle r^{*n/3} \tag{7.6}$$

where the superscript $^+$ is introduced to allow for a possible distinction between the new premultipliers and those in eq.(7.2).

The "new" main element in K62 is the so-called refined similarity hypothesis (RSH). Evidently, the difference between eq.(7.6) and eq.(7.2) will depend on the behaviour of $\langle\epsilon_r^{*n/3}\rangle$ in the IR. The possibility that $\langle\epsilon_r^{*n/3}\rangle$ will exhibit a power-law dependence in this range raises the issue of whether the scaling is "anomalous" in the sense that the exponent $\zeta_\alpha(n)$ in the relation

$$\langle(\delta\alpha^*)^n\rangle \sim r^{*\zeta_\alpha(n)} \tag{7.7}$$

may now differ from $n/3$. For the lognormal model for $\epsilon_r$ proposed in K62,

$$\langle\epsilon_r^n\rangle = D_n\langle\epsilon\rangle^n\left(\frac{L}{r}\right)^{\frac{\mu n(n-1)}{2}} \tag{7.8}$$

where $\mu$ is the intermittency parameter and $D_n$ are coefficients which may depend on the macrostructure of the flow. Many models — both cascade and non-cascade — have been proposed, some of which are arguably more reliable than the lognormal model. The limitations of the lognormal model are well known, especially when $n$ is large (e.g. [4]). This model is considered there because it has been found to be in reasonable agreement with both measured and numerical data for small $n$ and its extension to the scalar is relatively straightforward. Further, the DNS data, which provide complete information on $\epsilon$ (and $\chi$), have indicated that the pdfs of $\ln\epsilon_r$ (and $\ln\chi_r$) are reasonably well approximated by Gaussian distributions [479, 480], except perhaps when $\epsilon_r$ (and $\chi_r$) become large. Substitution of eq.(7.8) into eq.(7.6) yields a quadratic dependence on $n$ for the exponent of $r^*$, viz.

$$\langle(\delta\alpha^*)^n\rangle = C^+_{\alpha n}D_n L^{*\mu\frac{n}{3}\left(\frac{n-3}{6}\right)} r^{*\frac{n}{3}-\mu\frac{n}{3}\left(\frac{n-3}{6}\right)}\,. \tag{7.9}$$

Relation (7.9) implies an explicit dependence on $R_\lambda$ since, for isotropic turbulence,

$$L^* = C_\epsilon 15^{-3/4} R_\lambda^{3/2} . \tag{7.10}$$

Note that, in general, it may not be correct to assume that $C_\epsilon$ is equal to 1.

A number of authors (e.g. [258, 466, 16, 17]) have considered the extension of eq.(7.6) to a passive scalar[2] for $Pr \simeq 1$. The resulting expression for $\langle(\delta\theta^*)^n\rangle$ is

$$\langle(\delta\theta^*)^n\rangle = C_{\theta n}^+ \left\langle \epsilon_r^{*-n/6} \chi_r^{*n/2} \right\rangle r^{*n/3} . \tag{7.11}$$

Relation (7.11) takes into account fluctuations in $\epsilon_r$, $\chi_r$ and the possibility that these quantities may be correlated. The new element in eq.(7.11) is RSHP (or the refined similarity hypothesis for a passive scalar, e.g. [502, 453]). When the pdfs of $\epsilon_r$ and $\chi_r$ are lognormal and the joint pdf between $\epsilon_r$ and $\chi_r$ is bivariate lognormal (e.g. [17])

$$\langle \epsilon_r^{-n/6} \chi_r^{n/2} \rangle = \langle\epsilon\rangle^{-n/6} \langle\chi\rangle^{n/2} \exp(A\Omega_{0,n}) \left(\frac{L}{r}\right)^{\mu\Omega_{0,n}} \tag{7.12}$$

where $A$ is the additive (macrostructure dependent) constant in the expression for $\sigma^2$, the variance of $\ln \epsilon_r$, viz.

$$\sigma^2 = A + \mu \ln \frac{L}{r} \tag{7.13}$$

and

$$\Omega_{0,n} \equiv \frac{n}{12}\left(\frac{5n}{3} - n\rho - 2\right) \tag{7.14}$$

the correlation coefficient between $\ln \epsilon_r$ and $\ln \chi_r$ being designated by $\rho$ [note that the general definition of $\Omega_{m,n}$ is given later, Cf. eq.(7.21)]. Eq.(7.11) becomes

$$\langle(\delta\theta^*)^n\rangle = C_{\theta n}^+ \exp(A\Omega_{0,n}) L^{*\mu\Omega_{0,n}} r^{*\frac{n}{3}-\mu\Omega_{0,n}} \tag{7.15}$$

the dependence on the Reynolds number appearing explicitly through the term $L^{*\mu\Omega_{0,n}}$. For isotropic turbulence (and $Pr \simeq 1$), $L^*$ will depend on

[2] Van Atta [467] also considered the two cases $Pr \ll 1$ and $Pr \gg 1$, for which theoretical results were previously obtained by Batchelor et al. [36] and Batchelor [34]. For a more general discussion of the effect of $Pr$ on scalar structure functions, the reader should consult Gibson et al. [185] and Gibson [180]. Not much experimental information is available for $Pr \neq 1$, measurements becoming especially difficult when $Pr$ exceeds 1 since the smallest relevant scale, or Batchelor scale $\eta_B$, is equal to $\eta(Pr)^{-1/2}$. Useful insight into the statistics of the scalar, in particular the shape of the spectrum, is currently being gleaned from direct numerical simulations of both decaying and forced turbulence in a periodic box.

$R_\lambda$, $Pe$ and $C_\chi$ (e.g. [5]). Note that in general a distinction should be made between $L$ and $L_\theta$, the integral length scale for the scalar fluctuations. It has also been assumed, as in Van Atta [466], that the variance of $\ln \chi_r$ is indistinguishable from that of $\ln \epsilon_r$. This is unlikely to be valid; in particular, there is enough evidence to indicate that the magnitude of $\mu_\theta$, the intermittency of the scalar, is larger than that of $\mu$ (e.g. [447, 479, 480, 320, 343, 489, 324, 75])[3] notwithstanding the fact that the values of $\mu$ and especially $\mu_\theta$ depend on the particular method used for their determination (e.g. [480, 489, 451]). A few immediate observations can be made by comparing eq.(7.15) with eq.(7.9) when $n$ is equal to 2, i.e. for the lowest order moment of interest. These observations pertain both to differences with respect to the level of the scaling "anomaly" and the dependence on $R_\lambda$. For $n = 2$, eq.(7.9) indicates that the exponent of $r^*$ is always greater than 2/3, whereas eq.(7.15) suggests that the exponent of $r^*$ is likely to remain smaller than 2/3. A recent investigation into the magnitude of $\rho$ [18] concluded that, notwithstanding the uncertainties in replacing $\epsilon$ and $\chi$ by isotropic approximations, $\rho$ depends significantly on the type of flow. Further, the dependency of $\rho$ on $r^*$ and $R_\lambda$ cannot be ignored. Nonetheless, the magnitude of $\rho$ — typically in the range 0.1 to 0.3 — across the IR suggests that $\Omega_{0,n}$ is positive so that the exponent of $r^*$ in eq.(7.19) is likely to always be smaller than 2/3; the "anomalous" scaling, as predicted by RSH, RSHP and the lognormal model, is therefore inherently different between second-order velocity and scalar structure functions. Eq.(7.9) implies a negative $R_\lambda$ dependence when $n = 2$ in contrast to eq.(7.15), which indicates that the dependence on $R_\lambda$ is always likely to be positive. Note that relatively little attention was given to the case $n = 2$ in earlier assessments of different intermittency models, on the basis that predicted departures from "2/3" are too small to represent an effective test of the models. Although predictions by the lognormal model (as well as other models) seem to reproduce the $R_\lambda$ dependence of measured structure functions when $n$ is greater than 2 (but less than 8), e.g. Antonia et al. [13], it would be unwise to cursorily dismiss the previous observations for $n = 2$.

It is important to recall that theoretical expressions, derived from the Navier-Stokes and heat transport equations, are available for $\langle(\delta u)^3\rangle$ and $\langle\delta u(\delta\theta)^2\rangle$. The equation for $\langle(\delta u)^3\rangle$ was first given by Kolmogorov [254], viz.

$$\langle(\delta u)^3\rangle = 6\nu\frac{d}{dr}\langle(\delta u)^2\rangle - \frac{4}{5}\langle\epsilon\rangle r\ . \tag{7.16}$$

The assumptions needed to validate eq.(7.16) have been discussed by many authors (e.g. [333, 165, 298, 215]); it would appear that local homogeneity

[3]More general expressions for $\Omega_{m,n}$ (or $\Omega_{0,n}$), which distinguish between the variances of $\ln\epsilon_r$ and $\ln\chi_r$, have been given by Meneveau et al. [320] and Xu et al. [490].

and local isotropy are necessary requirements. The equation for $\langle \delta u (\delta\theta)^2 \rangle$, first written by Yaglom [491]

$$\langle \delta u (\delta\theta)^2 \rangle = 2\kappa \frac{d}{dr} \langle (\delta\theta)^2 \rangle - \frac{4}{3} \langle \chi \rangle r \tag{7.17}$$

also requires local homogeneity and local isotropy to be satisfied. In the inertial range, eq.(7.16) and eq.(7.17) reduce to

$$\langle (\delta u^*)^3 \rangle = -\frac{4}{5} r^* \tag{7.18}$$

and

$$\langle \delta u^* (\delta\theta^*)^2 \rangle = -\frac{4}{3} r^* \,. \tag{7.19}$$

Clearly, the linear behaviour with $r^*$ indicated by eq.(7.18) is satisfied by eq.(7.9) when $n = 3$. With the lognormal assumptions [16], the expression for the mixed velocity-temperature structure function is

$$\langle (\delta u^*)^m (\delta\theta^*)^n \rangle = C_{mn} r^{*\frac{m+n}{3}} \langle \epsilon_r^{*\frac{m}{2}-\frac{n}{6}} \chi_r^{*n/2} \rangle \,, \tag{7.20}$$

with

$$\langle \epsilon_r^{*\frac{m}{2}-\frac{n}{6}} \chi_r^{*\frac{n}{2}} \rangle = \exp(A\Omega_{m,n}) \left(\frac{L}{r}\right)^{\mu\Omega_{m,n}} \tag{7.21}$$

and

$$\Omega_{m,n} = \frac{1}{2}\left(\frac{m}{2} - \frac{n}{6}\right)\left(\frac{m}{2} - \frac{n}{6} - 1\right) + \frac{n}{4}\left(\frac{n}{2} - 1\right) + \left(\frac{m}{2} - \frac{n}{6}\right)\frac{n}{2}\rho$$

For $m = 1$ and $n = 2$,

$$\langle (\delta u^*)(\delta\theta^*)^2 \rangle = C_{12} \exp(A\Omega_{1,2}) L^{*\mu\Omega_{1,2}} r^{*1-\mu\Omega_{1,2}} \tag{7.22}$$

with $\Omega_{1,2} = (\rho - 5/12)/6$. It appears that the linear variation described in eq.(7.19) is recovered only if $\rho = 5/12$. As noted earlier, the magnitude of $\rho$ (allowing for possible dependencies with respect to $r$, $R_\lambda$ and the type of flow) is, in general, likely to be significantly smaller than 5/12.

## 7.3. A Few Questions which Arise from Section 2

It is clear that the analytical framework of Section 2, whether it is K41 or K62, is underpinned by assumptions which are only likely to be validated asymptotically, viz. at extremely large values of $R_\lambda$. This applies also to the deductions, eqs.(7.18) and (7.19), based on the equations of motion. In reality, for the turbulent flows that are considered and studied in the laboratory, $R_\lambda$ is only moderate and the conditions of isotropy and homogeneity are generally violated globally (at large scales) and often only imperfectly satisfied at small scales. Even for the atmospheric surface layer where $R_\lambda$ is typically of order $10^4$, the influence of the global anisotropy and inhomogeneity on the inertial range cannot be ignored (e.g. [448]).

The previous comments lead, in a natural way, to several not necessarily unrelated questions.

1. How do the structure functions $\langle(\delta\alpha^*)^n\rangle$ and $\langle(\delta\theta^*)^n\rangle$ in the DR and IR depend on $R_\lambda$ and on other parameters, such as the anisotropy and inhomogeneity associated with the large scales ? It is natural enough to inquire into the effect of the nature of the flow or of different initial conditions for nominally the same flow.
2. Does an inertial range, in the sense of K41, exist for laboratory type flows? Perhaps equivalently, is local isotropy satisfied in this range?
3. Is the scaling of scalar structure functions in the IR more (or less) anomalous than that of velocity structure functions? Perhaps more importantly, what is the most appropriate basis for comparing velocity and scalar statistics? And are there genuine differences in anisotropy between the velocity and scalar fields?

The following sections will address the previous issues. The objective is not necessarily to arrive at firm answers and conclusions, especially since the majority of these issues will continue to be the focus of research for some time in the future. As noted earlier, the intent is to provide a realistic appraisal of departures from the "asymptotic" framework described in Section 2.

### 7.4. $R_\lambda$ Dependence of $\langle(\delta\alpha^*)^2\rangle$ and $\langle(\delta\theta^*)^2\rangle$

We first consider here experimental data which allow the $R_\lambda$ dependence of $\langle(\delta\beta^*)^n\rangle$, with $\beta \equiv u,\ v,\ w,\ \theta$, to be quantified. The measurements were made in a number of flows (grid turbulence, wake, boundary layer, circular and plane jets, fully developed channel and pipe flows, atmospheric surface layer) over a significant range of $R_\lambda$ (30 to 4000) using an X-wire probe (either in a $u, v$ or $u, w$ combination) or a single cold wire (for $\theta$). Details of these measurements and appropriate references are given in [5, 360] (see also Pearson and Antonia [361]). A uniform treatment was applied to the data. This consisted of first correcting the spectra of $\beta$ for the attenuation at high frequencies caused by the imperfect spatial resolution of the probe, the noise contamination (also at high frequencies). The spectra were then extrapolated to sufficiently high frequencies (or wavenumbers, using Taylor's hypothesis) to provide adequate isotropic estimates of $\langle\epsilon\rangle$ and $\langle\chi\rangle$. A Fourier-transform was subsequently applied to the corrected and extrapolated spectrum, eq.(7.5), to estimate $\langle(\delta\beta)^2\rangle$.

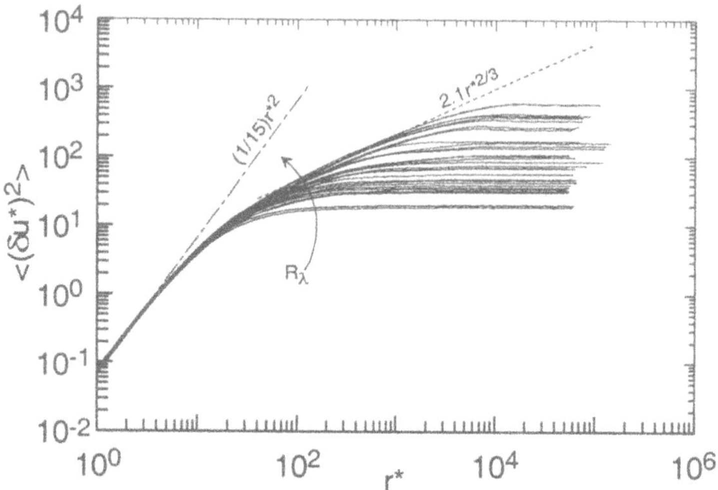


*Figure 7.1.* $R_\lambda$ dependence of Kolmogorov-normalized second-order moments of $\delta u$. No attempt is made to distinguish between the different curves but there is a systematic increase in $\langle(\delta u^*)^2\rangle$ in the direction of increasing $R_\lambda$, as indicated by the arrow. - - -, $2.1r^{*2/3}$ (K41 result for the IR with $C_u = 2.1$); — - —, $r^{*2}/15$ (limiting value, as $r^* \to 0$, of $\langle(\delta u^*)^2\rangle$ using local isotropy).

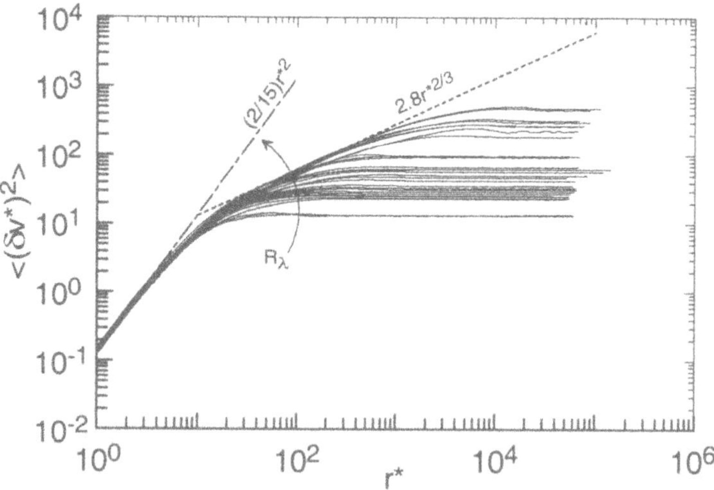


*Figure 7.2.* $R_\lambda$ dependence of Kolmogorov-normalized second-order moments of $\delta v$. No attempt is made to distinguish between the different curves but there is a systematic increase in $\langle(\delta v^*)^2\rangle$ in the direction of increasing $R_\lambda$, as indicated by the arrow. - - -, $2.8r^{*2/3}$ (K41 result for the IR with $C_v = 2.8$); — - —, $(2/15)r^{*2}$ (limiting value, as $r^* \to 0$, of $\langle(\delta v^*)^2\rangle$ using local isotropy).

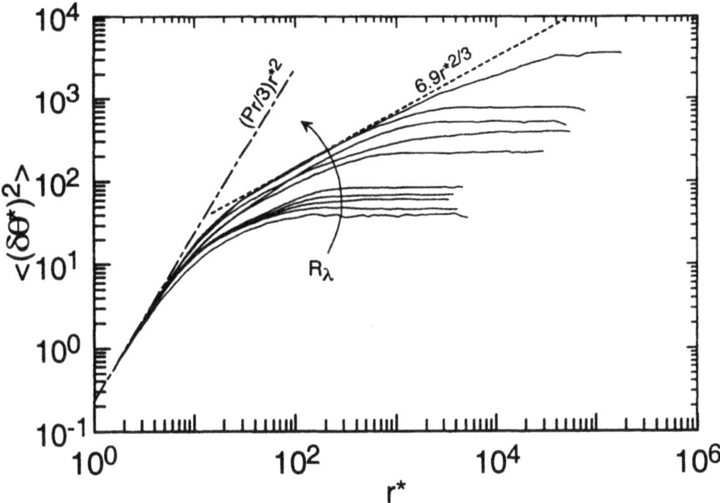


*Figure 7.3.* $R_\lambda$ dependence of Kolmogorov/Obukhov-normalized second-order moments of $\delta\theta$. No attempt is made to distinguish between the different curves but there is a systematic increase in $\langle(\delta\theta^*)^2\rangle$ in the direction of increasing $R_\lambda$, as indicated by the arrow. - - -, $6.9r^{*2/3}$ (O49 result for the IR with $C_\theta = 6.9$); — - —, $(Pr/3)r^{*2}$ (limiting value, as $r^* \to 0$, of $\langle(\delta\theta^*)^2\rangle$ using local isotropy).

The $R_\lambda$ behaviour of $\langle(\delta u^*)^2\rangle$, $\langle(\delta v^*)^2\rangle$ and $\langle(\delta\theta^*)^2\rangle$ is shown in Figures 7.1, 7.2 and 7.3 respectively. Several observations can be made in regard to the way both small and large scales depend on $R_\lambda$. There is seemingly good collapse for dissipative scales for both $\langle(\delta u^*)^2\rangle$ (Figure 7.1) and $\langle(\delta\theta^*)^2\rangle$ (Figure 7.3); this however mostly reflects the fact that the isotropic expressions $\langle\epsilon\rangle = 15\nu\langle(\partial u/\partial x)^2\rangle$ and $\langle\chi\rangle = 3\kappa\langle(\partial\theta/\partial x)^2\rangle$ were used to estimate $\langle\epsilon\rangle$ and $\langle\chi\rangle$. Consistently, there is less perfect collapse for $\langle(\delta v^*)^2\rangle$ at small $r^*$, indicating a non-negligible departure from local isotropy. There is little doubt however that there is a systematic increase with $R_\lambda$ of all three structure functions when $r^*$ is in the range 20–30 to 200–300, a range that would normally be identified with the IR; such an increase is in contradiction with the predictions of eqs.(7.9) and (7.10) since the exponent of $L^*$ is negative when $n = 2$. A closer examination of the variation of $\langle(\delta\beta^*)^2\rangle$ over this range of scales indicates that there is no rigorous power-law behaviour so that the concept of an inertial range, in the sense of K41 or K62, is not strictly tenable. The $R_\lambda$ trend displayed in Figures 7.1–7.3 does not rule out the possibility of an asymptotic approach towards power-law behaviours described by K41, O49 and K62, so that the framework introduced in K41 may be reasonable, provided the corresponding requirements are met. As expected, at sufficiently large $r^*$, $\langle(\delta\beta^*)^2\rangle$ becomes constant, the magnitude of this constant increasing linearly with $R_\lambda$ in

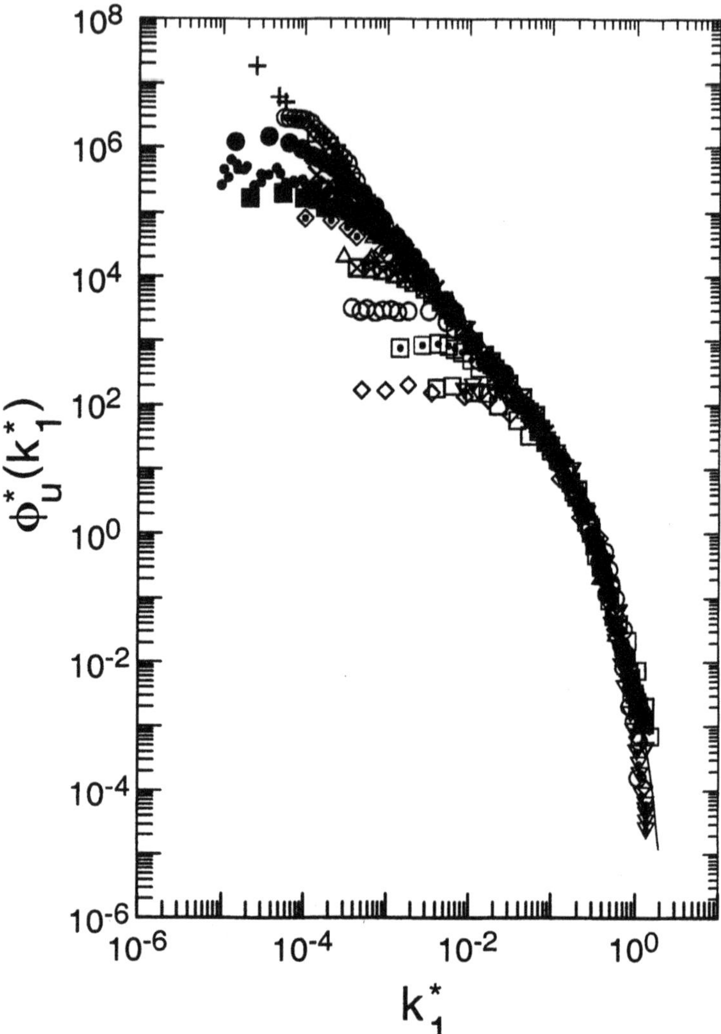


*Figure 7.4.* Kolmogorov's universal scaling for one-dimensional longitudinal power-spectra in different flows, after Saddoughi and Veeravalli [402]. No attempt is made to distinguish between the different symbols; the reader should refer to Figure 9 of Saddoughi and Veeravalli [402] from which this figure is taken. The magnitude of the spectral density at low wavenumbers increases with increasing $R_\lambda$. Note that here and in Figure 7.5, $\phi_u^*(k_1^*)$ is defined such that $\int_0^\infty \phi_u^*(k_1^*)dk_1^* = \langle u^{*2}\rangle$.

accordance with the relations

$$\underset{r^*\to\infty}{\mathcal{L}t}\langle(\delta u^*)^2\rangle = 2\langle u^{*2}\rangle = 2(15)^{-1/2}R_\lambda \tag{7.23}$$

$$\underset{r^*\to\infty}{\mathcal{L}t}\langle(\delta v^*)^2\rangle = 2\langle v^{*2}\rangle = 2\alpha_v(15)^{-1/2}R_\lambda \tag{7.24}$$

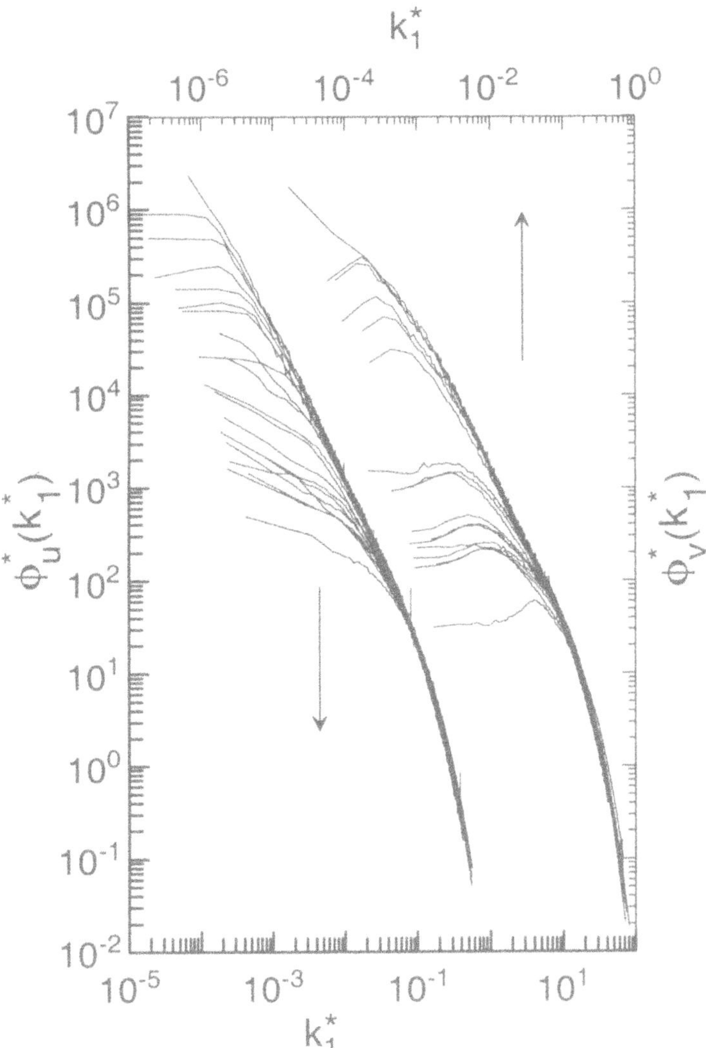


*Figure 7.5.* Kolmogorov-normalized one-dimensional longitudinal and transverse spectra, $40 \lesssim R_\lambda \lesssim 4250$, after Pearson [360]. Note that $\phi_u(k_1^*)$ is the left-hand collection of curves and $\phi_v(k_1^*)$ is the right-hand collection. No attempt is made to distinguish between the different curves but the magnitude of the spectral density at low wavenumbers increases with $R_\lambda$.

$$\mathcal{L}t_{r^* \to \infty} \langle (\delta\theta^*)^2 \rangle = 2\langle \theta^{*2} \rangle = 6R(15)^{-1/2} R_\lambda \,. \tag{7.25}$$

In obtaining eq.(7.23), (7.24), (7.25), the isotropic forms of $\langle \epsilon \rangle$ and $\langle \chi \rangle$ have been used; $\alpha_v$ denotes the ratio $\langle v^2 \rangle / \langle u^2 \rangle$, which is a measure of global anisotropy or anisotropy at large scales and $R \equiv (\langle \theta^2 \rangle / \langle \chi \rangle)/(\langle q^2 \rangle / \langle \epsilon \rangle)$ is the ratio of the time scales associated with $\theta^2$ and $q^2$ respectively. Note

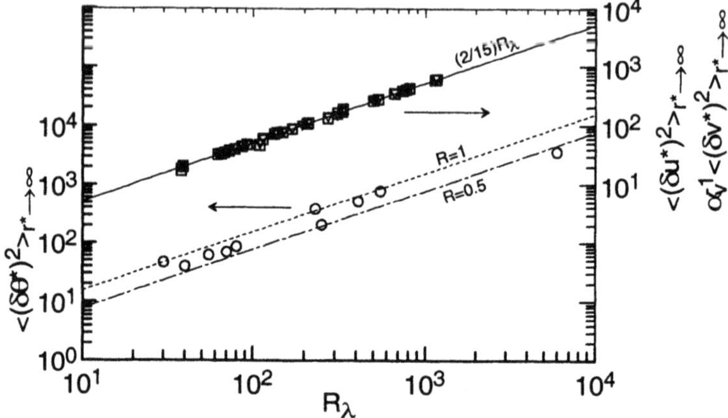


*Figure 7.6.* $R_\lambda$ dependence of limiting values, as $r^* \to \infty$, of $\langle(\delta u^*)^2\rangle$, $\langle(\delta v^*)^2\rangle$ and $\langle(\delta\theta^*)^2\rangle$. The data are the same as in Figures 7.1–7.3. ▽, $\langle(\delta u^*)^2\rangle$; □, $\alpha_v^{-1}\langle(\delta v^*)^2\rangle$; —, eq.(7.23) or eq.(7.24); ○, $\langle(\delta\theta)^2\rangle$; - - -, eq.(7.25) with $R = 1$; — - —, eq.(7.25) with $R = 0.5$.

that $R$ is proportional to the ratio $C_\epsilon/C_\chi$; for global isotropy, $\langle q^2\rangle \equiv 3\langle u^2\rangle$ and $R \equiv 1/3(C_\epsilon/C_\chi)$.

The observations made above with regard to Figure 7.1 apply equally well to published compilations of Kolmogorov-normalized $u$-spectra (e.g. [79, 402]). Figure 7.4, taken from the latter paper (their Figure 9), shows a reasona-ble collapse at large $k_1^*$ ($k_1 \equiv 2\pi f/\langle U\rangle$ is the one-dimensional wavenumber) as well as a systematic dilatation of a power-law range (the exponent is very nearly $-5/3$) as $R_\lambda$ increases (seemingly, this dilation is more extensive and convincing than that exhibited by $\langle(\delta u^*)^2\rangle$). The high wavenumber collapse is "forced" since the isotropic form of $\langle\epsilon\rangle$ has been used, as in Figure 7.1. At low wavenumbers, for which there is no collapse of $\phi_u^*(k_1^*)$, the $u$-spectra exhibit a lack of similarity between different flows. This is not surprising since some of the data are from wall-bounded flows for which a $k_1^{-1}$ behaviour [458] is observed. Such a behaviour is not seen in $\phi_v^*(k_1^*)$. Figure 7.5, taken from Pearson [360] (his Figure 6) shows much better similarity in shape at smaller wavenumbers by comparison to Figure 7.4; a similar observation could be inferred from the comparison of the $u$ and $v$ spectra obtained in grid turbulence by Mydlarski and Warhaft [342].

Figure 7.6 indicates that relations (7.23), (7.24) and (7.25) are reasona-bly well satisfied by the data. The limiting values ($r^* \to \infty$) of $\langle(\delta u^*)^2\rangle$ and $\langle(\delta v^*)^2\rangle$ are indistinguishable, when the latter are divided by $\alpha_v$. The scatter exhibited by the limiting values of $\langle(\delta\theta^*)^2\rangle$ reflects mainly the variation, both with $R_\lambda$ and between different flows, of the ratio $R$; the two dotted lines in Figure 7.6 correspond to relation (7.25) for $R = 0.5$ and

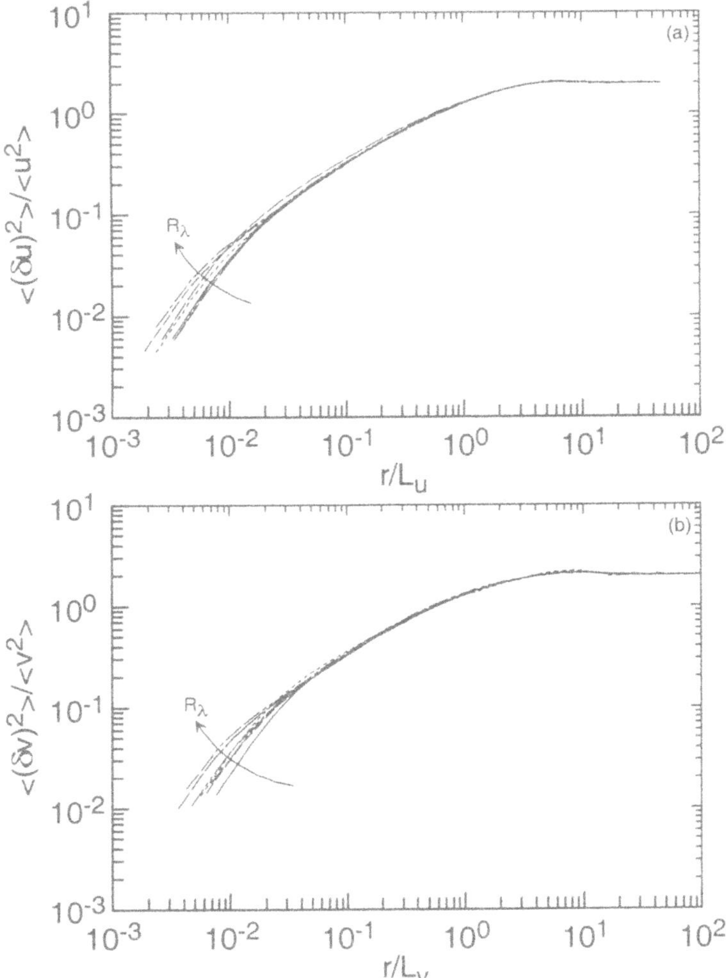


*Figure 7.7.* Second-order structure functions of $u$ and $v$ in the form $\langle(\delta\alpha)^2\rangle/\langle\alpha^2\rangle$ vs $r/L_\alpha$ ($\alpha \equiv u, v$). (a) $u$; (b) $v$. The data pertain to several different flows; the variation indicated by the arrow is in the direction of increasing $R_\lambda$.

$R = 1$.

Structure functions corresponding to energy-containing scales are expected to collapse when $\langle(\delta\beta)^2\rangle$ is normalized by $\langle\beta^2\rangle$ and $r$ is normalized by $L_\beta$. We have verified this for all three structure functions considered here; as an example, Figure 7.7 demonstrates the degree of collapse of $\langle(\delta u)^2\rangle$ (Figure 7.7a) and $\langle(\delta v)^2\rangle$ (Figure 7.7b) for data obtained on the axis/centreline of plane and round jets at different values of $R_\lambda$.

Structure functions of order higher than 2 also display a strong dependency on $R_\lambda$. They are not shown here; however, the pdfs of the increments

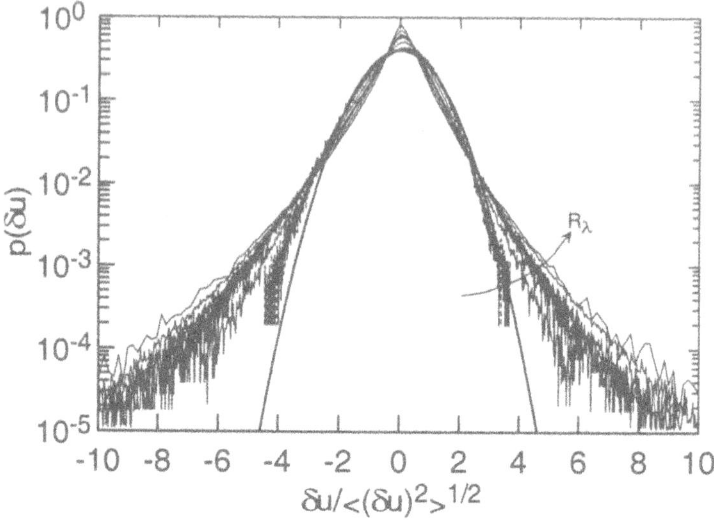


*Figure 7.8.* Probability density functions of $\delta u$ for $r^* \simeq 10$. The data pertain to several different flows; the variation indicated by the arrow is in the direction of increasing $R_\lambda$. The Gaussian distribution (thick solid curve) is shown as reference.

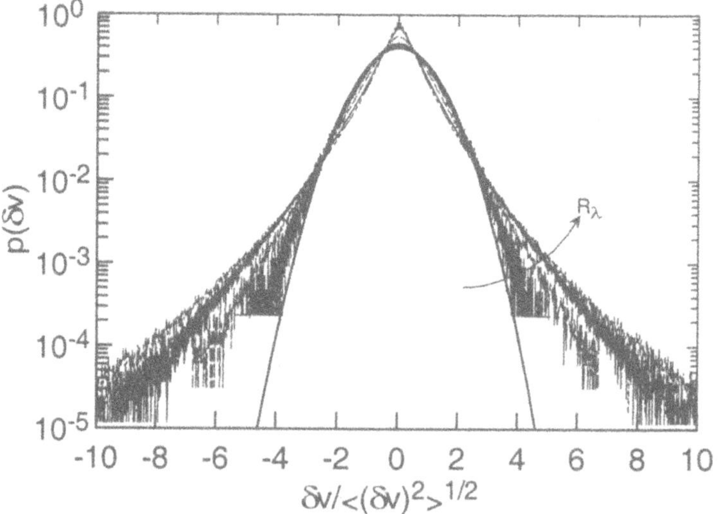


*Figure 7.9.* Probability density functions of $\delta v$ for $r^* \simeq 10$. The data pertain to several different flows; the variation indicated by the arrow is in the direction of increasing $R_\lambda$. The Gaussian distribution (thick solid curve) is shown as reference.

$\delta u$, $\delta v$ and $\delta \theta$, generated for $r^* = 10$, are shown in Figures 7.8, 7.9, 7.10 respectively. All three figures show a significant variation with $R_\lambda$, in particular, the peak of the pdf becomes narrower and more pronounced while

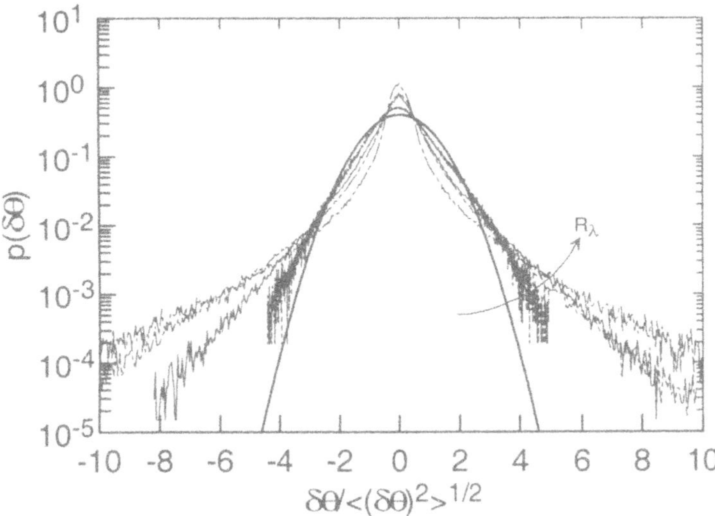


*Figure 7.10.* Probability density functions of $\delta\theta$ for $r^* \simeq 10$. The data pertain to several different flows; the variation indicated by the arrow is in the direction of increasing $R_\lambda$. The Gaussian distribution (thick solid curve) is shown as reference.

the tails of the pdf spread out to larger magnitudes as $R_\lambda$ increases. At the smallest $R_\lambda$, the pdf approaches a Gaussian distribution[4]. For large values of $|\delta\theta|$, the magnitudes of $p(\delta\theta)$ are larger than those of $p(\delta u)$ or $p(\delta v)$. Consequently, normalized higher-order even moments of $\delta\theta$ should be larger than those of $\delta u$ and $\delta v$. Some quantification of this is given in Figures 7.11 and 7.12. The flatness factor $F_{\delta\beta} \equiv \langle(\delta\beta)^4\rangle / \langle(\delta\beta)^2\rangle^2$ (Figure 7.11) is the largest for $\delta\theta$ and smallest for $\delta u$; also the rate of increase with $R_\lambda$ is biggest for $F_{\delta\theta}$. A possible inference is that small-scale fluctuations of $\delta\theta$ are more intermittent than those of $\delta v$ and, more especially, $\delta u$. The skewness of $\delta v$ (Figure 7.12) is nearly zero, independently of $R_\lambda$, reflecting the symmetry of the pdf of $\delta v$. The magnitude ($\simeq 0.4$) of $S_{\delta u}$ is comparable to that for the skewness of $(\partial u/\partial x)$ [447]. Figure 7.12 indicates that there is a significant variability in $S_{\delta\theta}$, including a change of sign, as was noted previously (e.g. [447, 183, 449, 450, 184, 323]) in the context of the temperature derivative skewness. The magnitude of $S_{\delta\theta}$ remains significant

[4]Attempts to calculate the evolution with $r^*$ and also $R_\lambda$ of $p(\delta\theta)$ or $p(\delta u)$ via transport equations for these pdfs have been proposed by, for example, Vaienti et al. [464], Friedrich and Peinke [164] and Marcq and Naert [315]. These approaches provide encouraging agreement with experimental (or DNS) data for the evolution of the pdf, although closures are needed for several unknown terms. At present, the closures are achieved on the basis of actual (measured or DNS) data. Obviously, this is not satisfactory and further work will be needed before the mechanisms responsible for the evolution of the pdfs can be explained either mathematically or physically.

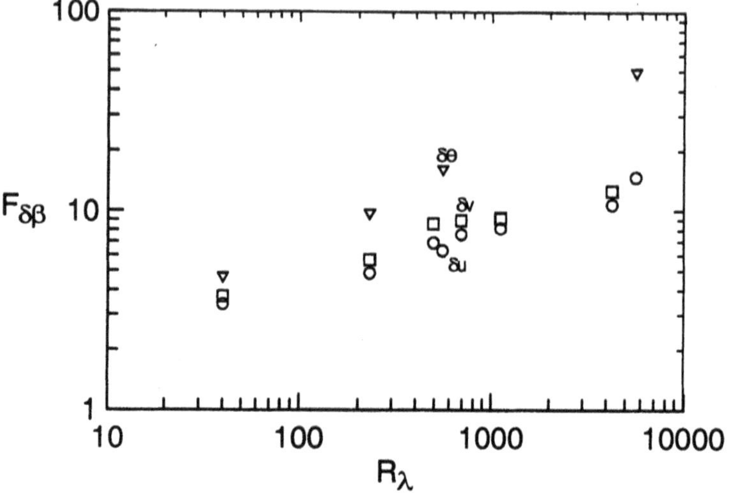


*Figure 7.11.* The flatness factor $F_{\delta\beta}$ for $r^* \simeq 10$. The data are the same as in Figures 7.8–7.10.

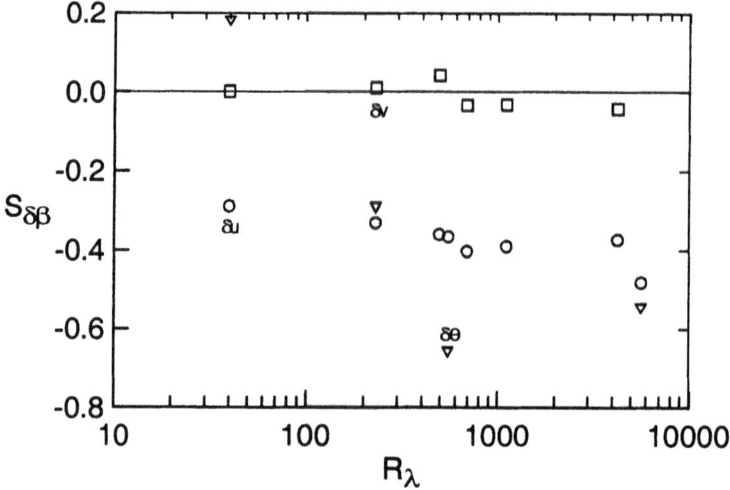


*Figure 7.12.* The skewness $S_{\delta\beta}$ for $r^* \simeq 10$. The data are the same as in Figures 7.8–7.10.

($\simeq 0.6$) on the axis of the circular jet ($R_\lambda \simeq 550$) and in the atmospheric surface layer ($R_\lambda \simeq 4000$), notwithstanding the relatively large values of $R_\lambda$ for these two flows.

It seems appropriate here to comment on the enhanced intermittency exhibited by the scalar field, relative to that of the velocity field. It has been known for some time that the scalar field exhibits features which are much more pronounced than in the velocity field. The presence of ramps in temperature signals was first observed by Taylor [457] in the unstable atmospheric surface layer; the earlier interpretation in the context of buoyant plumes [373] was subsequently revised [11, 90, 364] to recognise the ramps as the ubiquitous signature of the organized large scale motion. The sheet-like topology of the scalar has been highlighted both in experiments and numerical simulations; for example, Pumir [374, 375] has studied the steep scalar fronts in simulations of periodic box turbulence with an imposed mean scalar gradient. More recent theoretical and numerical analyses [261, 166] indicate that a passive scalar advected by a (non-intermittent) delta-correlated velocity field (which does not satisfy the Navier-Stokes equations) develops its own small-scale intermittency. The intermittency of the scalar field thus stems from two distinct sources. One is the intermittent nature of the advecting turbulent velocity field. The other results from the specific mathematical properties of the scalar advection-diffusion equation. These properties may explain, inter alia, the previously noted variability in the coefficient $\rho$.

## 7.5. A Framework for Comparing Velocity and Scalar Fields

The realisation that the advection of the instantaneous temperature in a turbulent flow is primarily controlled by the instantaneous velocity vector [101, 34, 309] — molecular effects tending to smooth out the smallest scale temperature fluctuations — provides a useful basis for a meaningful comparison between fluctuating velocity and scalar fields. Fulachier [168], see also Fulachier and Dumas [171], drew an analogy between the scalar correlation $\langle\theta(t)\theta(t+\tau)\rangle$ and the correlation $\langle \underset{\sim}{q}(t)\cdot\underset{\sim}{q}(t+\tau)\rangle$ where $\underset{\sim}{q}$ $(\equiv \underset{\sim}{i}\,u+\underset{\sim}{j}\,v+\underset{\sim}{k}\,w)$ is the fluctuating velocity vector ($\underset{\sim}{i}$, $\underset{\sim}{j}$, $\underset{\sim}{k}$ are unit vectors in the $x$, $y$, $z$ directions). For constant density flows, the transport equations for these correlations are reproduced below (using tensor notation)

$$
\begin{aligned}
&\langle U_i\rangle\frac{\partial}{\partial x_i}\langle\theta(t)\theta(t+\tau)\rangle+[\langle u_i(t)\theta(t+\tau)\rangle+\langle u_i(t+\tau)\theta(t)\rangle]\frac{\partial\langle T\rangle}{\partial x_i}\\
&+\left\langle\theta(t+\tau)\frac{\partial}{\partial x_i}[u_i(t)\theta(t)]\right\rangle+\left\langle\theta(t)\frac{\partial}{\partial x_i}[u_i(t+\tau)\theta(t+\tau)]\right\rangle\\
&-\kappa\left\langle\theta(t+\tau)\frac{\partial^2\theta(t)}{\partial x_i^2}+\theta(t)\frac{\partial^2\theta(t+\tau)}{\partial x_i^2}\right\rangle=0 \qquad (7.26)
\end{aligned}
$$

and

$$
\begin{aligned}
&\langle U_i\rangle\frac{\partial}{\partial x_i}\langle u_j(t)u_j(t+\tau)\rangle+[\langle u_i(t)u_j(t+\tau)\rangle+\langle u_i(t+\tau)u_j(t)\rangle]\frac{\partial\langle U_j\rangle}{\partial x_i}\\
+\;&\left\langle u_j(t+\tau)\frac{\partial}{\partial x_j}[u_i(t)u_j(t)]\right\rangle+\left\langle u_j(t)\frac{\partial}{\partial x_i}[u_i(t+\tau)u_j(t+\tau)]\right\rangle
\end{aligned}
$$

$$
\begin{aligned}
- & \quad \nu \left\{ \left\langle u_j(t+\tau) \left( \frac{\partial^2 u_j(t)}{\partial x_i^2} + \frac{\partial^2 u_i(t)}{\partial x_i \partial x_j} \right) \right\rangle \right. \\
+ & \quad \left. \left\langle u_j(t) \left( \frac{\partial^2 u_j(t+\tau)}{\partial x_i^2} + \frac{\partial^2 u_i(t+\tau)}{\partial x_i \partial x_j} \right) \right\rangle \right\} \\
+ & \quad \left\{ \left\langle u_j(t+\tau) \frac{\partial p(t)}{\partial x_j} \right\rangle + \left\langle u_j(t) \frac{\partial p(t+\tau)}{\partial x_j} \right\rangle \right\} = 0 \, . \qquad (7.27)
\end{aligned}
$$

When $Pr$ is 1, eqs.(7.26) and (7.27) are formally analogous except for the pressure containing terms in eq.(7.27) [$p$ is the kinematic pressure fluctuation]. These terms disappear for homogeneous turbulence. It follows that, for homogeneous turbulence, the equations for the temperature spectrum $\phi_\theta(f)$, the Fourier transform of $\langle \theta(t)\theta(t+\tau) \rangle$, and the spectrum $\phi_q(f)$, the Fourier transform of $\langle u_j(t)u_j(t+\tau) \rangle$, are also analogous. Fulachier [168] (also Fulachier and Dumas [171]) found that the distribution of $\phi_q(f)$, which can also be written as follows

$$
\langle q^2 \rangle \phi_q(f) = \langle u^2 \rangle \phi_u(f) + \langle v^2 \rangle \phi_v(f) + \langle w^2 \rangle \phi_w(f) \qquad (7.28)
$$

(with $\int_0^\infty \phi_\beta(f)df = 1$; note that this definition of $\phi_\beta(f)$ differs from that used in the context of Figures 7.4 and 7.5), followed closely that of $\phi_\theta(f)$ across almost the whole of a smooth wall boundary layer over a uniformly slightly heated surface, at least over a frequency range which accounts for most of $\langle \theta^2 \rangle$ and $\langle q^2 \rangle$. Figure 7.13, excerpted from Figure 7 of Fulachier and Dumas [171], clearly shows that $\phi_\theta(f)$ follows $\phi_q(f)$ much more closely than $\phi_u(f)$; the presentation is in the form $f\phi_\alpha(f)$ (on a linear scale) vs $\log f$ so as to highlight the range of frequencies which contribute most to the variances. This behaviour has since been confirmed, over a range of Reynolds numbers, both in the boundary layer, albeit with different types of surface conditions, or in different types of flows (e.g. [383, 170, 6, 10]. The spectral analogy is particularly good (Figure 7.14) for a nearly homogeneous turbulent flow with constant mean velocity and mean temperature gradients [455]. Although the non-homogeneity (in both $x$ and $y$ directions) is more important in the boundary layer than in the previous flow, the fact that the analogy works well seems to suggest that the maintained presence of production (or source) terms for $\langle q^2 \rangle$ and $\langle \theta^2 \rangle$ may play an important role in the context of this analogy. DNS data for boundary layer and fully developed pipe or channel flows indicate that the contribution from the pressure diffusion term in the transport equation of $\langle q^2 \rangle$ is small, even close to the wall. Fulachier and Antonia [170] noted that, for these wall flows, the strong link between velocity and temperature fluctuations close to the wall is present all along the flow and that such a link is absent in free shear flows. Indeed, the analogy between $\phi_q(f)$ and $\phi_\theta(f)$

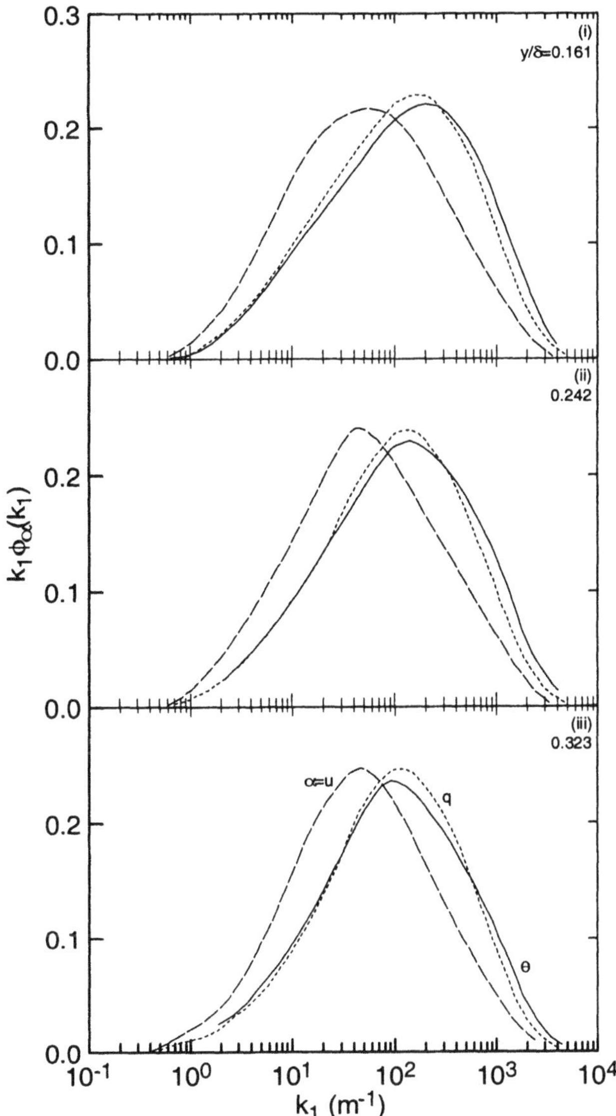


*Figure 7.13.* Comparison between the temperature spectrum and the energy spectrum in a smooth wall boundary layer with a uniform wall temperature. The $u$ spectrum is also shown for comparison. The curves are extracted from Figure 10 of Fulachier and Dumas [171]. —, $k_1\phi_\theta(k_1)$; - - -, $k_1\phi_q(k_1)$; – –, $k_1\phi_u(k_1)$. Note that here and for Figures 7.14 and 7.15, $\int_0^\infty \phi_\alpha(k_1)dk_1 - 1$.

is relatively poor on the axis of either jets or wakes, e.g. [14]. The analogy improves as the distance from the axis increases and the magnitudes of $\partial\langle U\rangle/\partial y$ and $\partial\langle T\rangle/\partial y$ increase. In view of the previous remarks, it is of

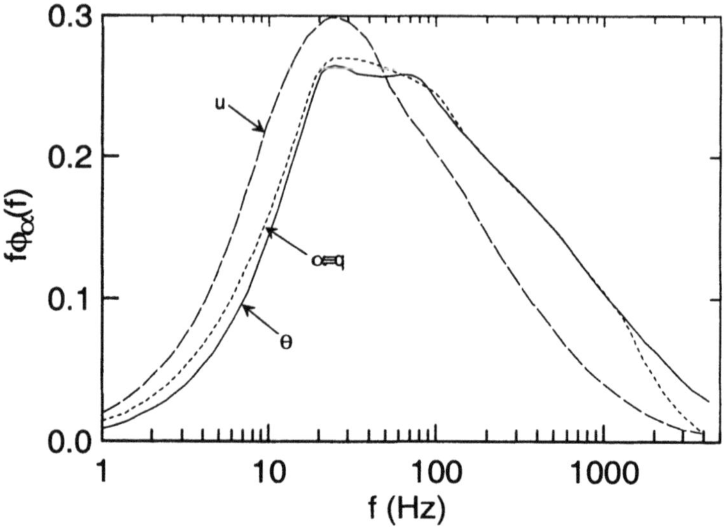


*Figure 7.14.* Comparison between the temperature spectrum and the energy spectrum in a uniform shear flow with a constant mean temperature gradient. The $u$ spectrum is also shown for comparison. The curves were calculated by Fulachier and Antonia [170] from the data of Tavoularis and Corrsin [455]. —, $f\phi_\theta(f)$; - - -, $f\phi_q(f)$; — —, $f\phi_u(f)$.

some interest to assess the validity of the analogy in decaying turbulence downstream of a grid. In this flow, there are no source terms, the rates of decay of $\langle q^2\rangle$ and $\langle\theta^2\rangle$ being set by $\langle\epsilon\rangle$ and $\langle\chi\rangle$ respectively, viz.

$$-\frac{\langle U\rangle}{2}\frac{d\langle q^2\rangle}{dx} = \langle\epsilon\rangle \tag{7.29}$$

$$-\frac{\langle U\rangle}{2}\frac{d\langle\theta^2\rangle}{dx} = \langle\chi\rangle\ . \tag{7.30}$$

Recently, simultaneous measurements were made of velocity and temperature fluctuations downstream of a grid-heated screen combination [501]. The distribution of $\phi_q(k_1)$, estimated from the spectra of $u$, $v$, $w$ as per relation (7.28), is compared with that of $\phi_\theta(k_1)$ in Figure 7.15. The two distributions are markedly different, $k_1\phi_\theta(k_1)$ peaking at a lower wavenumber than $k_1\phi_q(k_1)$; also, the former distribution has a smaller peak value than the latter. The shape of $k_1\phi_u(k_1)$ [included in Figure 7.15] is closer to that of $k_1\phi_\theta(k_1)$ than of $k_1\phi_q(k_1)$ [which, at higher wavenumbers, follows closely that of $k_1\phi_v(k_1)$ or $k_1\phi_w(k_1)$]. The relative behaviour of $\phi_q(k_1)$ and $\phi_\theta(k_1)$ [Figure 7.15] has also been observed in DNS data (Orlandi, private communication) obtained in decaying box turbulence. The difference between $\phi_q(k_1)$ and $\phi_\theta(k_1)$, which is in contrast with the analogous forms of (7.29) (7.30), suggests that the initial conditions for $q^2$ and $\theta^2$ are

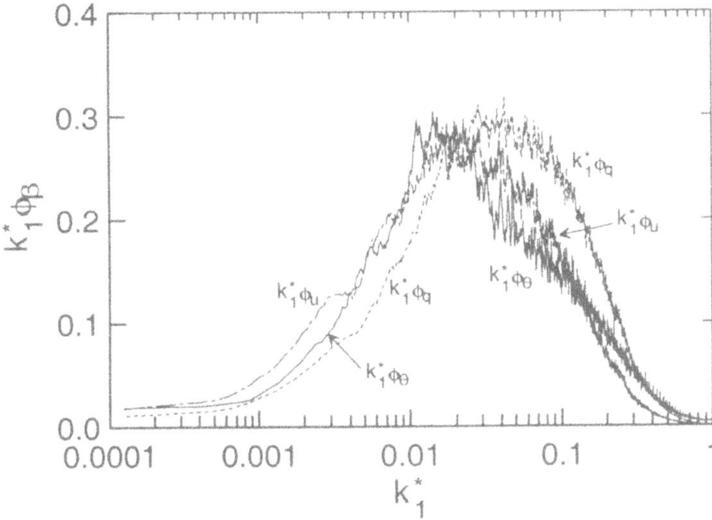


*Figure 7.15.* Comparison between the temperature spectrum and the energy spectrum in grid turbulence [501]. The measurements were made downstream of a biplane square mesh grid ($M$ = 24.76 mm with 4.76 mm × 4.76 mm square rod; solidity = 0.35). A screen (0.5 mm dia. Chromel wires) was placed at a distance $1.5M$ downstream of the grid to allow the introduction of heat into the flow. The $u$ spectrum is also shown for comparison. —, $k_1^*\phi_\theta(k_1^*)$; - - -, $k_1^*\phi_q(k_1^*)$; — - —, $k_1^*\phi_u(k_1^*)$.

also important in the context of achieving analogous forms for $\phi_q(k_1)$ and $\phi_\theta(k_1)$. It should be kept in mind that, even if eqs.(7.26) and (7.27) were formally similar, accepting that the effect of pressure could be totally ignored, similarity of solutions would also require the boundary and initial conditions to be identical. This latter requirement is unlikely to be met by the grid turbulence data of Zhou et al. [501, 500] (the grid and the mandoline have different rod/wire diameters and are not at the same $x$ location). Small differences between initial conditions (for the velocity and thermal fields) at the wake generator or the jet exit may be responsible for the lack of similarity between $\phi_\theta(k_1)$ and $\phi_q(k_1)$ observed on or near the axes in these two flows. Away from the axes, where $\partial\langle U\rangle/\partial y$ and $\partial\langle T\rangle/\partial y$ become important, the length scales associated with the production rates of $\langle q^2\rangle$ and $\langle\theta^2\rangle$ are more likely to be comparable.

Generally speaking, the scalar field remains strongly affected by the initial / boundary conditions, even far from the jet exit or from the grid / mandoline arrangement so that scalar small-scale properties continue to be dependent on the large-scale-flow-specific organization; the grid turbulence and experiment of Antonia et al. [12] and the flow between two counterrotating disks, considered by Danaila et al. [119], are two illustrations of this dependency. The orientation of the temperature ramps is directly connected to the relative orientations of the shear and the mean temperature gradient. Another connected observation (Auriault [25]) is the presence of an exponential tail in the temperature pdf measured on the axis of a heated jet, which appears on the side of the "cold" ambient temperature. This large-scale asymmetric feature for the temperature field "survives" very far from the jet exit, where one would expect the flow to be relatively well mixed, and directly affects the odd-order moments of the temperature increment. Therefore, for the scalar field to attain an asymptotic state in conformity with isotropy represents a formidable task, even when the Reynolds number is large, so that the correlation between $\delta u$ and $\delta\theta$ generally retains a value significantly different from zero, reflecting the correlation between $u$ and $\theta$.

It is also appropriate to comment on the relationship between the time-scale ratio $R$ and the status of the analogy between $\phi_q(k_1)$ and $\phi_\theta(k_1)$. It follows from eq.(7.28) that

$$\begin{aligned}\langle q^2\rangle \int_0^\infty k_1^2\phi_q(k_1)dk_1 &= \langle u^2\rangle \int_0^\infty k_1^2\phi_u(k_1)dk_1 + \langle v^2\rangle \int_0^\infty k_1^2\phi_v(k_1)dk_1 \\ &+ \langle w^2\rangle \int_0^\infty k_1^2\phi_w(k_1)dk_1\end{aligned}$$

or

$$\left\langle\left(\frac{\partial q}{\partial x}\right)^2\right\rangle = \left\langle\left(\frac{\partial u}{\partial x}\right)^2\right\rangle + \left\langle\left(\frac{\partial v}{\partial x}\right)^2\right\rangle + \left\langle\left(\frac{\partial w}{\partial x}\right)^2\right\rangle .$$

If local isotropy is assumed, $\langle\epsilon\rangle$ can be approximated by

$$\langle\epsilon\rangle = 3\nu\left\langle\left(\frac{\partial q}{\partial x}\right)^2\right\rangle \tag{7.31}$$

while

$$\langle\chi\rangle = 3\kappa\left\langle\left(\frac{\partial \theta}{\partial x}\right)^2\right\rangle . \tag{7.32}$$

The ratio $R$ can then be written as

$$R = Pr\,\frac{\int_0^\infty k_1^2\phi_q(k_1)dk_1}{\int_0^\infty k_1^2\phi_\theta(k_1)dk_1} . \tag{7.33}$$

If it is further assumed that $\phi_q(k_1) \simeq \phi_\theta(k_1)$, the magnitude of $R$ is equal to the Prandtl number. The time scale of the scalar fluctuation field would thus be equal to that of the velocity fluctuation field only when $Pr = 1$. For the grid turbulence data referred to earlier, $R \simeq 1$ [501] whereas $Pr \simeq 0.7$; we also recall that, in this flow, $\langle\epsilon\rangle$ and $\langle\chi\rangle$ are known accurately. Further, local isotropy is very closely approximated, but $\phi_q(k_1) \neq \phi_\theta(k_1)$ and the ratio of the two integral quantities on the right of eq.(7.33) is greater than 1, consistent with a measured value of $R$ greater than $Pr$. (Béguier et al. [41] also reported that $R \simeq 1$ for the heated grid turbulence data of Warhaft and Lumley [482]). For the slightly heated thin shear flows considered by [41], $R$ is reasonably uniform across each flow, with a value close to 0.5. Since $\phi_q(k_1)$ follows $\phi_\theta(k_1)$ closely across the boundary layer, the inequality $R < Pr$ ($\simeq 0.71$) in this flow is likely to be associated with different levels of departure from local isotropy of velocity and scalar fluctuations. Summaries of departures from local isotropy were reported in tabulated form, for a wide range of turbulent shear flows, by Browne et al. [62] and George and Hussein [178]. The value of 0.5 has yet to be established with confidence in most of the flows considered by Béguier et al. [41]. The major difficulty is of course the determination of $\langle \chi \rangle$ and more especially $\langle \epsilon \rangle$ which, in general, requires 12 terms to be estimated. Krishnamoorthy and Antonia [262] were able to measure $\langle \chi \rangle$ with reasonable accuracy in a zero pressure gradient boundary layer but had to infer $\tau_u$ ($\equiv \langle q^2 \rangle / \langle \epsilon \rangle$) from Laufer's [271] measurements in the near-wall region of a fully developed pipe flow. The resulting magnitude of $R$ was close to 1 in the near-wall region. For the quasi-homogeneous shear flow considered by Tavoularis and Corrsin [455], a value smaller than 0.5 was obtained. One would expect that DNS data should provide reliable estimates of $R$ since $\langle \epsilon \rangle$ and $\langle \chi \rangle$ can be obtained accurately. The channel flow DNS data of Kawamura et al. [242] suggest a constant value close to 0.5 (when $Pr = 0.71$) outside the wall region. It would of course be useful to have estimates, also for $Pr = 0.7$, in other simulated flows. It is relatively easy to show that the limiting wall value of $R$ is $Pr$. The channel flow simulations of Kawamura et al. [243] fully support this (a range of values of $R$, between 0.025 and 5, was considered). At the risk of digressing, it is also useful to recall that Antonia and Kim [7] observed that the limiting wall value of the turbulent Prandtl number is very close to 1, almost independently of $Pr$. This result, which may have implications with regard to scalar transport modelling, has since been confirmed in other flows and a wider range of $Pr$.

## 7.6. Further Implications of the Spectral Analogy

The similarity between $\phi_\theta$ and $\phi_q$ was re-examined in the context of second-order velocity and temperature structure functions by Antonia et al. [21]. Distributions of $D_q/\langle q^2 \rangle$, where $D_q \equiv \langle (\delta q)^2 \rangle = \langle (\delta u)^2 \rangle + \langle (\delta v)^2 \rangle + \langle (\delta w)^2 \rangle$ followed closely those of $D_\theta / \langle \theta^2 \rangle$ in both wakes and boundary layers. The distributions in Figure 7.16 (reproduced from Figure 7 of their paper; this figure uses data from Mestayer [321, 322] and Saddoughi and Veeravalli [402]) underline the quality of the similarity. Note that in each flow, the

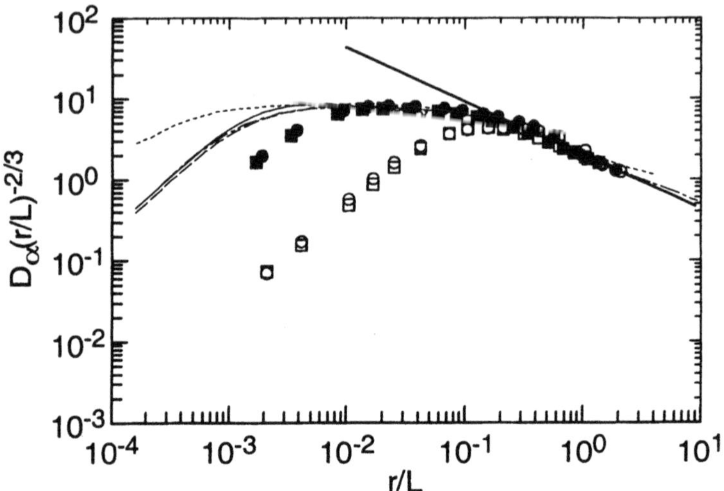


*Figure 7.16.* Comparison between $D_q/\langle q^2\rangle$ and $D_\theta/\langle\theta^2\rangle$ in several flows, after Antonia et al. [21]. Wake ($R_\lambda = 40$): □, $D_q$; ○, $D_\theta$. Wake ($R_\lambda = 230$): ■, $D_q$; ●, $D_\theta$. Boundary layer (Mestayer [321, 322], $R_\lambda = 500$): —, $D_q$; — - - —, $D_\theta$. Boundary Layer (Saddoughi and Veeravalli [402]): —,$D_q$ ($R_\lambda = 600$); - - -, $D_q$ ($R_\lambda = 1450$). The thick solid line represents $D_\alpha/\langle\alpha^2\rangle = 2$, i.e. the limiting value at large $r/L$. Note that, for the data in this figure, $L$ was defined by $L = \langle q^2\rangle^{3/2}/\langle\epsilon\rangle$.

comparison was made at nominally the same $R_\lambda$ [5].

When $Pr \simeq 1$, the similarity between spectra or structure functions implies a simple connection between the inertial range behaviours of $q^2$ and $\theta^2$ and, more particularly, a connection between the inertial range premultipliers $C_\theta$ and $C_q$ (the subscript 2 has been dropped here since only second-order structure functions are under consideration). For convenience, Antonia et al. [21] remained strictly within the framework of K41 and O49 and showed that

$$C_\theta = C_q\frac{\langle\theta^{*2}\rangle}{\langle q^{*2}\rangle} = C_q R\ ,$$

with

$$C_q \equiv D_q r^{*-2/3}$$

and

$$C_\theta \equiv D_\theta r^{*-2/3}\ .$$

[5]Corrsin [102] suggested that the use of $\langle u^2\rangle^{1/2}$ and $\lambda_u$ is ambiguous in non-isotropic turbulence because of the directional dependence of these quantities. Fulachier and Antonia [169] suggested that more appropriate definitions for the turbulence Reynolds and Péclet numbers are $R_{\lambda_q} \equiv \langle q^2\rangle^{1/2}\lambda_q/\nu$ and $Pe_q \equiv \langle q^2\rangle^{1/2}\lambda_\theta/\kappa$ where $\lambda_q \equiv \langle q^2\rangle^{1/2}/\langle(\partial q/\partial x)^2\rangle^{1/2}$.

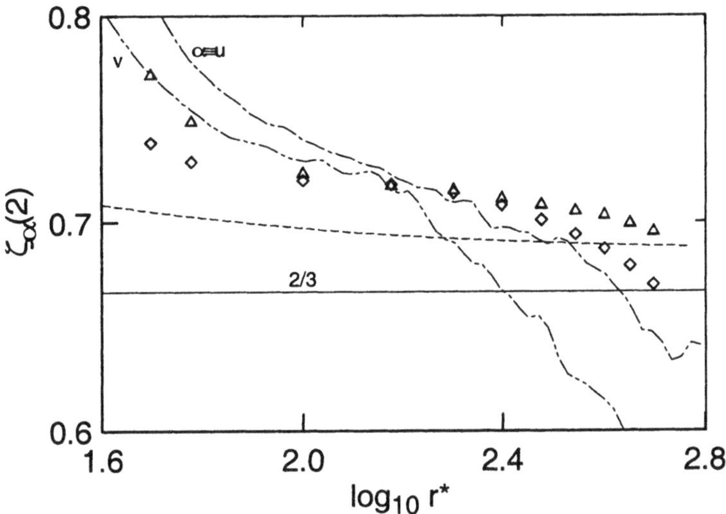


*Figure 7.17.* Distribution of exponent $\zeta_\alpha(2)$ in a region ($1.6 \lesssim \log_{10} r^* \lesssim 2.8$) which would probably be identified as the IR in a log-log plot of $\langle(\delta\alpha^*)^2\rangle$ vs $r^*$. The data were obtained in the atmospheric surface layer [9]. $\triangle$, $\zeta_u(2) = 2 - 2c_u(2)$; $\diamond$, $\zeta_v(2) = 2 - 2c_v(2)$; — - —, $\zeta_u(2) = d[\log_{10}\langle(\delta u^*)^2\rangle]/d(\log_{10} r^*)$; — - - —, $\zeta_v(n) = d[\log_{10}\langle(\delta v^*)^2\rangle]/d(\log_{10} r^*)$; - - -, $\zeta_u(2) = 2/3 + d(\log_{10}\langle\epsilon_r^{*2/3}\rangle)/d(\log_{10} r^*)$; —, K41 value : $\zeta_u(2) = \zeta_v(2) = 2/3$.

The relative magnitudes of $C_q$ and $C_\theta$ thus depend on the value of the timescale ratio. In the case of a wake, for which $R$ is known with some degree of confidence, Antonia et al. [21] proposed values of about 4 and 8 for $C_\theta$ and $C_q$ respectively. A more general connection between $C_q$ and $C_\theta$ should of course take into account the intermittencies of $\epsilon$ and $\chi$. Scaling exponents for $q$ and $\theta$ should first be estimated before evaluating the corresponding premultipliers.

Many different methods are available for estimating the scaling exponents $\zeta_\beta(n)$. It is not our intention to review these methods. It is important however to note that methods such as the extended self-similarity (ESS) technique [43] can only yield "relative", instead of "absolute", magnitudes and hence caution should be applied when interpreting results from such methods. Of more serious concern is whether it is indeed possible to obtain unique estimates of these exponents, at least for the Reynolds numbers typically encountered in the laboratory. This concern is only heightened when one enquires into the "constancy" with respect to $r$ of the structure function exponents. This can be done directly by examining the $r$ dependency of the exponent $\zeta_\beta(n)$, obtained by differentiating $\log\langle(\delta\beta^*)^n\rangle$ with respect to $\log r$. Antonia and Smalley [9] examined data in the atmospheric surface layer ($R_\lambda \simeq 4000$) for $u$ and $v$ and plotted $\zeta_\beta(n)$ [for $n = 2$, 4, 6] as a function of $\log r$. The results for $n = 2$, shown in Figure 7.17

(taken from Figure 2 of their paper), clearly indicate that the magnitudes of $\zeta_u(2)$ and $\zeta_v(2)$ continue to decrease with $\log_{10} r^*$ over a range which would normally be identified with the IR and in which the exponents ought to be constant. It is also evident that $\zeta_u(2)$ remains somewhat larger than $\zeta_v(2)$, in violation of local isotropy; recall that, for isotropy,

$$\langle(\delta v)^2\rangle = \langle(\delta w)^2\rangle = \left(1 + \frac{r}{2}\frac{d}{dr}\right)\langle(\delta u)^2\rangle$$

so that $\zeta_u(2) = \zeta_v(2) = \zeta_w(2)$ follows after substituting eq.(7.7) into the above relation. Whilst it may be argued that the proximity to the ground (the X-probe was located at a height of 1.72 m) may have had some effect on the distributions of Figure 7.17, similar trends are evident in the atmospheric surface layer data at larger $R_\lambda$ ($\simeq 10^4$) and larger heights. There seems little doubt that $\zeta_\theta(2)$ would exhibit the same trend as either $\zeta_u(2)$ or $\zeta_v(2)$, so that the concept of an IR or an inertial-convective range appears meaningful only when the stipulations in K41 (or K62) are satisfied strictly.

Another, relatively attractive, means of estimating $\zeta_\beta(n)$ is to start with a parametric description for $\langle(\delta\beta)^2\rangle$. A plausible description when $\beta \equiv u$ is given by [453, 433, 190, 319]

$$\langle(\delta u)^2\rangle = \frac{\langle\epsilon\rangle}{15\nu}\frac{r^2}{\left[1 + \left(\frac{r}{r_{cu}}\right)^2\right]^{c_u}} \tag{7.34}$$

where $r_{cu}$ can be identified with the cross-over between the dissipative and inertial ranges and $c_\beta \equiv [2 - \zeta_\beta(2)]/2$. Expressions similar to eq.(7.34) can be written for $\langle(\delta v)^2\rangle$, $\langle(\delta w)^2\rangle$ and hence $\langle(\delta q)^2\rangle$ [20]. The latter authors also proposed the following descriptions for $\langle(\delta\theta)^2\rangle$

$$\langle(\delta\theta)^2\rangle = \frac{\langle\chi\rangle}{3\kappa}\frac{r^2}{\left[1 + \left(\frac{r}{r_{c\theta}}\right)^2\right]^{c_\theta}} \tag{7.35}$$

where $r_{c\theta}$ may, in general, be different from $r_{cu}$ or, for that matter, $r_{cv}$ or $r_{cw}$. One may expect $r_{c\theta}$ and $r_{cq}$ to be close to each other when $Pr \simeq 1$. Expressions (7.34) and (7.35) provide high quality fits to available data in range of flows and values of $R_\lambda$ across both the DR and IR. A more general description for $\langle(\delta u)^2\rangle$ extending to $r \simeq L$ has been considered by Dhruva, see Kurien and Sreenivasan [264]. For consistency with the previously observed trends of $\langle(\delta\beta^*)^2\rangle$, both in terms of $r^*$ and $R_\lambda$, one would expect that expressions (7.34) and (7.35) ought to yield values of

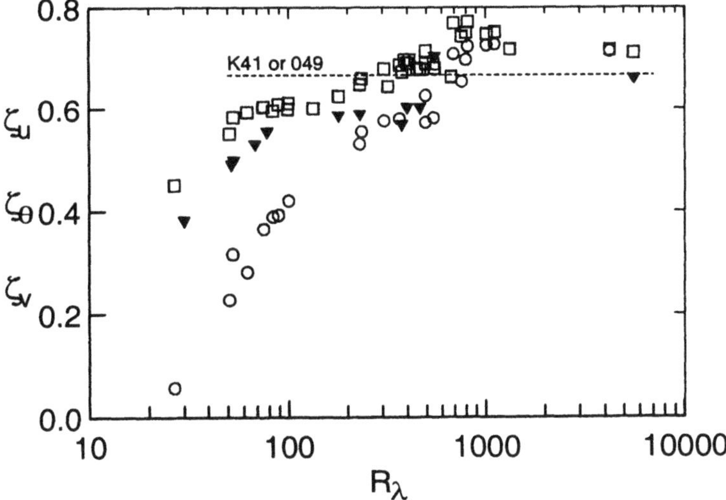


*Figure 7.18.* Dependence of $\zeta_u(2)$, $\zeta_v(2)$ and $\zeta_\theta(2)$ on $R_\lambda$ using the flows considered in Figures 7.1–7.3. □, $\zeta_u(2)$; ○, $\zeta_v(2)$; ▼, $\zeta_\theta(2)$. The broken line corresponds to the K41 (or O49) value ("2/3").

$\zeta_\beta(2)$ which, for a given $r^*$, increase with $R_\lambda$ possibly asymptoting to constant values at sufficiently large $R_\lambda$. The application of expressions such as eq.(7.34) to the ASL data of Figure 7.17 does indeed result in a slight decrease with $r^*$ of the magnitudes of $\zeta_u(2)$ and $\zeta_v(2)$ across the IR (strictly, a scaling range). This is as it ought to be since a reliable description of $\langle(\delta\beta)^2\rangle$ must "mimic" the $r$ dependence of the local slope, as discussed previously. It is axiomatic that when "unique" (absolute) values of $\zeta_\beta(2)$ are quoted, as is usually the case in the literature, they can only at best represent "averages" across the IR.

Antonia et al. [19] presented values of $\zeta_u(2)$, $\zeta_v(2)$ and $\zeta_\theta(2)$ which corresponded to a particular value of $r^*$ (that for which the magnitudes of $\langle(\delta u^*)^3\rangle/r^*$ or $\langle(\delta u^*)(\delta\theta^*)^2\rangle/r^*$ are maximum) for three types of flows (grid turbulence, plane wakes and circular jets). Figure 7.18 combines these results with those estimated for some of the flows, including the ASL used in earlier figures. Despite the scatter, the plotted data exhibit unmistakable trends. The magnitudes of all three exponents increase with $R_\lambda$, arguably approaching constant values when $R_\lambda$ is of order $10^4$. The increase is most rapid for $\zeta_v(2)$ and slowest for $\zeta_u(2)$ (the use of ESS, with $\langle(\delta u)^3\rangle$ used as "reference", would result in $\zeta_u(2)$ remaining essentially independent of $R_\lambda$, as reported for example by Arneodo et al. [22]). It should be stressed that the magnitudes of the exponents are not uniquely determined by $R_\lambda$, even when fully developed turbulence is considered. They depend on other parameters, such as the mean shear, the proximity to the wall, the nature of

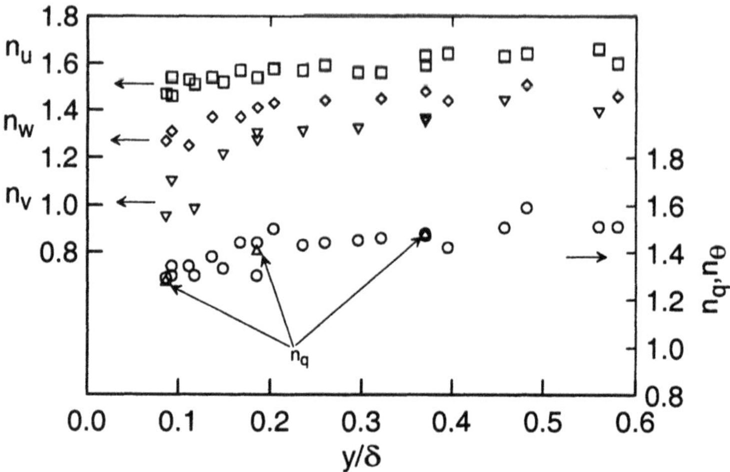


*Figure 7.19.* Variation of spectral exponents across a rough wall boundary layer. The data are taken from Antonia and Smalley [9]. □, $n_u$; ▽, $n_v$; ◇, $n_w$; ○, $n_\theta$; △, $n_q$. To avoid confusion, $n_q$ and $n_\theta$ are plotted on the right coordinate.

the flow and, in particular, the effect of initial conditions. What the previous influences have in common may be their effect on the degree of anisotropy of the large scale motion. It is fair to say that there is an urgent need for systematic investigations into the effects of these parameters, either in isolation or judicious combinations, before published values of the difference $[\zeta_u(2) - \zeta_v(2)]$ and its dependence on $R_\lambda$ obtained from both simulations and experiments, can be interpreted unambiguously. Clearly, when the turbulence is not fully developed, the influence of other parameters such as the intermittency associated with a turbulent/non-turbulent interface, must be taken into account (e.g. [265]); it is of course important that this effect is isolated from that of the mean shear (this should not be difficult since the mean shear is small towards the edge of a boundary layer or a wake; the Kolmogorov-normalized mean shear has been observed [324] to remain constant across a significant cross-section in jets and wakes).

It is pertinent also to stress the need for experiments to include measurements (preferably taken simultaneously) of all three velocity fluctuations, so that $\zeta_q$ can be estimated and compared with $\zeta_\theta$. The velocity fluctuation $w$ was not measured in a number of the flows used in Figure 7.1. For the grid turbulence data, the magnitudes of $\zeta_q$ ($\simeq 0.41$) and $\zeta_\theta$ ($\simeq 0.53$) are different, reflecting the failure of the spectral analogy (see Section 5, Figure 7.15) in this flow. For the boundary layer, these exponents are, as expected, nearly equal. In a recent study of a rough wall boundary layer [10], the spectral exponents $n_\beta$ ($\phi_\beta(k_1) \sim k_1^{-n_\beta}$ in the IR) were estimated for $\beta \equiv u$, $v$, $w$, $q$ and $\theta$ across a significant portion of the layer. It was noted that it was "easier" to identify a power-law range in $\phi_\beta(k_1)$ than in $\langle(\delta\beta)^2\rangle$, especially when a trial and error method (e.g. [446, 342]) is used. In theory, the exponents $n_\beta$ and $\zeta_\beta$ should be related ($n_\beta = \zeta_\beta + 1$) but, in reality, the correspondence between $n_\beta$ and $\zeta_\beta$ may deviate from this relation (see Hou et al. [219] for a recent discussion of this). This was indeed observed by Antonia and Smalley [10], although, in general, good consistency was obtained between the estimates of $n_\beta$ and the values of $\zeta_\beta$ inferred from the parameterized fits to $\langle(\delta\beta)^2\rangle$. The magnitude of $n_\beta$ increase with $y/\delta$ (Figure 7.19) remaining smaller than 2/3 (K41 or O49); the magnitudes of $n_q$ and $n_\theta$ are in reasonable agreements. The variability of the exponents $n_\beta$ (or $\zeta_\beta$) implies, in turn, that the Kolmogorov-Obukhov premultipliers will also vary. This has indeed been verified (e.g. [19]), the $R_\lambda$ dependence of $C_\beta$ tending to reflect that of $\zeta_\beta$ (or $n_\beta$).

A final comment in connection with the current "status" of RSH and RSHP seems worthwhile. It is relatively straightforward to investigate directly the IR behaviour of $\langle x_\beta^{*2}\rangle$ (where $\beta \equiv u$, $v$, $w$ or $q$) and $\langle x_\theta^{*2}\rangle$, i.e. the quantities which feature in K62 (or its extension for a passive scalar), viz.

$$\langle(\delta\beta^*)^2\rangle = C_\beta\langle x_\beta^{*2}\rangle \tag{7.36}$$

with $x_\beta = (r\epsilon_r)^{1/3}$ for $\beta \equiv u$, $v$, $w$, $q$ and $x_\theta \equiv r^{1/3}\epsilon_r^{-1/6}\chi_r^{1/2}$. The implementation, in the context of experimental data, has to contend with the "surrogacy" issue since the true values of $\epsilon$ and $\chi$ are not available in general. Nonetheless, recent tests [10] indicate that the use of isotropic approximations for $\epsilon$ and $\chi$ yields opposite trends for the velocity and scalar power-law exponents. While the velocity exponent (included in Figure 7.17) is qualitatively similar to that inferred directly from $\langle(\delta\beta)^2\rangle$, its value is always larger than 2/3, the scalar exponent (not shown here) tends to increase with $r$ and its magnitude is always smaller than 2/3. It would of course be useful to seek confirmation for these trends from DNS databases.

## 7.7. Effect of Large Scale Anisotropy in the Inertial Range : Some Recent Developments

As noted in Section 2, eqs.(7.16) and (7.17) assume local isotropy. Cautious interpretation is therefore needed if they are to be used in evaluating the behaviour of measured structure functions. A first step towards understanding how the energy input at large scales may affect the small scale motions was recently taken by Danaila et al. [120]. More general forms of eqs.(7.16) and (7.17), in the context of decaying grid turbulence, are given by

$$-\langle(\delta u)^3\rangle = \frac{4}{5}\langle\epsilon\rangle r - 6\nu\frac{d}{dr}\langle(\delta u)^2\rangle + 3\frac{\langle U\rangle}{r^4}\int_0^r y^4\frac{\partial}{\partial x}\langle(\delta u)^2\rangle dy \tag{7.37}$$

$$-\langle\delta u(\delta\theta)^2\rangle = \frac{4}{3}\langle\chi\rangle r - 2\kappa\frac{d}{dr}\langle(\delta\theta)^2\rangle + \frac{\langle U\rangle}{r^2}\int_0^r y^2\frac{\partial}{\partial x}\langle(\delta\theta)^2\rangle dy\,, \qquad (7.38)$$

where $y$ is a dummy variable (replacing $r$). These equations correctly reflect the decay of $\langle q^2\rangle$ and $\langle\theta^2\rangle$ at very large scales. They also correctly represent the decay of $\langle\epsilon\rangle$ and $\langle\chi\rangle$ at small scales [20]. More importantly, the magnitudes of the integral or "source" terms in eqs.(7.37) and (7.38) become comparable to those of $\langle(\delta u)^3\rangle$[6] and $\langle\delta u(\delta\theta)^2\rangle$ in the inertial range. An important implication is that the usually reported departures from the "4/5" and "4/3" behaviours need to be reassessed in the context of eqs. (7.37) and (7.38). These equations are closely satisfied by grid turbulence data at both small $R_\lambda$ [120] and relatively large $R_\lambda$ [343, 342]. They can also be used to provide more meaningful estimates of $\langle\epsilon\rangle$ and $\langle\chi\rangle$ than via relations (7.18) and (7.19). According to eqs.(7.37) and (7.38), $-\langle(\delta u)^3\rangle$ and $-\langle\delta u(\delta\theta)^2\rangle$ must be smaller than $(4/5)\langle\epsilon\rangle r$ and $(4/3)\langle\chi\rangle r$ respectively since the second and third terms on the right sides of eqs.(7.37) and (7.38) are both negative. Published data which violate these inequalities most likely suffer from inaccurate estimates of $\langle\epsilon\rangle$ and $\langle\chi\rangle$. While the accurate estimation of $\langle\epsilon\rangle$ and $\langle\chi\rangle$ is relatively straightforward in grid turbulence, it remains a formidable challenge in the wider context of non-homogeneous (sheared) turbulence. Using the parameterizations of eqs.(7.34) and (7.35) in conjunction with eqs.(7.37) and (7.38), Zhou et al. [501, 500] showed that a value of $R_\lambda$ of about $4\times10^4$ was needed before the 4/5 and 4/3 behaviours are satisfied (to 1% accuracy) in grid turbulence.

A major future challenge will be to identify the versions of eqs.(7.37) and (7.38) which apply to more complex flows than grid turbulence. To date, encouraging results have been obtained by Danaila et al. [118] for the central region of an isothermal fully developed channel flow; in this case, there is no streamwise decay of $\langle q^2\rangle$ but $\langle\epsilon\rangle$ is balanced by a large-scale diffusion of turbulent energy away from the walls. Appropriate "source" terms will need to be identified, by adopting a structured approach, in different types of flows, e.g. those, like the boundary layer, where there is continuous injection of energy and others, such as jets and wakes, which are naturally decaying.

[6]Strictly, eq.(7.37) should be replaced by an equation for $\langle\delta u(\delta q)^2\rangle$, in view of the analogy between $\langle(\delta q)^2\rangle$ and $\langle(\delta\theta)^2\rangle$ discussed in Section 6. Such an equation can be easily written. The homogeneous-isotropic form of this equation was presented and discussed in [15].

## 7.8. Concluding Remarks

Several points and conclusions emerge from issues, addressed in this chapter, concerning mainly the characteristics of the small-scale motion but also the manner in which these characteristics may be affected by the large-scale motion and its inherent anisotropy.

Firstly, it must be recalled that the assumptions underpinning the similarity hypotheses that make up K41 and K62 are only likely to be satisfied at extremely large Reynolds numbers. This is also true for the hypotheses contained in O49 and its later extensions to account for the spatio-temporal intermittencies in the fluctuating energy and scalar dissipation rates. In the majority of experimental and numerical investigations, the magnitude of $R_\lambda$ is only moderate and, to further compound the difficulty, the anisotropy of the large-scale motion is not negligible. Although the combined effect of the latter two factors may be difficult to discern on the dissipative scales it is only too apparent on scales normally associated with the inertial range. For these scales, there is a gradual increase in the magnitude of the corresponding velocity and scalar structure functions as $R_\lambda$ increases, with, arguably, an asymptotic approach to a power-law behaviour, which is in the spirit of both the original and revised similarity hypotheses. This gradual evolution towards an asymptotic state reflects in essence a relatively slow approach towards local isotropy, illustrated for example by the manner in which the magnitude of $\zeta_v(2)$ approaches that of $\zeta_u(2)$. Several important consequences accompany this asymptotic behaviour :

- Like the scaling exponents, the magnitudes of the IR premultipliers also evolve with $R_\lambda$. There is sufficient evidence to indicate that they depend on the type of flow and, in a given flow, on the initial conditions.
- The dimensionless energy dissipation rate and scalar dissipation rate constants $C_\epsilon$ and $C_\chi$ also vary with $R_\lambda$, flow type and initial conditions. This variation needs to be recognised when relating $R_\lambda$ to the global flow Reynolds number, for example the Reynolds number based on the jet exit velocity and nozzle diameter. It is not difficult to appreciate why the ratio $C_\epsilon/C_\chi$, which is proportional to the dissipation time scale ratio $R$, is an important indicator of the relative behaviour of the scalar and energy fluctuations in the context of different flows and different initial conditions.
- The theoretical IR predictions from the equations developed by Kolmogorov [254] and Yaglom [491] are also approached in an asymptotic fashion. Attempts to reformulate these equations to include the local non-homogeneity, such as was done by Danaila et al. [120] in the case of grid turbulence, provides an adequate description of the manner in which the "4/5" and "4/3" behaviours are approached [501, 500].

Although the majority of the results presented have related to the structure function, similar comments and conclusions could be made with regard to the spectrum, notwithstanding the subtleness of interpreting the translation between physical and spectral domains. The spectral energy-scalar variance analogy, first established by Fulachier [168], has provided a robust platform for comparing scalar and velocity fields. This analogy is most effective in flows with continuous injections of energy and scalar variance, such as a boundary layer over a heated wall or when the local mean velocity and temperature gradients are important. Its validity in flows without such source terms, for example decaying grid turbulence, appears to intimately depend on how closely matched the initial conditions for the energy and scalar fields are. The spectral analogy or, equivalently the analogy between $\langle(\delta q)^2\rangle$ and $\langle(\delta\theta)^2\rangle$, has useful implications for the inertial range when $Pr \simeq 1$.

CHAPTER 8

# THE STRUCTURE OF SOME VARIABLE-DENSITY LOW-SPEED SHEAR FLOWS

*This chapter focuses on the influence of density contrasts on the development of some basic low-speed shear flows. The specific features of these variable-density flows are best accounted for as seen from their vorticity dynamics. The baroclinic torque, connecting misaligned pressure and density gradients, reorganizes the vorticity field according to the fluid inertia. It is introduced after a short literature survey. Then the particular cases of the mixing layer and the jet are examined. The two-dimensional and some three-dimensional aspects are documented, based on temporally and spatially developing numerical simulations.*

## 8.1. Introduction

In the paper of Brown & Roshko [61], often cited because of its striking spark shadow pictures of coherent structures, one can find a clear introduction to the questions risen at that time upon density and compressibility effects on the self-preserving development of turbulent mixing layers. The main concern was then on the alteration of their downstream spreading rate. That paper demonstrated that the density ratio between the two streams had a significant influence on the spreading rate even if milder than the drastic reduction resulting from increasing Mach numbers. It was followed by numerous studies of fully turbulent compressible mixing-layers connected with new proposals for the turbulence closure-schemes in the compressible context, see the successful attempts by Sarkar [408] or Zeman [495] and the comprehensive review by Lele [291]. Later studies, Vreman et al. [477] and Sarkar & Pantano [414] have benefited from the direct numerical simulations and pointed toward the reduction of pressure fluctuations and the shifting of the maximum Reynolds stress towards the lighter stream. These studies, motivated by a renewed interest in high speed civil aircraft propulsion, focused on the fully turbulent statistical prediction of the shear layer. The density ratio and the connected mean and fluctuating inertia effects were merely considered as resulting from the upstream supersonic conditions.

Other industrial applications in low-speed combustion and mixing are nevertheless demanding the analysis and prediction of inertia effects alone. Provided these findings are uncorrelated to compressibility, this may also provide a simpler context to elucidate the density ratio effect and transpose the conclusions to compressible flows. This approach is clearly supported by the discussion developed by Lele [288] upon the vorticity dynamics of the supersonic mixing layer in the phase of transition to turbulence. The density ratio and the convective Mach number are shown to have separate effects on the spreading rate. The convective Mach number is seen to trigger the growth rate of the rollups in their moving frame while the density ratio changes their convection speed. The latter effect is finally attributed to a significant contribution of the baroclinic torque to the vorticity budget, in agreement with the incompressible analysis by Soteriou and Ghoniem [438]. The baroclinic torque is seen to yield an asymmetric vorticity field and the standard elliptic eddy is turned into a coma-shaped layer standing on the light side of the rotation center. The implications of such a redistribution of vorticity in inertia dominated low-speed flows stand as the focus of the present chapter.

Thus, the present contribution deals with the baroclinic effects beyond the Boussinesq approximation but uncorrelated with compressibility effects. This issue will be discussed in Chapter 9. The baroclinic torque results from the inertial component of the pressure gradient only and the vorticity is seen to evolve in a quasi-solenoidal velocity field. This purely inertial influence of density variations is likely to occur in high Reynolds number and high Froude number mixing of fluids of different densities or in thermal mixing. Such variable-density shear-layers were previously considered by Davey & Roshko [122] who concluded that the situation where the lighter fluid is moving faster leads to an increase in the amplification rate of instability oscillations. More recently a step further was achieved in the numerical analysis of spatially evolving two-dimensional shear layers by Soteriou & Ghoniem [438]. They confirmed that an asymmetric entrainment resulting from the baroclinic torque is responsible for the shifting of the center of the main structures into the lighter fluid, a different convection velocity and, as stated before, a modified spreading rate. The three-dimen-sional study of Knio & Ghoniem [253] sets the frame for the stability analysis of the baroclinically-altered roller. However, this issue remains unanswered and it stands as one crucial target in the present research.

The relevancy of such a question also emerges when considering the specific features of the cousin prototype flow, the variable-density jet. From the early work of Corrsin & Uberoi [104] in 1949 to the review by List [301] in 1982 and the late papers by Richards & Pitts [392] and Panchapakesan & Lumley [353] in 1993, the sensitivity of the mean fields and statistical moments of fully turbulent jets to the density ratio has been extensively documented. The Reynolds shear-stress budget is demonstrated to be complemented with mean-density gradient terms that enhance the diffusion of light jets ending in higher spreading and mixing rates. Correlations with density fluctuations, i.e., turbulent mass fluxes, may be also invoked, together with mean inertia effects, to analyze the behaviour of turbulent variable-density jets, see Chassaing et al. [86].

Besides, the experiments of Crow & Champagne [116] have asserted the idea that even at high Reynolds numbers the dynamics and downstream evolution of the jet strongly depend on the development of deterministic vortical structures associated with amplified instability modes. The linear stability analysis reviewed by Michalke [325] in 1984, has been very successful in predicting the preferred modes of the jet and stressed the influence of the radius to momentum-thickness ratio $R/\theta$. The particular instability of low-density jets was also noticed at that time and analyzed later on by Monkewitz & Sohn [337] to be an absolute instability occurring at a local density ratio below of 0.72 in the axisymmetric case. These studies were followed by both theoretical and experimental efforts that converged toward a quite unified picture of the "exotic" character of the low-density jet.

From the experiments reported by Monkewitz et al. [334] in 1990 and the similar ones by Kyle & Sreenivasan [266] in 1993, it is found that low-density jets, below a critical density ratio, exhibit both a powerful self-excited column mode and typical lateral ejections of light fluid at right angle of the jet axis. Both features are spectacularly recovered in the helium round jet by Hermouche [213] and in plane jets by Raynal et al. [378]. The latter study underlined the influence of the distance between the inflection points of velocity and density profiles.

In the previously quoted linear stability analysis, the column mode is merely acknowledged as a result of the deductive formalism of the stability analysis. It seems, though, that inertia effects embedded in the linearized description of the flow, are not well identified and that there is still a lack of cause-to-effect explanation that may extend in the non-linear regime. The numerical experiments by Grinstein et al. [189] are a noticeable attempt in such a direction. This may well be crucial to design efficient control strategies to switch on the absolute mode on homogeneous jets, see for instance Chapin et al. [78].

As for the onset of side-jets, it is out of the scope of the linearized approaches and is often described as a by-product of the low-density jet self-excitation, see Huerre & Monkewitz [225]. In the experimental investigation performed by Monkewitz & Pfizenmaier [335], it is concluded that pairs of counter-rotating streamwise vortices, analogous to those encountered in the three-dimensional mixing-layer, are triggering radial ejections roughly periodically spaced in the azimuthal direction. The numerical study of temporally-evolving corrugated homogeneous jets by Brancher *et al.* [55] confirms this scenario stressing the persistence of Biot-Savart induced side-jets lying in the braid region between the wavy vortex rings. In both attempts it is assumed that, with respect to the side-jet mechanism, the externally forced homogeneous jets reproduce the features of the self-excited low-density jets. The comprehensive exploration of the three-dimensional stability of round jets, performed by Brancher [54], is also based on this assumption. The last part of the present chapter aims at gathering new elements on both the onset of the column mode and the consequence of the baroclinic torque on the spontaneous occurrence of side ejections in light jets.

## 8.2. The baroclinic torques

The generation-destruction of vorticity by the baroclinic torque is not a specific feature of industrial flows. It occurs in any non-barotropic flow and is a well identified process that also affects the development of geophysical flows under the Boussinesq approximation, see Turner [463]. In this situation the baroclinic torque has been demonstrated to have a strong influence on the two-dimensional and three-dimensional stability characteristics of the shear flow. It allows secondary modes to occur in the saddle point region of two-dimensional stratified shear-layers, as described by Staquet [452]. It has a strong influence on the vorticity distribution of the Kelvin-Helmholtz billows that bias the spanwise stability of the three-dimensional stratified mixing layer, e.g. Klaassen & Peltier [250], Cortesi *et al.* [105] and the experimental investigation by Schowalter et al. [423].

While inertia dominated flows exhibit a significant internal acceleration due to the flow itself, these buoyancy dominated flows are immersed in the external gravity field, constant in direction and intensity, that combines with slight density gradients. Such externally accelerated variable-density flows are also encountered in internal combustion engines. Borée [50] and Bury [64] have discussed the relative competition between the time scales of the unsteadiness and the baroclinic torque in a light jet issuing in a highly pulsed co-flow. Whereas unsteadiness dominates the close-to-outlet development of the jet, a transition to a baroclinically driven behaviour occurs further downstream.

Thus it turns out that the nature of the baroclinic torque depends on the nature of the acceleration field, hence the use of the plural in the title of this section which aims at clarifying the different situations where a significant baroclinic torque has to be accounted for.

It is argued here that a relevant view of the changes in flows carrying inhomogeneous density fields may be sought from the analysis of the vorticity dynamics. The compressibility and inertia effects are to be considered from the general form[1] of the vorticity transport equation:

$$\frac{d\omega}{dt} = \underbrace{(\omega \cdot \nabla)\mathbf{u}}_{(i)} - \underbrace{\frac{1}{\rho^2}\nabla P \times \nabla \rho}_{(ii)} - \underbrace{\omega\, d}_{(iii)} + \underbrace{\nu \Delta \omega}_{(iv)} , \tag{8.1}$$

where $(i)$ is associated with the vortex-stretching mechanism $(ii)$ with the baroclinic torque, $(iii)$ is the dilatation term with $d = \nabla{\cdot}\mathbf{u}$ and $(iv)$ the diffusion term. This equation results from taking the curl of the momentum equation within the Navier-Stokes model:

$$\frac{d\rho}{dt} = -\rho\, d , \tag{8.2}$$

$$\frac{d\mathbf{u}}{dt} = \mathbf{g} - \dot{\mathbf{u}}_r - \frac{1}{\rho}\nabla p + \nu \Delta \mathbf{u} + \frac{\nu}{3}\nabla d . \tag{8.3}$$

Here, the velocity field is expressed in a possibly accelerated reference frame, which velocity is denoted by $\mathbf{u}_r$. The gravity field is $\mathbf{g}$. In the limit of an infinite Reynolds number or in the inviscid situation, the pressure gradient is seen to result from the following generalized acceleration field $\mathbf{a}^*$:

$$\mathbf{a}^* = \frac{d\mathbf{u}}{dt} - \mathbf{g} + \dot{\mathbf{u}}_r , \tag{8.4}$$

so that the inviscid baroclinic torque $\mathbf{b}$ may be recast in the form:

$$\mathbf{b} = \mathbf{a}^* \times \frac{1}{\rho}\nabla \rho . \tag{8.5}$$

The equation governing the enstrophy $z = \frac{1}{2}\omega^2$ exhibits then a corresponding source/sink term that may compete with the vortex stretching, $\mathbf{s} = (\omega \cdot \nabla)\mathbf{u}$, and dilatational enstrophy production terms:

$$\frac{dz}{dt} = \mathbf{s} \cdot \omega + \mathbf{b} \cdot \omega - 2d\, z + \nu \Delta z - \nu \mathrm{tr}(\nabla \omega)^2 , \tag{8.6}$$

where tr denotes the trace of the matrix.

That comprehensive view of the different origins of the baroclinic torque may be illustrated in situations where only one of its components is active. In the pioneering work of Thorpe [460] in 1968 or in the recent numerical

[1] The extra terms, coming from the viscosity dependence to the fluid composition or temperature, are discarded from the general expression of the equation (see Chapter 6.

simulation by Andreassen *et al.* [2], the gravitational baroclinic torque is seen to be the dominant vorticity source. Flows that are considered in the present chapter are inertia-dominated and the acceleration field results from unsteadiness only.

## 8.3. The two-dimensional mixing layer

As sketched in figure 8.1, the mixing layer is a downstream developing flow where a fast stream merges with a parallel and slower one. Here the density gradient and the velocity gradient are pointing upward in a so-called co-gradient mixing-layer. When the density and the streamwise-velocity gradients are in opposite directions, the situation is said to be of counter-gradient type. The spatially evolving flow is often associated with its time-evolving equivalent which, though suffering from different symmetry properties, e.g. Corcos & Sherman [99], is a convenient model for both theoretical and numerical studies. The temporal model aims at following the time development of a streamwise portion of the flow in a frame moving downstream at the convection velocity $u_c$. The time may be converted back to the streamwise distance to the splitter plate, assuming that $x = u_c(t - t_0)$. The approximation is also made that, provided the length of the temporal frame is a multiple of a streamwise periodic pattern, the flow is assumed to be streamwise periodic.

The shear layer is highly unstable and develops well organized spanwise flow patterns as those observed by Brown & Roshko [61] at high Reynolds numbers. The growth rate and the streamwise wavelength of this Kelvin-Helmholtz instability are only weakly affected by the density ratio between the two-streams, see Landau & Lifchitz [267], page 154, Maslowe & Kelly [317] and Soteriou & Ghoniem [438]. Meanwhile, the flow spreading rate, the entrainment rate and the convective velocity of the eddies are known to be affected by the density ratio, as reviewed by Dimotakis [131].

The two-dimensional development of this flow is considered first with the purpose of extracting the specific features due to the baroclinic torque, as seen from the vorticity and strain fields of a temporally evolving model. The occurrence of a two-dimensional baroclinic secondary instability mode is then discussed before a short description of spatially developing mixing-layers.

### 8.3.1. THE TWO-DIMENSIONAL ROLL-UP

The temporal evolution of a streamwise wavelength $\lambda_x$ of the Kelvin-Helmholtz instability is considered. The streamwise velocity is $-U + u_c$ in the lower free-stream and $+U + u_c$ in the upper one. The initial vorticity thickness of the error function velocity profile is quoted $\delta_\omega^0$ and is chosen to

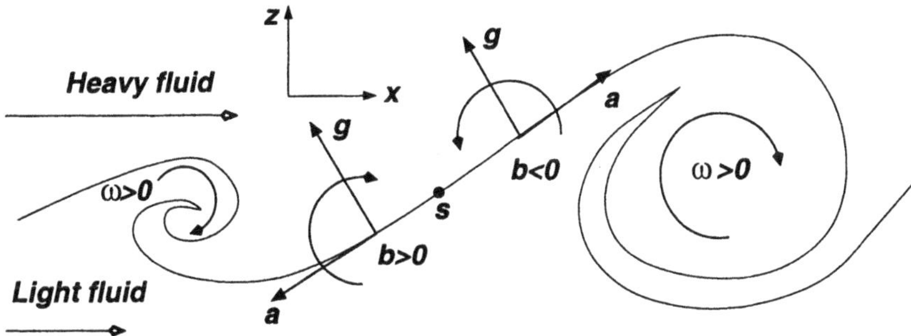


*Figure 8.1.* Sketch of the co-gradient mixing-layer between a fast heavy stream and a slow lighter one. The density gradient is $\mathbf{g} = \nabla\rho$. Source and sink contributions of the baroclinic torque to the vorticity field are outlined.

fit the most unstable mode by setting $\lambda_x = 7.3\ \delta_\omega^0$. The flow is perturbed using the eigenfunctions for $\mathbf{u}$ and $\rho$ given by the linear stability analysis of the passive scalar situation. This will ensure the relevancy of the comparisons between passive and active scalar situations. The phase speed of the velocity perturbation $u_c$, which is non zero in the variable density case, is superimposed on the perturbed profile in order to keep the main structure centered in the domain.

Within the temporal approximation, it is demonstrated that no differences are expected between the co-gradient and the counter-gradient configurations. The galilean transformation $(x, z) \rightarrow (x* = -x, z* = -z)$ turns one situation into the other such that replacing the density ratio $s_\rho$ by $1/s_\rho$, where $s_\rho = \rho_{\text{upper}}/\rho_{\text{lower}}$, has no effect on the flow development. In the followings, the upper free stream is then arbitrarily chosen to be associated with the heavier fluid. The Reynolds number is defined as $R_e = U\delta_\omega^0/\nu$ and the Prandtl-Schmidt number is set to unity. Normalization is performed using the scales $U$, $\delta_\omega^0$ and $\tau = \delta_\omega^0/U$.

In the flow configuration of figure 8.1, and due to the mean streamwise velocity field, upward shifted fluid lumps are experiencing a downstream acceleration while downward shifted particles are decelerated. In the temporal frame, moving at the eddy convection speed, both fluid elements are advected apart from the stationary saddle point, marked $\mathbf{s}$ in the figure.

This acceleration field combined with an upward density gradient gives rise to opposite sign contributions of the baroclinic torque to the vorticity field.

Such a source-sink association led Soteriou and Ghoniem [438] to consider the added baroclinic vorticity as a superimposed counter-rotative dipole. This dipole interpretation reveals very successful in explaining the asymmetric entrainment rate and the density effect on the convection speed, both expected from experimental evidence. The density field in figure 8.5(b)

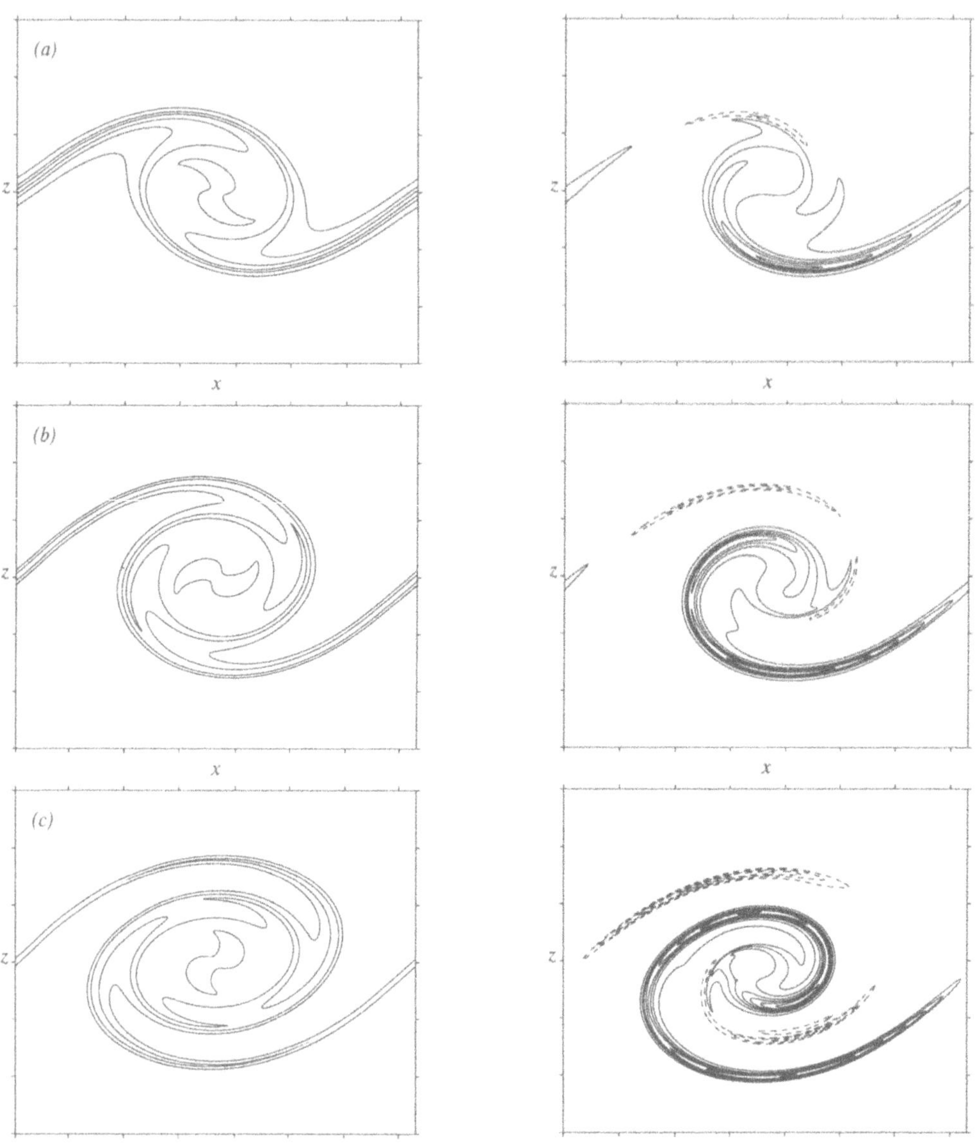


*Figure 8.2.* Vorticity $\omega_y$ contours in the two-dimensional variable density mixing-layer at $R_e = 3000$. Left column: passive-scalar case, right column : variable-density situation with density ratio of $s_\rho = 3$. The time normalized by $\tau$, is (a) $t = 8$, (b) $t = 10$, (c) $t = 12$. Contour increment is $1/2\tau$ for the passive-scalar case and twice that value for the variable-density case. The negative contours are drawn with dashed lines; tic marks along $x$ and $z$ coordinates are every $\delta_\omega^0$.

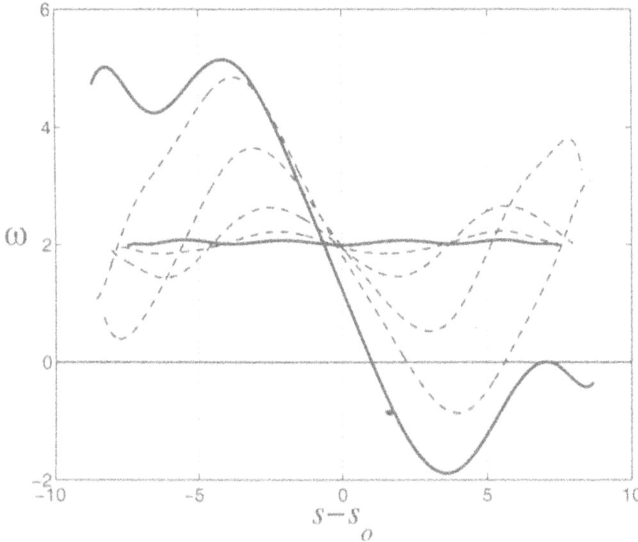


*Figure 8.3.* Normalized vorticity $\omega \times \tau$ along the central material line (where $\rho = (\rho_{\text{upper}} + \rho_{\text{lower}})/2$) of the variable density mixing layer with $R_e = 1500$ and density ratio $s_\rho = 3$. The solid curves are the initial vorticity level and the one at t=10. Dashed lines are every $\Delta t = 2$ and the dot-dashed line marks the zero threshold. The abscissa is given along the normalized curvilinear coordinate $s/\delta_\omega^0$ associated to the material line, $s_o$ being the origin at the saddle point.

illustrates the preferential entrainment of light fluid into the core. The shifting of the center of the eddy towards the light-fluid stream also results from entrainment asymmetry and continuity, see Soteriou & Ghoniem[438].

This source-sink system yields first a strong asymmetry of the vorticity field, as illustrated in figure 8.2 and later induces a strong concentration of the circulation on thinning vorticity sheets of opposite signs, see figure 8.2. The vorticity is redistributed in favor of the vorticity braid-end pointing toward the light side (left-side of figure 8.3), the other end being vorticity depleted in a first stage and then fed with a vorticity source of sign opposite to the one of the initial layer. These two opposite-sign vorticity sheets wrap around an asymmetric primary structure core located at the center of the streamlines pattern, see figure 8.5(d). Due to (i) the alternate sign of the acceleration field along the vorticity sheets and (ii) the folding of the density field changing the direction of the density-gradient, the baroclinic torque exhibits consecutive regions of positive and negative contributions to the local vorticity. Figure 8.5(c) shows how the spatial distribution of the baroclinic torque has evolved from the simple counter-rotative dipole proposed by Soteriou and Ghoniem [438] toward an intricate multipole after the saturation time.

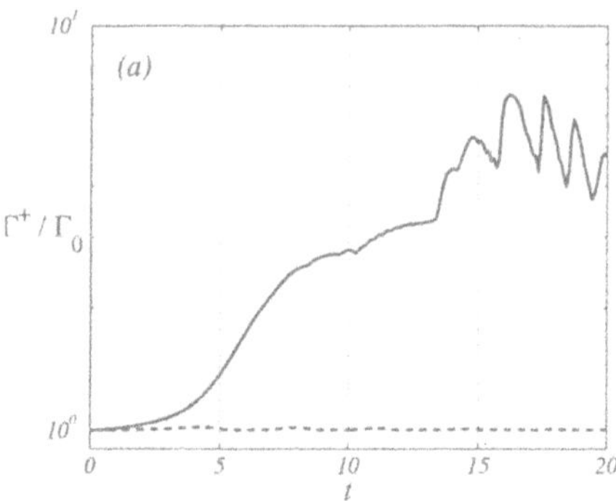

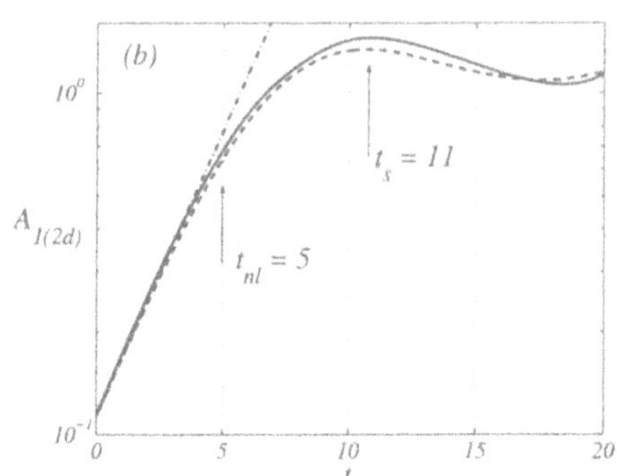


*Figure 8.4.* Time evolution of the circulation $\Gamma^+$ (a) and the kinetic energy of the primary mode $A_1(2d)$ (b).
Left: increase of the circulation $\Gamma^+$ associated with baroclinically enhanced positive regions of vorticity, normalized by the initial circulation $\Gamma_0$.
Right: before $t_{nl} \approx 5$, the linear analysis (dot-dashed line) predicts an exponential growth of the kinetic energy of the primary mode $A_1(2d)$, associated to the streamwise wavenumber $k_x = 2\pi/\lambda_x$. Then the shear-layer enters a non-linear stage leading to the saturation of the K-H mode at $t_s \approx 11$. The dashed lines give the passive scalar references.

Though the growth of the primary energy mode is only weakly affected by the baroclinic torque, — see figure 8.4(b) —, significant amounts of negative and positive circulations are generated at both ends of the braid and within the core. These contributions have to cancel in order to fulfill the invariance of the global circulation over the period: $\Gamma_0 = 2U\lambda_x$. Figure 8.4(a) illustrates the departure of the positive circulation $\Gamma^+$ from that reference level. By the saturation time $t_s$, the positive circulation has grown up to three times the initial value. A corresponding negative circulation has been generated meanwhile at the other end of the braid lying on the heavy-side of the main structure.

In contrast with the passive scalar case, where a smooth elliptic core structure is receiving advected vorticity from the strained braids (left column of figure 8.2), the filamentary nature of the vorticity field is more pronounced and yields a wider spectrum of energetic wavenumbers. The trend to develop this filamentary structure, since limited by viscous diffusion, intensifies with increasing Reynolds numbers (compare figure 8.5(a) and 8.2 bottom-right). The departure of this vortical structure from the often invoked prototype Stuart vortex is striking. The streamline pattern has the same global features, see figure 8.5(d), excepted for the symmetries of the Stuart vortex. As noted first by Knio & Ghoniem [252], no such steady approximation of the variable density roller is suited for the three-

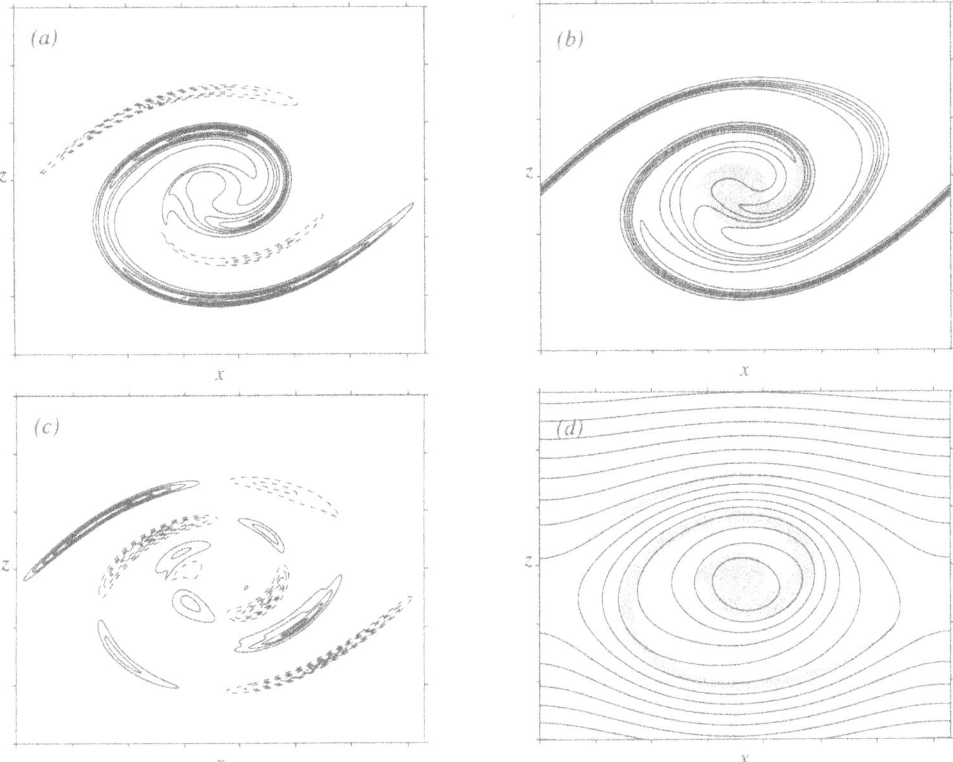


*Figure 8.5.* Description of the flow fields of the two-dimensional variable density mixing-layer at $R_e = 1500$ at $t = 12$. (a) vorticity contours (increment $U/\delta_\omega^0$), (b) density contours (increment $\Delta_\rho/6$), (c) baroclinic torque (increment $0.5\delta_\omega^{0\,2}$), (d) streamlines. The negative contours are drawn with dashed lines; in figures (b) and (d), the shaded region corresponds to absolute vorticity levels above $2U/\delta_\omega^0$; tic marks along $x$ and $z$ coordinates are every $\delta_\omega^0$.

dimensional stability analysis of the variable-density mixing layer, a point discussed in section 4.

### 8.3.2. THE VARIABLE-DENSITY SHEAR AND STRAIN FIELDS

As far as the mixing properties or the three-dimensionalization process of this flow are concerned, a description of the *2-D* strain field is needed. Given the strain tensor $\mathcal{D} = (\nabla \mathbf{u} + \nabla \mathbf{u}^t)/2$ the question arises regarding the proper direction of examination of the strain. In a pure shear flow with $2\lambda = \mathrm{d}u/\mathrm{d}y$, eq.(8.7.a), the eigen vector of the strain tensor is tilted at $\pi/4$ of the vorticity sheet, whereas in the pure strain flow, eq.(8.7.b), the maximum stretch is obviously aligned with the streamwise direction, viz.

$$(a) \quad \mathcal{D} = \begin{pmatrix} 0 & \lambda \\ \lambda & 0 \end{pmatrix} \quad , \quad (b) \quad \mathcal{D} = \begin{pmatrix} \gamma & 0 \\ 0 & -\gamma \end{pmatrix} . \tag{8.7}$$

In the general case, the strain tensor has to be projected in the direction of interest to give the stretching rate in that direction. Klaassen and Peltier [250] used simplified expressions of the shearing and stretching deformation that are valid for an horizontal line only. We choose the less ambiguous projection onto the material line, everywhere normal to the density gradient $\mathbf{g} = \nabla \rho$.

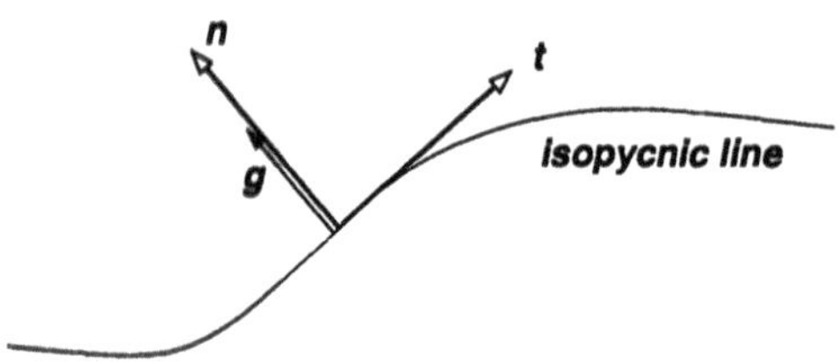


*Figure 8.6.* The local frame of reference $(t, n)$ for the analysis of the strain rate.

Within the local frame of reference $(t, n)$ sketched in figure 8.6 and designed for the braid region analysis by Corcos & Sherman [98], the shear rate $\lambda$ and the strain rate $\gamma$ of the material line are defined as:

$$\lambda = \frac{\partial u_t}{\partial n} \quad , \quad \gamma = \frac{\partial u_t}{\partial t} . \tag{8.8}$$

They can also be readily given by the following tensorial expressions:

$$\lambda = \mathbf{n} \cdot \nabla \mathbf{u} \cdot \mathbf{t} \quad , \quad \gamma = \mathbf{t} \cdot \nabla \mathbf{u} \cdot \mathbf{t} \equiv \mathcal{D} : \mathbf{tt} . \tag{8.9}$$

Here $\mathbf{t}$ and $\mathbf{n}$ are the unitary vectors of the local reference frame $(t, n)$.

The evolution of the strain rate at the saddle point $\gamma_s$ is given in figure 8.7. It is weakly affected by the redistribution of vorticity in the main structure. According to Corcos & Sherman [99], the strain rate at the saddle point depends on the circulation accumulated within the main structure core and reaches a plateau after the saturation of the primary instability mode, i.e., after $t_s$. In the variable-density situation, the circulation is still globally invariant in the main structure, with cancellation of baroclinic source and sink terms, so that the strain rate at the saddle point is expected to be independent of density effects.

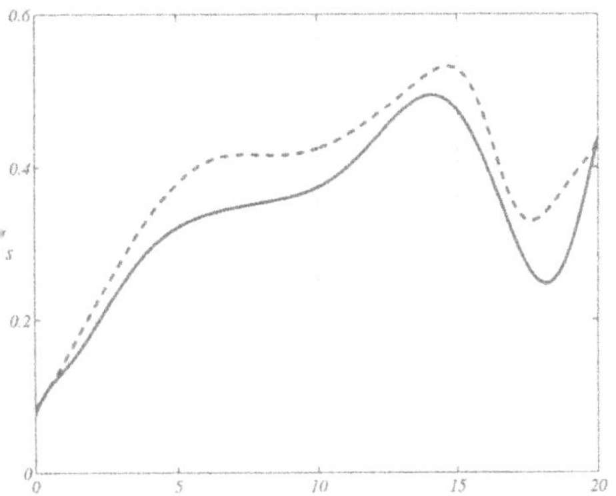


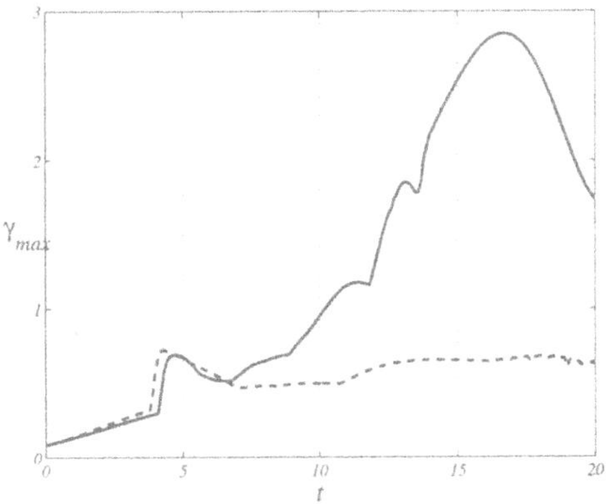


*Figure 8.7.* Left: evolution of the strain rate $\gamma_s$ normal to the density gradient at the saddle point between the main *2-D* structures. Right: evolution of the overall maximum of the strain rate $\gamma_{\max}$. Solid line: variable density mixing-layer at $s_\rho = 3$; dashed line: passive scalar mixing-layer.

The invariance of the circulation of the main structure does not extend to the local amounts of the circulation and a more detailed investigation of the stretching field is needed. In figure 8.7 the maximum value of the strain rate over the whole field is shown to be drastically enhanced due to the new vorticity distribution. By the end of the primary mode saturation, levels of strain twice as much as their passive scalar counterparts can be found in the strain field. After $t_s \approx 11$ this level even peaks toward normalized strain rates of 3, which is four times as much as in the passive scalar situation.

The question of the location of these high strain rates is answered in figure 8.8, where the symmetric strain field of the passive scalar rollup at $t = 10$ and the baroclinically modified one are compared.

In figure 8.8(a) the shear of the passive scalar rollup is seen to be concentrated, with vorticity, in the core region. The higher strain rates (above $3/8\tau$) are found on the braid, in region $S\mathbf{1}$, centered around the saddle point and spreading on both sides of the central material line. Secondary regions of high stain (locally reaching $3/8\tau$), marked $S\mathbf{2}$, are located near the center of the structure core and concern the light or heavy fluid lumps entrained into the eddy.

Figure 8.8(b) reveals a clear asymmetric development of the shear and strain fields. Due to the intense baroclinic torque, the left-end of the braid is submitted to a shear rate four times as high as in the passive scalar rollup. Besides, three regions of high strain rates are clearly identified. The regions $S\mathbf{1}$ and $S\mathbf{2}$ in figure 8.8(a) are recovered, but region $S\mathbf{1}$ is shifted

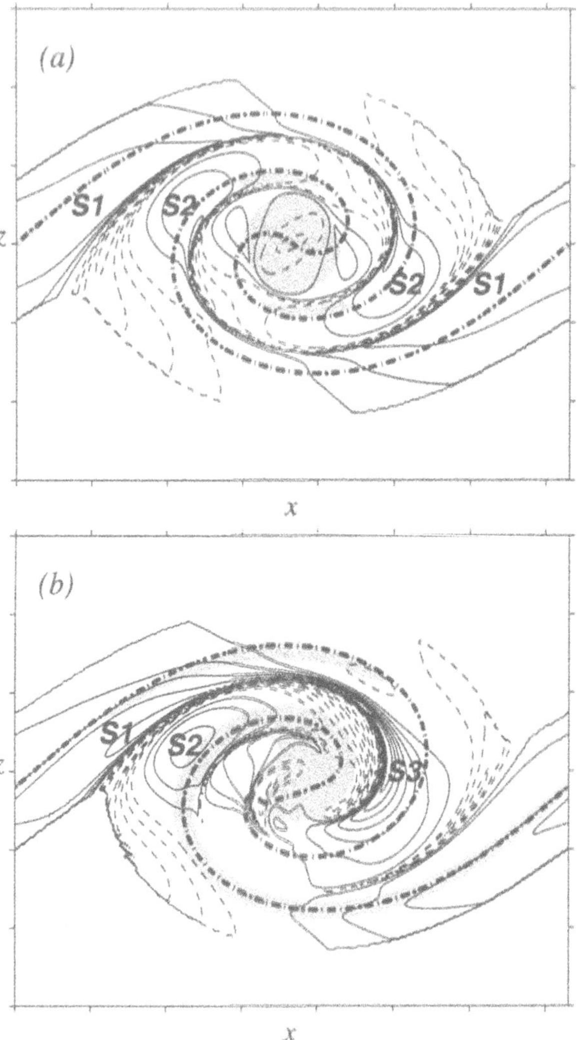


*Figure 8.8.* Contours at $t = 10$ of the strain rate $\gamma$ (increment is $1/8\tau$, solid lines for positively stretched region, dashed line for negative strain rate or compression), and the shaded region corresponds to shear rates $|\lambda|$ above $1/\tau$, the thick dot-dashed line is the central material line: (a) passive scalar case, (b) variable density mixing layer with $s_\rho = 3$.

downward into the lighter fluid region, where the strain rate reaches values above $5/8\tau$. The region marked $S\mathbf{3}$ holds the absolute maximum of the strain rate at $t = 10$ above $7/8\tau$ in agreement with the observations in figure 8.7.

The baroclinic torque is seen to enhance the shear rates and reorganize the strain field. The lighter part of the mixing layer is then submitted to much higher strain at the center and at the ends of the braid. Such a detailed description gives the first elements that may lead to a better understanding of the development of three-dimensional streamwise structures through the vortex stretching mechanism.

### 8.3.3. THE TWO-DIMENSIONAL BAROCLINIC SECONDARY INSTABILITY

At infinite Reynolds numbers Reinaud *et al.* [382] showed that the vorticity-enhanced side of the braid is likely to break-up into two-dimensional secondary rollups, as illustrated in figure 8.9. Using an inviscid vortex element method, Reinaud [380] identified the mechanism of this two-dimensional secondary instability. The secondary structures, of Kelvin-Helmholtz type, are growing spatio-temporally from amplified disturbances issuing from the saddle point region. In figure 8.10 the crosswise oscillations of the central isopycnic line $(s, z)$ are given with reference to the position of a best-fit curve $(s, z_r)$ obtained from a sixth order polynomial regression of $(s, z)$. The oscillations are more amplified on the left-end of the braid. The conditions of this amplification can be discussed with the aid of a simplified model for the strained variable-density braid.

The frame of the stability analysis of a strained vorticity strip is given by Dritschel *et al.*[135]. The strain is seen to have a two-fold stabilizing effect. It compresses the flow in the crosswise direction and stretches the wavelength of upcoming perturbations far beyond the most amplified wavelength to strip-thickness ratio. In the case of a uniformly strained constant vorticity strip, a strain to vorticity ratio of 0.25 is sufficient to prevent any disturbance amplification. Below that threshold, the growth is still very slow at 0.065, but time intervals of positive amplification rate are observed that favor the break-up of the strip into vortices. This criterion has been usefully recast by Staquet [452] for the stability analysis of the baroclinic layer in the stratified mixing layer under the Boussinesq approximation. It is reconsidered here — see also Reinaud *et al.* [382] — for the stability analysis of the braid region of the inertia-dominated variable-density mixing layer.

A simplified model of the variable density braid is proposed, as sketched in figure 8.11, which consists in a uniformly strained vorticity and density-gradient strip. The straining velocity field is $\mathbf{u}_\gamma = (\gamma x, -\gamma z)$. The initial vorticity on the strip is $\omega_0 > 0$, the uniform density is $\rho_0$ and the transverse positive density gradient $g_0 > 0$. Due to the strain-induced exponential growth of the density gradient $g_z = g_0 \exp(\gamma t)$ and to the acceleration associated with $\mathbf{u}_\gamma$, the temporal evolution of the vorticity on the strip is

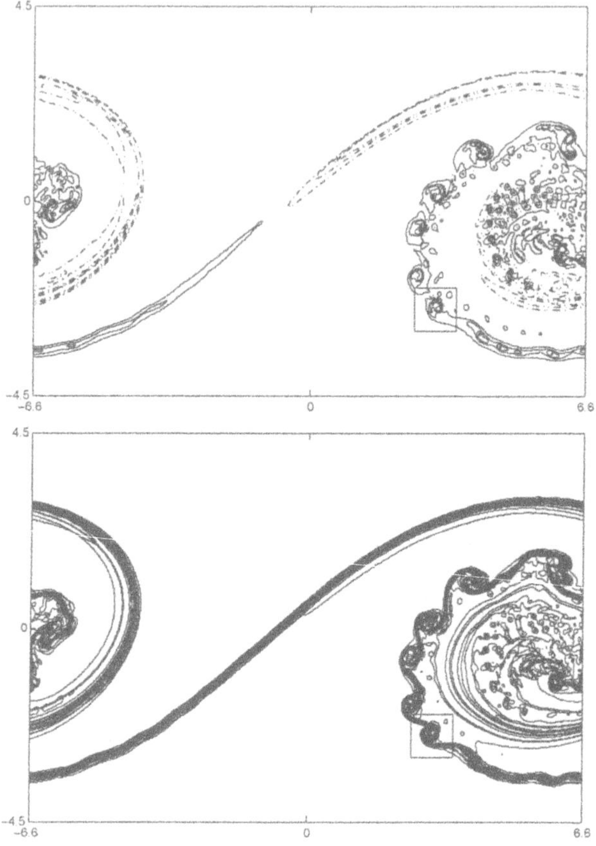


*Figure 8.9.* Secondary rollup of the baroclinically enhanced vorticity braid at $t = 12\delta_{\omega 0}/U$ from [380]. (Top) Vorticity contours: solid line for positive vorticity, dashed lines for negative ones. (Bottom) Density contours.

given by:

$$\omega(x,t) = -\frac{\gamma g_0}{\rho_0} x \sinh(\gamma t) + \omega_0 . \tag{8.10}$$

Note that the $x$-linear evolution of the vorticity is recovered along the curvilinear coordinate $s$ on the braid of the mixing layer, as illustrated in figure 8.3. The model predicts an exponentially growing source of positive vorticity for negative abscissa and a corresponding negative vorticity supply at far positive locations. Near the origin, a transient decrease from the initial vorticity level is expected. The strain to vorticity ratio evolves with an inverse trend, going to zero at far ends of the vorticity strip, thus yielding perturbation-sensitive behaviours there. Moreover, given a strain

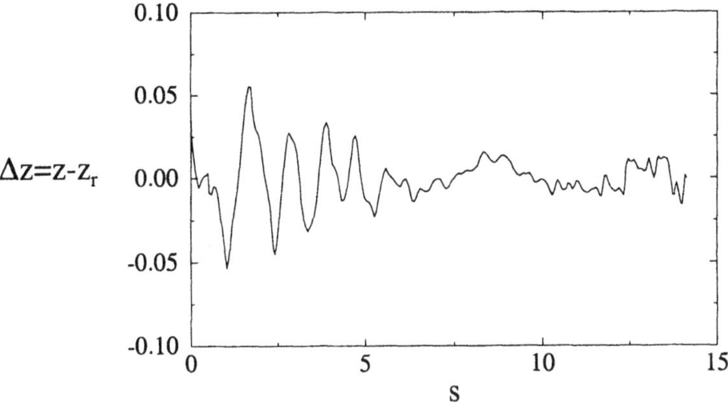


*Figure 8.10.* Crosswise oscillations of the central isopycnic line along the curvilinear coordinate $s$ from [380].

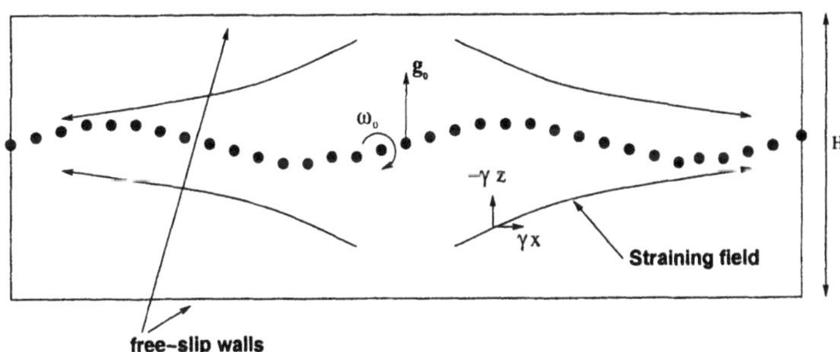


*Figure 8.11.* A simplified model of the variable-density vorticity strip submitted to a parallel uniform strain, from [380].

to vorticity ratio threshold, the positions at which that threshold is reached are getting closer to the origin with increasing time.

This prediction is confirmed by a vortex-method simulation of a truncated version of the simplified model of figure 8.11. The analysis of the results is performed based on the following Helmholtz decomposition of the cross-wise velocity: $v = v_\gamma + v_\omega$. The potential component of the transverse velocity $v_\gamma$ is due to the strain field, and $v_\omega$ is the solenoidal component induced by the vorticity distribution. Time is normalized by $1/\omega_0$ and length scales by $\lambda_0$, the initial wavelength of the perturbation.

At $t = 2$ in figure 8.12(left) the transverse velocity in section A is

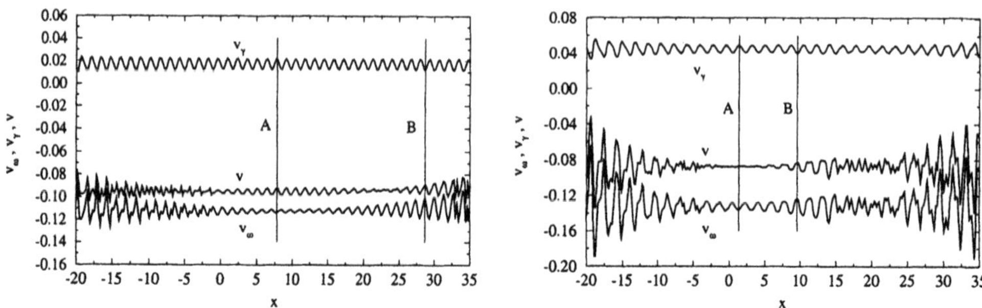


*Figure 8.12.* Distribution of the decomposed transverse velocity on the truncated vorticity/density-gradient strip at $t = 2$ (left) and $t = 4$ (right), from [380].

dominated by its stabilizing potential component $v_\gamma$, whereas in section B the dominant rotational component $v_\omega$ works at amplifying the undulations of the strip. A neutral region stands in-between where the whole transverse velocity $v$ collapses, indicating a neutral behavior against the stability analysis. Later on, at $t = 4$, region A becomes neutrally stable and B is clearly dominated by the crosswise oscillation due to $v_\omega$.

It is concluded that the baroclinic torque, favored by the acceleration and the higher density-gradient, both produced by the strain field, significantly increase the receptivity of the vorticity layers to perturbations, possibly yielding secondary rollups of the vorticity braids. This inviscid analysis has to be moderated when dealing with finite Reynolds number flows. The damping of density gradients by molecular diffusion is responsible for the smoother development of the variable density braid observed in section 8.3.1.

### 8.3.4. THE SPATIALLY EVOLVING VARIABLE-DENSITY MIXING-LAYER

In figure 8.13 the shape of the forced spatially developing mixing layer is presented from the position of vortex elements carrying the Lagrangian description of the vorticity and density-gradient. The mixing layer is bounded crosswise by two free-slip walls, separated by the channel height $h$, in order to suppress the rotation due to the truncation operated at the outgoing boundary conditions, see Reinaud [380] for the assessment of the implemented boundary conditions. The phase of the given snapshots, relative to the harmonic splitter plate oscillations, is the same, so that the sensitivity of the convection velocity to the density ratio is effectively restituted. As expected the counter-gradient case (a) exhibits a lower convection velocity than the co-gradient case (c), the passive scalar mixing layer (b)

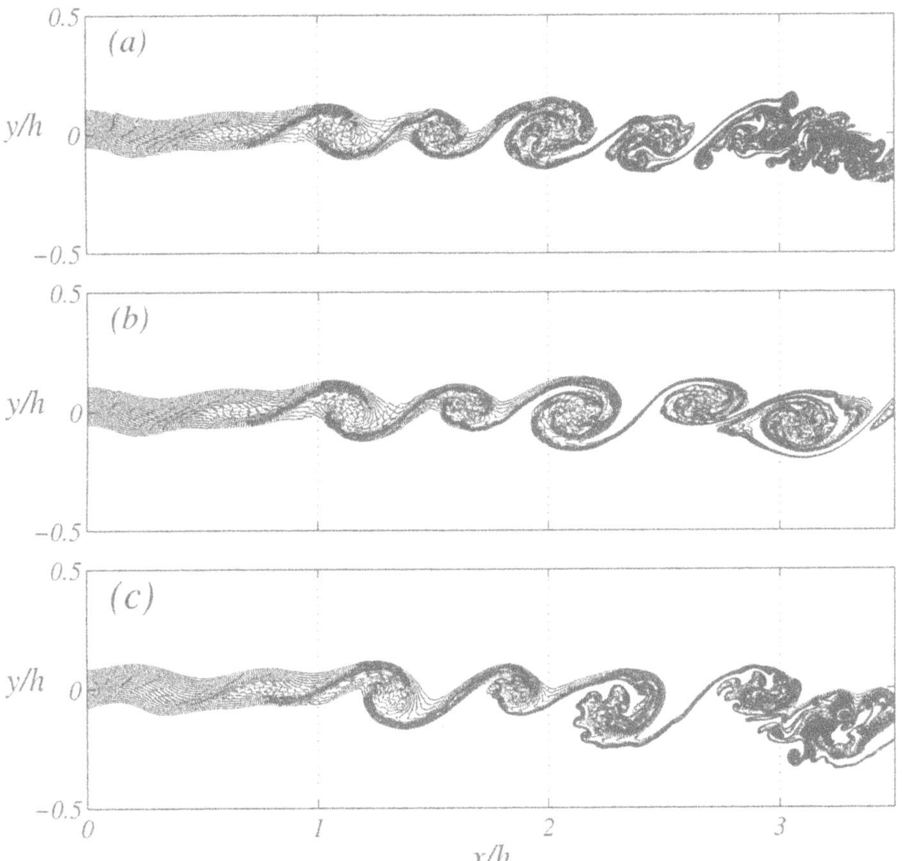


*Figure 8.13.* Spatial development of the mixing layer from the positions of vortex elements: (a) counter-gradient with $s_\rho = 1/3$, (b) constant density case, (c) co-gradient with $s_\rho = 3$, from [380].

being an intermediate situation. The entrainment rate, measured from the mean density profiles is also affected according to the experimental data analyzed by Dimotakis [131] and the previous lagrangian simulations by Soteriou & Ghoniem [438].

The second influence of the density ratio is observed in the spatial development of the pairing process. As already stated in [438] the counter-gradient pairing occurs on a different mode when compared to the constant density situation and the co-gradient pairing. In the latter case, two primary structures spiral toward one another at a section $x = 3h$ and merge downstream their circulations to form a larger and wider structure. In the former counter-gradient case, the primary structures are getting closer when moving downstream and amalgamate into a composite new structure of a wider streamwise extent but with the same crosswise thickness as the original eddies.

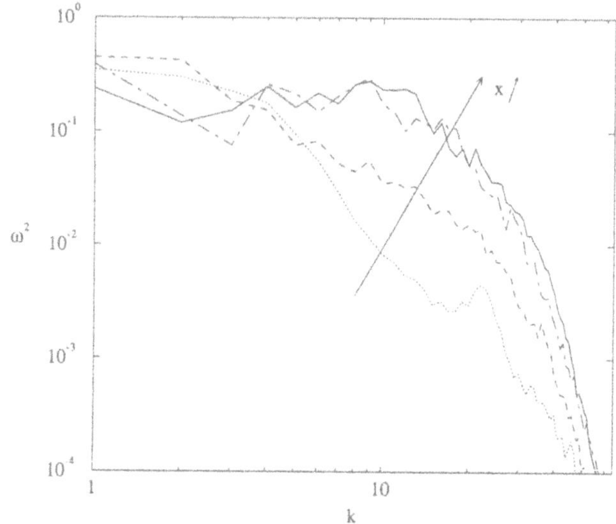


*Figure 8.14.* Downstream evolution of the enstrophy spectrum, from [380].

Here, such differences in the spatial development of the pairing process are being blurred by the other obvious and original feature of these nearly-inviscid simulations, namely the quick occurrence of the secondary roll-ups predicted by the temporal simulations. In both counter-gradient and co-gradient cases, the pairing process has to occur on primary structures that have been modified by intense baroclinic vorticity sources, yielding sub-structures less regularly organized, but similar in nature to the secondary roll-ups of the temporal simulations. Near the outlet section of the counter-gradient mixing-layer, small grained rollers can be observed even on both sides of the long structure resulting from the amalgamation of the primary ones.

The spreading of the enstrophy spectrum with the downstream distance to the splitter plate is given in figure 8.14 in the co-gradient case. Due to the distribution of vorticity on the thinning sheets and to the secondary instability, the transition to turbulence is expected to be much quicker in variable-density mixing layers, a point favorable to an efficient mixing in reacting flows.

## 8.4. The structure of the *3-D* shear-layer

In three-dimensional flows the vorticity dynamics is affected by the vortex stretching mechanism that enables enstrophy to travel among vorticity components through *3-D* instability modes. The consequences of the baroclinic redistribution of the spanwise vorticity on the development of three-dimensional modes are the focus point of this section. The interference with the pairing process and the emergence of further subharmonics are not yet considered.

Experimental evidence, e.g. Bernal & Roshko [44], stability analysis by Pierrehumbert & Widnall [365], Corcos et Lin [97], and direct numerical simulations, e.g. Rogers & Moser [396], all converge toward a similar route to three-dimensionality, leading to streamwise vortices lying in the braid region as a result of both an instability located in the Kelvin-Helmholtz (KH) billow called the translative instability (TI) and one located in the vorticity-depleted braid, hereafter called the shear instability (SI). Knio & Ghoniem [252] in 1992 contributed to the first analysis of these co-working mechanisms under non-symmetric vorticity conditions resulting from a weak baroclinic torque. They focused on symmetry losses and acknowledged for uneven intensification and weakening of the streamwise vorticity.

As stated by these authors the baroclinic torque is responsible for such a different two-dimensional structure that the results on the spanwise stability of Stuart vortices or even the KH billow are irrelevant to the three-dimensional stability properties of the variable-density situation. Though this case demands a currently unavailable stability study, a step further has been attempted in intensifying the density variation and refining the crosswise description of the layer in order to get a full baroclinic torque effect, i.e., opposite-sign vorticity sheets as in Reinaud *et al.* [381]. The vortex core being vorticity depleted in favor of surrounding vorticity cups, the translative instability mechanism is expected to weaken, leaving vorticity cups submitted to the braid instability. This scenario is examined here.

Two simulations are analyzed, one solving the passive scalar (PS) equations and the other (VD) with full variable density effects. The parameters of the PS and VD cases are reported in table 8.1.

| Name | $\rho_{up}/\rho_{\ell}$ | $R_e$ | $N_x$ | $u_c$ | $\Gamma_x/\Gamma_y^0$ |
|---|---|---|---|---|---|
| PS | / | 500 | 128 | 0. | 0.023 |
| VD | 3 | 500 | 192 | 0.28 | 0.023 |

TABLE 8.1. Global parameters of the passive scalar and baroclinically modified three-dimensional simulations.

Throughout this chapter, time is normalized by $\tau = \delta_\omega^0/U$, vorticity by the initial peak value $2U/\delta_\omega^0$ and strain by $1/\tau$. Unless quoted, the vorticity

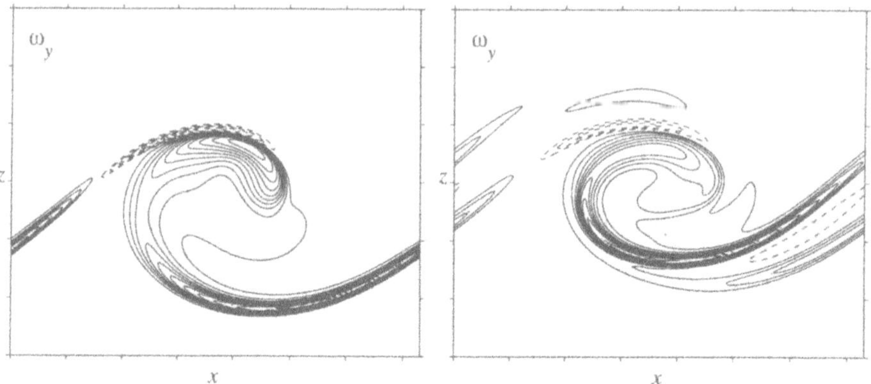


*Figure 8.15.* Contours of spanwise vorticity $\omega_y$ of the three-dimensional variable density mixing-layer at $t = 8$. Top: Base Plane cross section, bottom: Rib Plane cross section.

contours increment is always $U/\delta_\omega^0$, starting from zero. The positive vorticity contours are sketched by solid lines and negative contours by dashed ones. The tic marks along the spatial coordinates are distributed every $\delta_\omega^0$.

The structure of the spanwise vorticity cross-section is derived directly from the folded distribution established in *2-D* situation. In both the rib plane and the "off-rib" one, spanwise vorticity is redistributed in thin sheets of alternate signs. The contour maps given in figure 8.15, taken at $t = 8$, are still simple. In the "off-rib" base plane, cutting the center of the upper spanwise mushroom structure, two counter-rotative thin vorticity layers are brought closer to each other than in the *2-D* case, locally defining a jet flow of light fluid moving towards the heavy side. In the rib plane, the sketch is quite similar, though no more symmetric, to the one given by Rogers & Moser [396], Fig. 19(a). The analysis of subsequent spanwise vorticity maps is more difficult, due to the complex structure of the core region as seen from the streamwise vorticity contours in figure 8.18.

Streamwise vorticity is collapsing into rib vortices as in the constant density case. At $t = 8$, figure 8.16 shows that the streamwise structure is growing more rapidly on its right side lying above the main structure. This can be clearly associated with the favorable effect of the additional strain in that region, as mentioned in the *2-D* analysis. A weak region of negative streamwise vorticity is also noted at the center of the core and is the signature of the still active translative instability mechanism. At $t = 12$ the main contribution to the streamwise circulation comes from the dominant rib vortices, see figure 8.17.

The mechanism acting on the spanwise vorticity in *2-D* flows, namely the baroclinic source on the light side and sink on the heavy side, is recovered in the streamwise direction. From figure 8.18 (right views) the

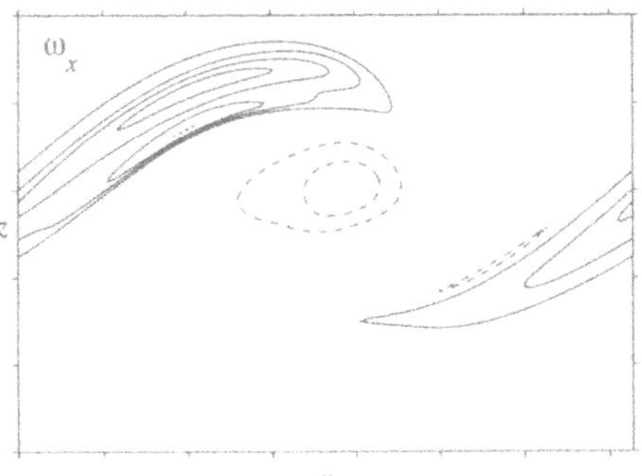


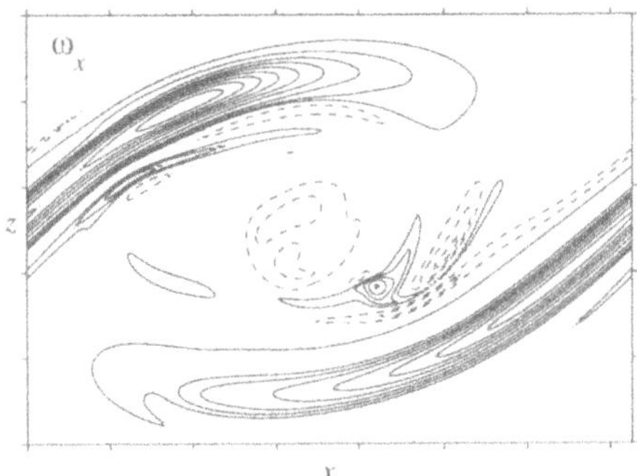


*Figure 8.16.* Contours of streamwise vorticity $\omega_x$ at $t = 8$ (left) and $t = 12$ (right). Contour increments is $U/\delta_\omega^0$, tic marks are at $\delta_\omega^0$.

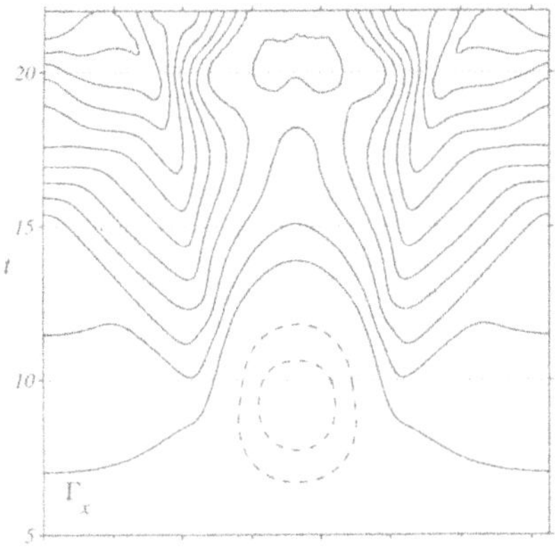


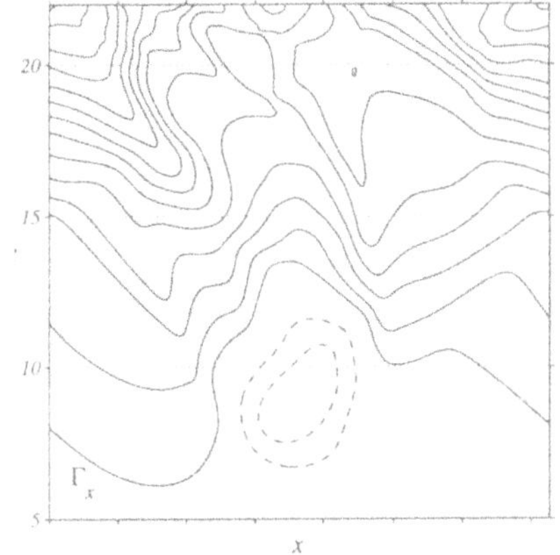


*Figure 8.17.* Time evolution of the streamwise circulation $\Gamma_x(x, t)$. Left: passive scalar, right: variable density.

rib vortices are developing with a counter-rotative companion layer on the heavy side. This association yields again a higher entrainment of light fluid into the heavy medium, as seen from the density half maps.

The spatial structure of the rib vortices is seen to be affected by the modified strain field and by the streamwise baroclinic torque. The sensitivity of the amplification rate of the three-dimensional modes to these changes is discussed now. Figure 8.19 gives the time evolution of the energy content of the pure two-dimensional mode $A_{10}$, corresponding to the streamwise and spanwise wavenumbers $(k_x, k_y) = (2\pi/\lambda_x, 0)$, and of the cumulative three-dimensional modes, i.e., those which satisfy $k_y \neq 0$. In this semilog plots the slopes of linear portions are equivalent to the exponential amplification rates of the small perturbations. Both the PS and VD cases undergo a two-stage evolution for $A_{3D}$, the first one is connected with the *2-D* growing

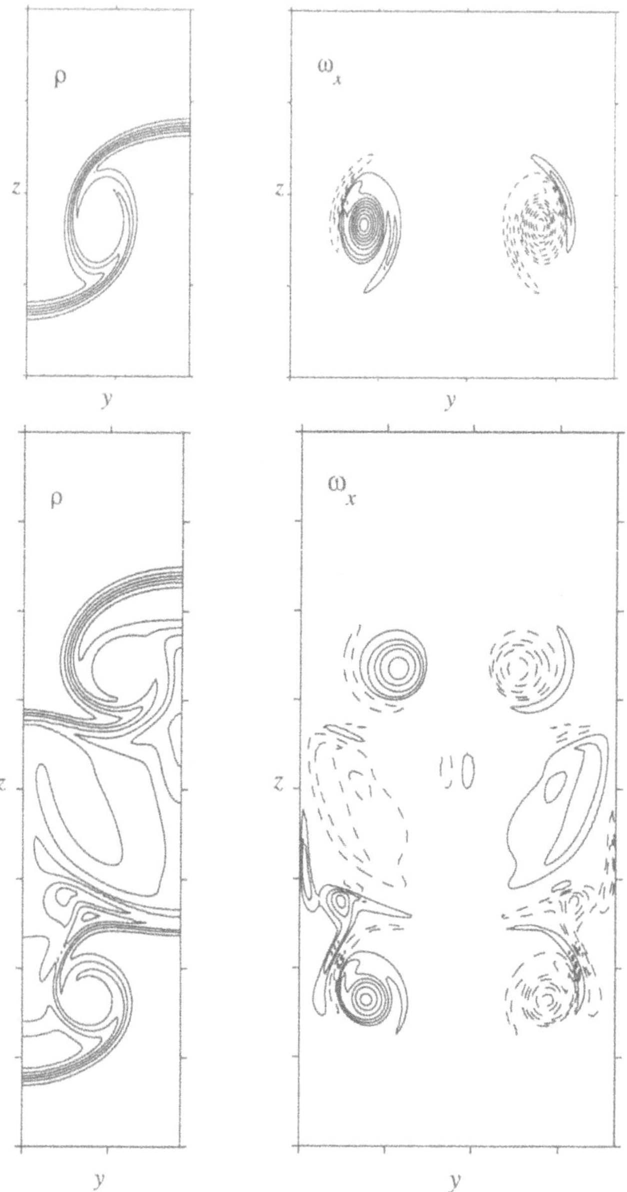


*Figure 8.18.* Density Contours and streamwise vorticity of the three-dimensional variable density mixing-layer at $t = 12$. Contour increments for the density (left) are $\Delta\rho/6$. Top: Mid-braid cross section, bottom: Core plane cross section.

mode and the second one is specific of the three-dimentionalization of the flow. The VD exhibits a delayed transition at $t = 12$ (9 for PS) and a lower amplification rate $\sigma_{3D-VD} = \sigma_{3D-PS}/2$. This first conclusion stands for

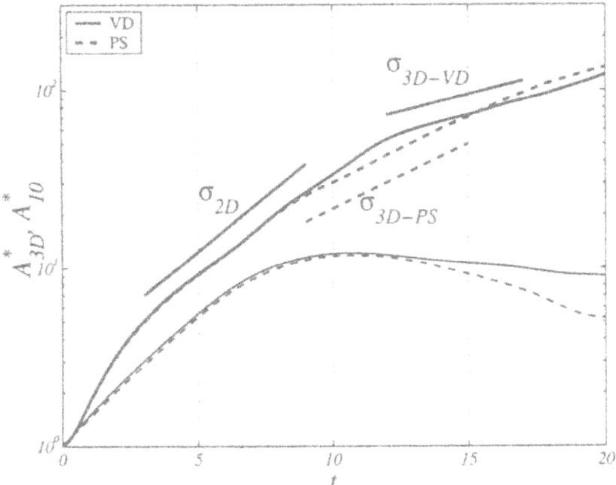


*Figure 8.19.* Time evolution of the normalized energy in all three-dimensional modes $A_{3D}$ and in the two-dimensional one $A_{10}$.

the linear response of the layer to the presently small perturbations. The non-linear regime, as well as the higher Reynolds number domain, have to be further explored.

## 8.5. The structure of variable-density jets

### 8.5.1. THE TWO-DIMENSIONAL JET

The temporal evolution of a light jet is considered with respect to a passive scalar jet, similar in all respects but the inhibition of the baroclinic torque in the vorticity equation. Both are developing jet instabilities over perturbed mean velocity and mean density profiles of the hyperbolic tangent family:

$$F(z) = \frac{1}{2}[1 + \tanh\{\frac{R}{4\theta}(\frac{1}{z} - z)\}] , \qquad (8.11)$$

$$U(z) = U_\infty + U_s F(z) \quad \text{and} \quad \rho(z) = \rho_\infty + \rho_s F(z) , \qquad (8.12)$$

where $U_\infty$ and $\rho_\infty$ are the velocity and the density of the surroundings. $U_s$ and $\rho_s$ are the jet excess velocity $U_{\text{jet}} - U_\infty$ and the density excess $\rho_{\text{jet}} - \rho_\infty$. The velocity ratio is denoted by $r = U_\infty/U_s$, and the density ratio by $s_\rho = \rho_{\text{jet}}/\rho_\infty$, which can be recovered from $1 + \rho_s/\rho_\infty$. Note that the velocity ratio used in Huerre & Monkewitz [225] is $\Lambda = 1/(2r+1)$. The velocity is normalized by $U_s$, the spatial coordinates by the radius or half height $R = D/2$ and the time by $\tau = R/U_s$. The Reynolds number is given by $R_e = U_s D/\nu$ and the Schmidt number is set to unity.

The perturbation is the most unstable mode resulting from the linear stability of the homogeneous two-dimensional jet. According to Brancher

[54], for a radius to momentum thickness ratio of $R/\theta = 11$, this mode is characterized by the streamwise wavelength $\lambda_x = R/0.41$ and the expected group velocity of the perturbation is $c_r = U_\infty + U_s/2$. The temporal analysis may be transformed back to the spatial development of the co-flowing jet by the transformation $x = c_r(t - t_0)$. The mode frequency of the spatial jet $f = c_r/\lambda_x$ building the Strouhal number $S_t = f\,D/U_{\text{jet}}$, may hence be given by $S_t = R/\lambda_x$ for the jet in quiescent surroundings for which $r = 0$. Because of the global convection by the co-flowing velocity $U_\infty$ and due to the density effect on the convection velocity $c_r$, a more general form for the temporal equivalent of the Strouhal number is:

$$S_t = G(r, s_\rho)\, R/\lambda_x \,. \tag{8.13}$$

Since the group velocity is seen to increase with increasing the density ratio, the low-density jet ($s_\rho < 1, r = 0$) exhibits then a lower convection velocity: $c_r < U_s/2$ and a larger spreading rate. As for the mixing layer simulation, the mean velocity profile is here added a negative co-flowing velocity that ensures a steady domain-centered position of the main structure.

Figure 8.20 illustrates the comparison between the vorticity fields of the two-dimensional primary mode of a passive scalar jet, with $(r, s_\rho) = (-0.5, 1.)$, and the vorticity fields in a variable-density jet, with $(r, s_\rho) = (-0.4, 0.33)$. Due to a double increment between vorticity contours and the increased number of contours in the variable density jet, it is clearly seen that the vorticity field is highly contrasted in response to significant baroclinic sources and sinks. At first sight the primary structure is similar to the one encountered in the two-dimensional mixing layer and exhibits a similar streamline pattern as the one observed in figures 8.5 and 8.8. Any difference may be inferred from the symmetry constraint on the axis, a constraint which increases with decreasing values of $R/\theta$.

Due to the streamwise periodicity, the initial circulation over, say, the lower half of the jet is $\Gamma_o = U_s\lambda_x$. Since the far velocity field is uniform at $z \to -\infty$, the departure of the circulation from this initial value may be due to an evolving contribution of the streamwise velocity along the axis. It is readily given by:

$$\Gamma(t) - \Gamma_0 = \int_0^{\lambda_x} \{u_x(x, 0; t) - U_{\text{jet}}\}\, \mathrm{d}x \,. \tag{8.14}$$

The present simulations result in a perfect cancellation of vorticity sources and sinks such that the integral on the right of equation (8.14) is zero and the circulation over the half of a temporally-evolving variable-density jet is found to be time-invariant. The integral in equation (8.14) must then vanish and the positive and negative fluctuations around the mean axial velocity

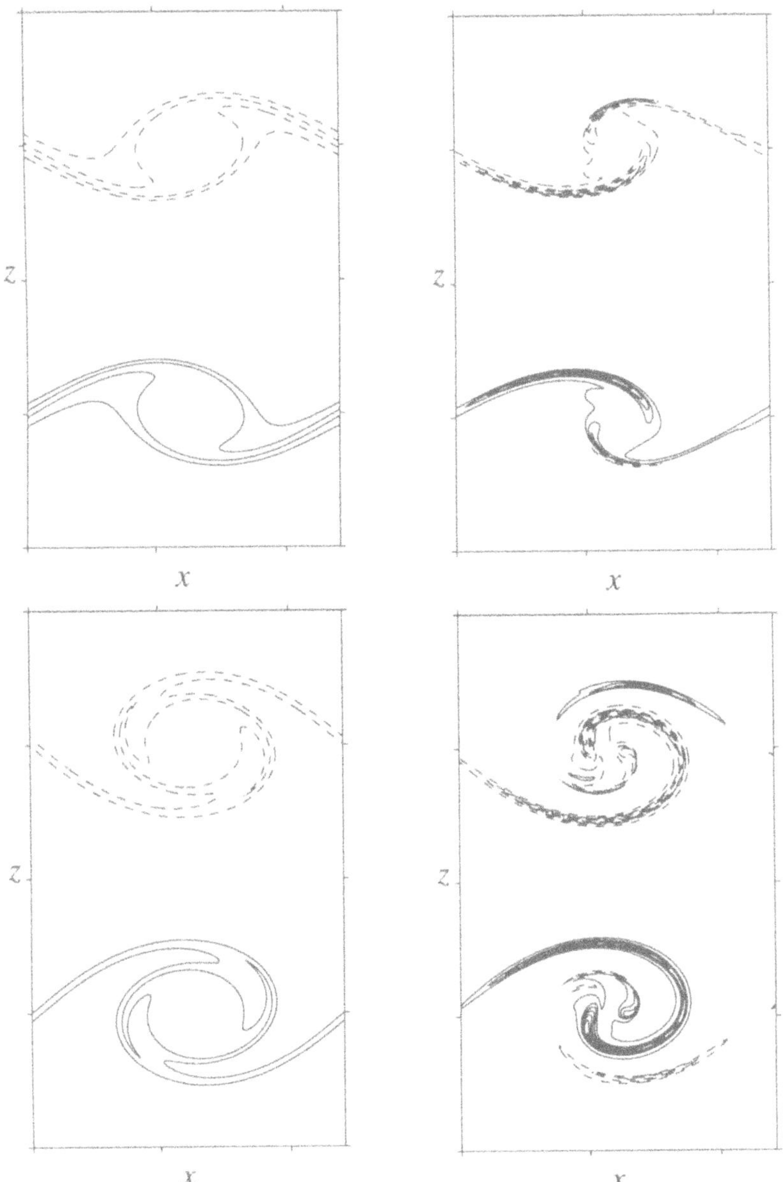


*Figure 8.20.* Vorticity $\omega_y$ contours of the two-dimensional variable density jet at $R_e = 5000$. Left column: passive-scalar case, right column : variable-density situation with density ratio of $s_\rho = 3$. The time, normalized by $\tau$, is $t = 4.5$ (top), $t = 6.8$ (bottom). Contour increment is $1/\tau$ for the passive scalar case and $2/\tau$ for the low-density jet. The negative contours of $\omega_y$ are dashed and tic marks along $x$ and $z$ coordinates are every jet radius.

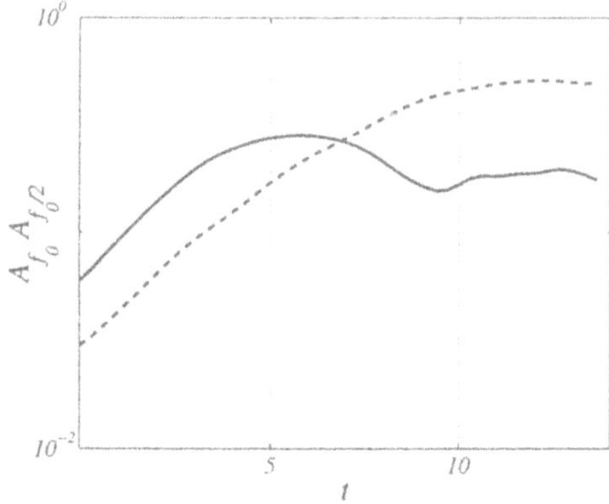


*Figure 8.21.* Modal energy of the primary mode $A_{f_0}$ (solid line) and of the subharmonic mode $A_{f_0/2}$ (dashed line) in the case of the low-density jet with $s_\rho = 1/3$, $R_e = 5000$ and $R/\theta = 11$. The primary-mode saturation time is $t_s(f_0) = 6\tau$ and the subharmonic one is $t_s(f_0/2) = 12\tau$.

cancel. However, due to the radically different distribution of vorticity in the structure, the symmetry axis does not see the same streamwise velocity. Considering the Biot-Savart law, it is expected that the contribution from the closer and baroclinically enhanced vorticity sheets are yielding a higher induced velocity at sections coincident with the streamwise position of the main structure in the low-density jet.

This conjecture is further examined in the more complete configuration where the first pairing mode is excited and develops over the primary one. The chronology of the flow development is given in figure 8.21. According to the standard findings, the primary mode saturates first and, with the pairing onset, the subharmonic mode increases until it saturates too.

Figure 8.22 describes the vorticity field of the low-density jet at the saturation times of the fundamental and subharmonic modes, together with the associated axis profiles of the streamwise velocity. The baroclinic torque has redistributed vorticity, enhancing the contribution close to the axis. During the pairing event, the axis region is being pinched between intense vorticity sheets with the rapid engulfment of the upstream structure into the downstream one. This has to be related with experimental facts, such as the flow visualizations given in figure 9 in Monkewitz *et al.* [334] or on the movie frames in figure 6 of the paper by Kyle & Sreenivasan [266]. Both of them are clearly supporting such a "pair-and-pinch" scenario of the low-density jet as it is seen also from the shaded density field in figure 8.22. Since the vorticity fields are not easily measured, the former paper merely concludes that the vorticity is concentrated in the streamwise direction in oscillating light jets. It can be confirmed here that due to the baroclinic

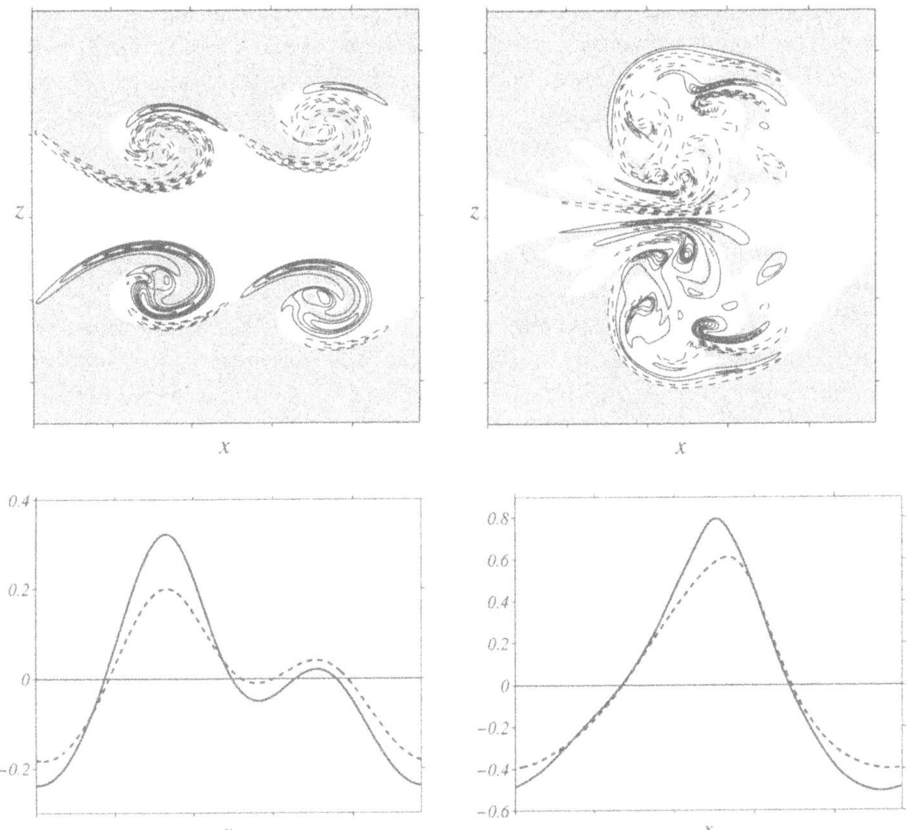


*Figure 8.22.* Vorticity contours in the two-dimensional low-density jet at $R_e = 5000$ and $s_\rho = 1/3$. Left column: $t = 6$, contour increment is $1/\tau$. Right: $t = 12$, contour increment is $2/\tau$. The negative contours of $\omega_y$ are dashed and tic marks along $x$ and $z$ are every jet radius. The shaded area corresponds to heavy fluid of density above the mean density $\rho_{\rm jet} + \rho_s/2$. Bottom views are the corresponding normalized on-axis streamwise velocity $(u(x, 0) - U_{\rm jet})/U_s$ compared to axis velocity profiles of the passive scalar jet (dashed lines).

torque, the vorticity is much more concentrated in the crosswise direction.

The two previously cited papers also focus on the spectral and statistical content of centerline velocity fluctuations and near-field pressure fluctuations. Both are indicating higher fluctuations in low-density jets with a peak intensity located at the pairing downstream location. Here the axial velocity profile is seen to be accordingly more contrasted for the light jet than for the homogeneous jet. During the primary rollup the peak value is 50% higher and close to 30% higher during the pairing event. Besides, the velocity defect is seen to be very close to -0.5 near the head of the pairing structures, indicating a close relation to the stationary region in the equivalent spatial jet and a slight reversal flow in the moving temporal frame. The higher extrema centerline velocities contributes to the observed powerful fluctuation spectra. It is inferred here that this baroclinically-modified induction mechanism on the centerline of the jet is connected with the observed column mode of low-density jets. If any pressure feedback effect is to be invoked in a compressible context, it must be noted also that the radiated noise from the baroclinically modified main structure is different from the one radiated by the smooth mono-pole of the homogeneous or passive scalar jet.

The profile of the streamwise velocity on the jet centerline also reveals much higher strain rates, $\gamma \approx \partial u_x / \partial x$, in the core region located before the main rollup. Such a difference may give a clue for the enhancement of the streamwise vorticity in the braid region of round jets increasing thus the induced radial velocities responsible for side jets ejection, according to the scheme proposed by Brancher *et al.* [55]. Then, a sounder explanation may be proposed that links the vorticity dynamics of the fundamental axisymmetric mode with the generation of side jets. This issue is illustrated on the following three-dimensional simulations of temporally evolving corrugated jets.

### 8.5.2. CORRUGATED AXISYMMETRIC LIGHT JETS

The corrugation of a nozzle has been widely considered to be an efficient device to promote mixing in industrial applications. When the corrugation is slight, it is also considered to be a relevant case for testing the onset of three-dimensionality and the transition to turbulence in axisymmetric jets, see Martin & Meiburg [316]. We present here an illustration of the baroclinic torque influence on the development of a corrugated temporally evolving low-density jet. The base flow is similar to the one defined by equation (8.12) and the axial perturbation is found by linear stability analysis on the same Strouhal number $S_t = R/\lambda_x = 0.4$ as in the two-dimensional case. The corrugation is obtained by a radial displacement of the radius of the inflection point of the base flow, say $r(\phi)$, in every point similar to the one used in Brancher *et al.* [55]:

$$r(\phi) = R\,[1 + 0.05\cos(k_a\,\phi)]\,. \tag{8.15}$$

For consistency reasons with the configuration of Brancher, the azimuthal wave number is set at $k_a = 3$ which is found to be the selected mode when the homogeneous temporal jet is submitted to a white-noise azimuthal perturbation. A passive-scalar jet and a low-density with $s_\rho = 1/3$ jets are simulated at $R_e = U_s\,D/\nu = 1900$ and $R_{e\theta} = U_s\,\theta/\nu = 170$. The time scale is $\tau = R/U_s$.

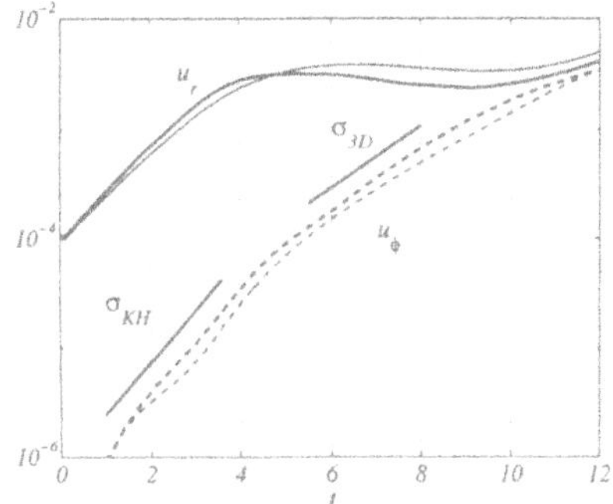


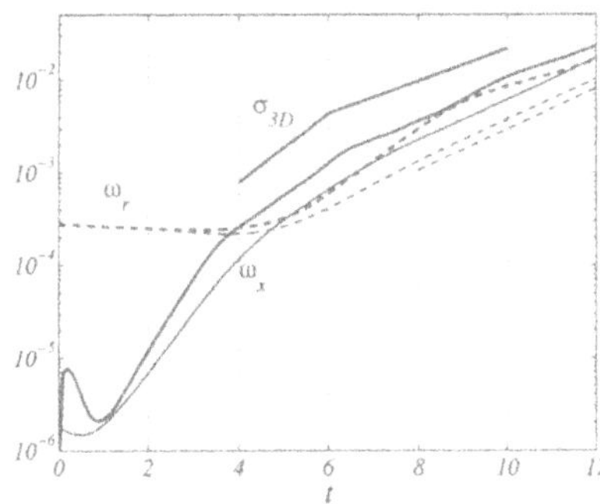


*Figure 8.23.* Left: Mean squared radial ($u_r$ solid lines) and azimuthal ($u_\phi$ dashed lines) velocities of the corrugated low-density jet with $s_\rho = 1/3$ (thick lines) and the passive scalar jet (light lines). Right: streamwise ($\omega_x$ solid lines) and radial ($\omega_r$ dashed lines) enstrophy components.

The mean quadratic velocities in the radial and azimuthal directions are given in figure 8.23 (left). It is seen that both jets experience a primary mode saturation at $t \approx 6$ and similar growth rates of the Kelvin Helmholtz instability (though higher in the light jet case). The time period between $t = 5$ and $t = 8$ corresponds to a linear growing of three-dimensional modes associated with non-zero azimuthal velocities (radial velocities are also associated with two dimensional modes). The growth of the azimuthal kinetic energy is slightly higher for the low-density jet. The marked amplification slope gives $\sigma_{3D} = 0.66$ for the light jet and $\sigma_{3D} = 0.56$ gives the best fit for the passive jet.

A significant difference is also observed on the enstrophy components relative to the three-dimensional activity in the jet. It is shown in figure 8.23 (right). From $t = 4$ the low-density jet exhibits a much higher increase of streamwise and radial vorticity, as compared with the long range amplification of the three-dimensional modes in the passive scalar jet (slope in dashed line). At distinct times, both streamwise and radial enstrophy components experience a sudden decrease of their growth rates, preventing them to reach much higher levels. Coarser resolutions of the low-density jet allow to argue that, due to the baroclinic torque redistribution of enstrophy toward higher wave numbers, the diffusive effects are damping the growth of the three-dimensional mode sooner in the low-density jet than in the standard one. However it is confirmed that the higher strain rates observed at the saturation time in the two-dimensional simulations are favoring the intensity of the streamwise and radial components of the vorticity field. Higher Reynolds numbers are needed to extend the influence of a full baroclinic vorticity source later in time.

In figure 8.24 the late development of both flows are compared. Despite

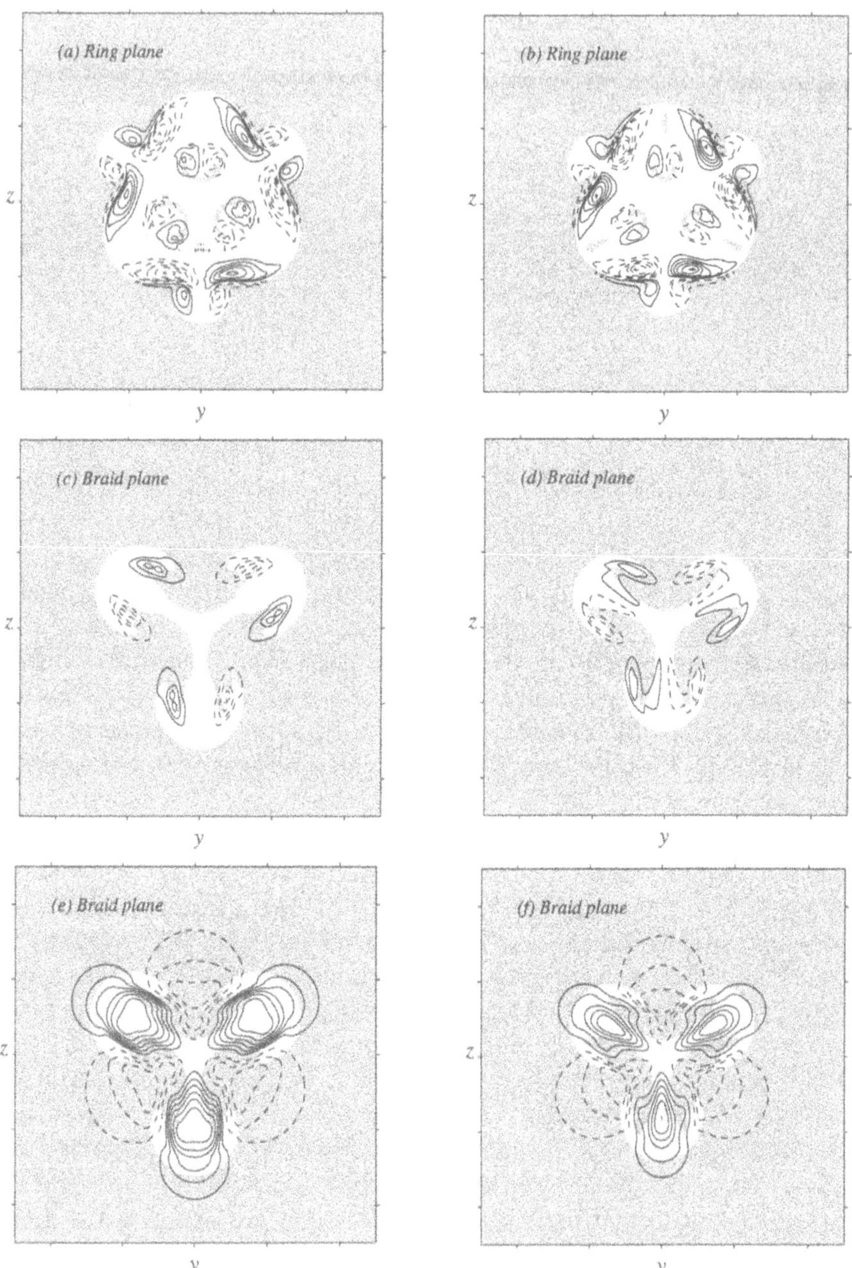


*Figure 8.24.* (a) to (d): cross sections of the streamwise vorticity at $t = 12\tau$. Contours increment is $1/\tau$ and the shaded area correspond to heavy fluid (density greater than the average $\rho_{\text{jet}} + \rho_s/2$). Views (a) and (c) are the ring and braid planes of the passive scalar jet, (b) and (d) are the ring and braid planes of the low density jet with $s_\rho = 1/3$. Views (e) and (f) are the radial velocity fields corresponding to vorticity field (b) and (d), contour increment is $U_s/20$.

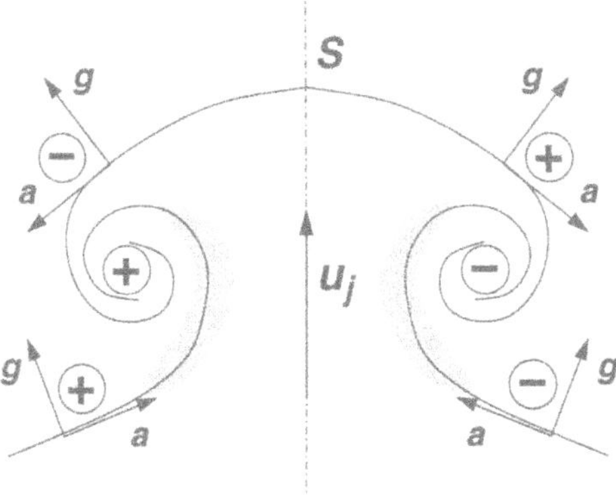


*Figure 8.25.* Sketch of the baroclinic contribution on a low-density mushroom structure possibly generated by streamwise ribs in the braid region of variable density mixing-layer or a light jet. The density gradient **g** points upward and the acceleration **a** is oriented according to the advection away from symmetry planes and toward the streamwise rib. The shaded area are baroclinically enhanced vorticity layers promoting the induction of the centerline velocity $u_j$.

the strong influence of viscous diffusion at that time, several observations can be put forward. Long after the primary mode saturation, the extra strain rates observed on the centerline of the two-dimensional low-density jets do not seem to enhance the strength of streamwise vortices in the braid region. Note that the streamwise ribs are not located on the centerline, at least during the rollup of the fundamental mode. The pinching of the jet due to the pairing process may be favorable to the stretching of streamwise vortices by forcing them toward the centerline.

The extensional strain field of the baroclinically modified vortex ring is not the only difference between the homogeneous and the light jet. The baroclinic torque has also a streamwise component that redistributes vorticity in the crosswise planes. In figure 8.25 the sketch of the baroclinic sources and sinks is given on a streamwise view of an outward mushroom structure, resulting from a rib-pair induction on a low-density jet. It is seen that streamwise vortices are supplied by a baroclinic source on the shaded region near the symmetry plane $S$. The Biot-Savart induced radial velocity $u_j$ benefits directly from that vorticity enhancement near the symmetry plane, just as the centerline velocity of the two-dimensional jet in figure 8.22. This mechanism acts also at the enhancement of the acceleration, and its associated strain field, in the entrainment region below the mushroom pattern. Thus, a divergent sequence where the baroclinically enhanced acceleration feeds back the baroclinic torque in the shaded region of increasing vorticity, may be a valuable scenario for the occurrence of spontaneous radial ejection on the braid of light jets. The stability of a pair of counter-rotating vortices in the variable-density braid region is being examined as in Lin & Corcos [297] to answer whether this divergent mechanism exhibits the very high amplification rates associated with side jets.

CHAPTER 9

# THE HIGH-SPEED TURBULENT SHEAR LAYER

*In high speed flows, significant density fluctuations can be generated by the flow itself. In this Chapter, compressibility effects in free shear flows with uniform and non uniform density are discussed in detail. The free shear layer is choosen as a useful benchmark for evaluating such effects, since, unlike the jet, neither the Mach number nor the density ratio are decreasing with distance. Recent DNS results are used to illustrate important aspects of the compressible free shear layer, including Mach number effects and variable density effects.*

## 9.1. Introduction to compressible turbulence

Turbulence with large density changes is important in high-speed applications such as supersonic aviation, missiles and high-energy devices. In the realm of nature, flows on planetary and galactic scales involve compressibility effects. A distinguishing feature of compressibility effects in the high-speed regime is that, unlike the low-speed situation, significant density fluctuations can be generated by the flow itself if the Mach number, $M = U/c$ is sufficiently large. Density changes arise from the velocity field in the following ways. First, velocity fluctuations generate pressure fluctuations which, in a compressible fluid, leads to density fluctuations. Second, viscous heating leads to entropy (temperature) fluctuations that have associated density fluctuations. Both of these sources become important when an appropriately defined Mach number becomes sufficiently large. The terms, 'acoustic' mode and 'entropy' mode, are often used to distinguish between the pressure-related and entropy-related generation of density fluctuations while the term 'vorticity' mode is used to denote vortical fluctuations. Kovasznay [259] was probably the first to introduce such a modal decomposition to compressible turbulence and to consider the linear coupling between the modes. It should be noted that, in classical acoustics, the 'acoustic' mode involves a linearly-evolving velocity field; this is not the case in high-speed compressible turbulence. Indeed, it is the coupling of the acoustic/entropy modes with the vorticity mode that is the essence of the problem of compressible turbulence.

In high-speed flows, apart from density changes introduced by the flow, there may be, in addition, density fluctuations generated by mechanisms that are operative in the low-speed regime. For example, in supersonic combustion, density fluctuations are associated not only with high Mach number, but also by mechanisms operative at low speeds such as heterogeneous fluids, as well as reaction-associated phenomena, namely, heat release and compositional changes. Unravelling and predicting the combined effect of these notably distinct phenomena is a challenge.

If the density is uniform in a flow, its effect is through the Reynolds number, $Re = \rho UL/\mu$. The mechanisms that determine the influence of a nonuniform density field on the flow depend on the flow speed (see Chapter 4). At low speeds, more precisely low Froude number, $Fr = \Delta\rho U^2/\rho gl$, the differential gravitational acceleration of fluid elements with different densities becomes important. There are many situations in which gravitational acceleration dominates momentum transport so that the variation in density drives the flow, for instance in an avalanche or a fire. At somewhat higher speeds, the influence of density is manifest through the inertia term in the momentum transport equation. Thus, the response of a lighter lump of fluid to an applied force is faster than that of a heavier lump of fluid. Also, at these speeds, vortex dynamics is affected by density variation through the baroclinic term, $-\nabla p \times \nabla\rho/\rho^2$. At still higher speeds, when the Mach number is not small, the 'acoustic' mode affects the turbulence. The finite speed of propagation, namely the speed of sound, limits the space-time region of influence of a local pressure fluctuation. The consequent change in the contribution of the pressure-gradient term to the momentum balance in turbulent flows may be significant. At even higher Mach numbers, the formation of shocks and shocklets affects the turbulence. Shock-induced separation, shock-enhanced mixing as well as the Richtmyer-Meshkov instability induced by a shock accelerating through a density interface are all examples unique to compressible turbulence.

To summarize, at low speeds, density variation influences the unsteady turbulent flow through gravity, inertia and, in reactive flows, perhaps additionally through viscous terms. At high speeds, the influence of density changes on momentum transport through inertia and viscous terms remains important and, in addition, the influence on the pressure gradient term through the pressure-density coupling gains significance.

Diverse applications in engineering, astrophysics and geophysical flows that involve fundamentally distinct couplings between density fluctuations and turbulence have motivated the study of compressible turbulence. Most of the work has been experimental and theoretical in nature. Computational studies of compressible flows with direct numerical simulation (DNS) or large eddy simulation (LES) have been more recent. Bradshaw [52] and, more recently, Lele [291], and Smits and Dussauge [434] have reviewed progress in high-speed turbulent flows. Studies of variable-density effects in the low-speed regime can be found in Fulachier *et al.* [173], and Fulachier *et al.* [172]. The present contribution is limited to high-speed, nonreacting free shear flows with the objective of giving a self-contained exposition of important compressibility effects in these flows. The reader is referred to Chapter 2 as well as the aforementioned reviews for discussions of other important examples of high-speed turbulent flows such as the supersonic or hypersonic boundary layer, shock/boundary layer interaction, and the Richtmyer-Meshkov instability.

The free shear layer is discussed in detail. Unlike the jet, the shear layer retains strong compressibility effects in both near and far-fields since the Mach number, $M_c$ and density ratio, $s_\rho = \rho_2/\rho_1$ do not decrease with distance. Therefore, the shear layer is a useful benchmark flow for evaluating compressibility effects in turbulent free shear flows. Section 9.2 summarizes the mathematical definitions and formalism used later. Section 9.3 introduces the compressible free shear layer problem. Section 9.4 introduces our recent DNS study whose results are used later to illustrate important aspects of the compressible free shear layer. Section 9.5 discusses the Mach number effect when the free-stream densities are equal, while section 9.6 discusses the variable density effect at a constant Mach number, $M_c = 0.7$. Finally, section 9.7 concludes the chapter by identifying future avenues of research in the compressible free shear layer and indicating connections with other compressible flows.

## 9.2. Mathematical preliminaries

The evolution of the instantaneous fields in a high-speed turbulent flow is governed by the compressible Navier-Stokes equations. The equations governing the mean fields and second-order statistics (Reynolds stresses, variance and cross-correlation of thermodynamic variables, mass flux and heat flux) can be derived by appropriate averaging (see Chapter 5). In the case ofcompressible flows, it is customary to use Favre-averages. The Favre-average, $\phi$, is defined by

$$\tilde{\phi} = \frac{\overline{\rho\phi}}{\bar{\rho}}, \tag{9.1}$$

where $\overline{\phi}$ denotes the Reynolds average. Fluctuations can be defined with respect to the Favre-average or Reynolds-average. Here, $\phi'$ denotes Reynolds fluctuations and $\phi''$ denotes Favre fluctuations.

The turbulent stress tensor $R_{ij}$, is defined by,

$$R_{ij} = \frac{\overline{\rho u_i'' u_j''}}{\bar{\rho}} \,. \tag{9.2}$$

Consider a single-component compressible fluid. Averaging the equations for conservation of mass and momentum gives the following system of equations for the mean density and mean velocity,

$$\begin{aligned} \frac{\partial \bar{\rho}}{\partial t} + \frac{\partial (\bar{\rho} \widetilde{U}_k)}{\partial x_k} &= 0\,, & (9.3)\\ \frac{\partial (\bar{\rho} \widetilde{U}_i)}{\partial t} + \frac{\partial (\bar{\rho} \widetilde{U}_k \widetilde{U}_i)}{\partial x_k} &= -\frac{\partial \bar{P}}{\partial x_i} + \frac{\partial}{\partial x_k}(\bar{\sigma}_{ik} - \bar{\rho} R_{ik})\,. & (9.4) \end{aligned}$$

The energy equation can be written in various forms. One choice leads to the following equation for the mean pressure in an ideal gas,

$$\frac{\partial \bar{P}}{\partial t} + \overline{U}_k \frac{\partial \overline{P}}{\partial x_k} = -(\gamma - 1)(\Pi - \overline{\phi}) - \gamma \overline{P} \frac{\partial \overline{U}_k}{\partial x_k} - \frac{\partial}{\partial x_k}\left( \overline{p' u_k'} - \overline{\frac{\gamma \kappa^*}{Re Pr} \frac{\partial T}{\partial x_k}} \right), \tag{9.5}$$

where $\Pi = \overline{p'd'}$ represents the pressure-dilatation correlation, and $\phi = \sigma_{ij}\, \partial U_i / \partial x_j$ denotes the viscous dissipation. Note that $d'$ denotes the fluctuating component of dilatation, $\nabla \cdot \mathbf{u}$. The mean value, $\overline{\phi}$, represents average viscous loss from mechanical energy to internal energy, while the terms $\Pi$ and $\overline{P}\, \partial \overline{U}_k / \partial x_k$ arise from averaging the pressure-work which, unlike the viscous term, is a potentially reversible coupling between mechanical and internal energy (see Chapter 5).

The Reynolds stress equation in a compressible flow is given by — Cf. Chapter 5, eq.(5.36) —,

$$\frac{\partial (\overline{\rho} R_{ij})}{\partial t} + \frac{\partial (\overline{\rho} \widetilde{U}_k R_{ij})}{\partial x_k} = \overline{\rho}(P_{ij} - \epsilon_{ij}) - \frac{\partial T_{ijk}}{\partial x_k} + \Pi_{ij} + \Sigma_{ij} \,. \tag{9.6}$$

The contributors to the right-hand-side of the Reynolds stress equation, eq. (9.6) are as follows:

Turbulent production, $P_{ij} = -\left( R_{ik}\frac{\partial \tilde{U}_j}{\partial x_k} + R_{jk}\frac{\partial \tilde{U}_i}{\partial x_k} \right)$,

Turbulent dissipation, $\epsilon_{ij} = \frac{1}{\bar{\rho}}\overline{\left( \tau'_{jk}\frac{\partial u''_i}{\partial x_k} + \tau'_{ik}\frac{\partial u''_j}{\partial x_k} \right)}$,

Pressure strain, $\Pi_{ij} = \overline{p'\left( \frac{\partial u''_i}{\partial x_j} + \frac{\partial u''_j}{\partial x_i} \right)}$,

Transport,

$$T_{ijk} = \overline{\rho u''_i u''_j u''_k} + \overline{p' u'_i}\delta_{jk} + \overline{p' u'_j}\delta_{ik} - (\overline{\tau'_{jk}u''_i} + \overline{\tau'_{ik}u''_j}),$$

Mass flux coupling, $\Sigma_{ij} = \overline{u''_i}\left( \frac{\partial \bar{\tau}_{jk}}{\partial x_k} - \frac{\partial \bar{P}}{\partial x_j} \right) + \overline{u''_j}\left( \frac{\partial \bar{\tau}_{ik}}{\partial x_k} - \frac{\partial \bar{P}}{\partial x_i} \right)$.

With respect to the incompressible counterpart, the mass flux term, $\Sigma_{ij}$, is an explicitly new term. But the other terms that carry over from the incompressible case can be *implicitly* affected by compressibility.

The following transport equation for the turbulence kinetic energy, $\tilde{k} = R_{ii}/2$, is obtained from eq. (9.6) by contracting the indices,

$$\frac{\partial(\bar{\rho}\tilde{k})}{\partial t} + \frac{\partial(\bar{\rho}\tilde{u}_k\tilde{k})}{\partial x_k} = \bar{\rho}(P_{rod} - \epsilon) - \frac{\partial T_k}{\partial x_k} + \Pi + \Sigma_{kk}/2\,, \qquad (9.7)$$

with $P_{rod}$, $\epsilon$, and $\partial T_k/\partial x_k$ representing the usual contributions: turbulent production, turbulent dissipation, and transport, respectively. The pressure dilatation,

$$\Pi = \frac{\Pi_{kk}}{2} = \overline{p'\frac{\partial u'_k}{\partial x_k}},$$

exchanges energy between the fluctuating velocity and the fluctuating pressure. The turbulent dissipation rate, $\epsilon$, can be rewritten, as demonstrated by Sarkar *et al.* [416] and recalled in Chapter 6, to show that, in the case of compressible flow, there is an additional term, the compressible dissipation,

$$\bar{\rho}\epsilon_c = \frac{4}{3}\overline{\mu\frac{\partial u'_k}{\partial x_k}\frac{\partial u'_k}{\partial x_k}},$$

which is always positive and acts as a sink for the turbulent kinetic energy. The pressure dilatation and compressible dissipation (also called the dilatational dissipation) are explicit compressibility terms in the turbulence kinetic energy budget that disappear in the limit of incompressibility.

## 9.3. Introduction to the compressible shear layer

The compressible shear layer between two parallel-flowing streams of fluid with a relative velocity shows a remarkable decrease of thickness growth rate with increasing Mach number. In addition to being a phenomenon of fundamental interest, such a compressibility effect is of technological interest. For example, reduction in high-speed mixing has a negative impact on supersonic propulsion concepts such as the scramjet, while supersonic jet noise phenomena and mitigation concepts are affected by any changes in mixing properties at high speeds. During the 1980's and early 1990's, when high-speed civilian transport programs such as the National Aerospace Plane (NASP) and the supersonic transport (SST) in the USA, as well as similar programs in Europe were in vogue, there was a especially strong interest in the compressible shear layer. Examples of experimental studies carried out during that period include Chinzei *et al.* [95], Papamoschou and Roshko [358], Elliott and Samimy [146], Goebel and Dutton [186], Hall, Dimotakis and Rosemann [200], Barre, Quine and Dussauge [30], Clemens and Mungal [96], and Chambres, Barre and Bonnet [76]. Earlier experimental studies are summarized by Birch and Eggers [45].

The convective Mach number, $M_c$, introduced by [49] has become popular as the global parameter that determines compressibility effects in the free shear layer. Denoting the velocity, density and speed of sound in the high-speed stream by $U_1$, $\rho_1$, $c_1$ and corresponding quantities in the low-speed stream by $U_2$, $\rho_2$, $c_2$ and, assuming equal specific heats, gives $M_c = (U_1 - U_2)/(c_1 + c_2)$. From experimental data, see Fig. 9.3 for example, it is clear that there is a general trend of decreasing thickness growth rate with increasing values of $M_c$. There is a significant scatter in the data pointing towards the possible importance of parameters other than $M_c$, for example the density ratio, $s_\rho = \rho_2/\rho_1$, in the problem.

Although the convective Mach number, $M_c$, is the overall global Mach number that is widely used, there are other important compressibility parameters. The turbulence Mach number, $M_t = \sqrt{2\widetilde{k}}/\bar{c}$ with $\widetilde{k}$ representing the turbulence kinetic energy estimates compressibility effects due to fluctuating motion distinct from the mean velocity field. The gradient Mach number, $M_g = Sl/\bar{c}$ where $S$ is the mean shear and $l$ the length scale in the direction of shear is a 'field Mach number' in shear flows. The parameter, $M_g$ can be viewed as the ratio of an acoustic time scale, $l/c$, for a large eddy to the mean distortion time scale, $1/S$. If $S$ is generalized to denote $|\nabla \mathbf{u}|$, the gradient Mach number is applicable to flows more general than free shear flow. It is possible that in some high-speed flows, for example multi-component or reacting flows, the density and temperature fluctuations may not be determined by Mach number alone. If so, other density and temperature-related nondimensional parameters may need to be introduced to describe observed compressibility effects.

In addition to the overall growth rate, measurements of the turbulence structure and mixing efficiency have been obtained. There is agreement that the Reynolds shear stress as well as cross-stream turbulence intensity decreases with increasing value of $M_c$. However, as discussed later in section 9.5.2, there is some controversy regarding the effect on streamwise turbulence intensity. The mixing efficiency, usually defined with respect to a scalar, is the ratio of the amount that is molecularly mixed to the amount that is entrained and is useful to distinguish between effects on 'large-scale' entrainment and 'small-scale' mixing. The general conclusion is that the compressibility effect on the mixing efficiency is small relative to that on the thickness (defined using either mean scalar or mean velocity profiles).

The direct numerical simulation (DNS) approach had become popular in the 1980s as a method for investigating fundamental physics of turbulent flows in a detailed fashion without the uncertain influence of turbulence models. Therefore, it is not surprising that DNS was tried in the arena of compressible turbulent flows. The simplest problem, homogeneous isotropic turbulence was simulated by Passot and Pouquet [359], Erlebacher *et al.* [149], Sarkar *et al.* [416], and Lee *et al.* [284]. It was found that there is augmented dissipation due to the so-called compressible or dilatational dissipation which increased the decay rate of turbulence kinetic energy. However, it became apparent that, due to the absence of mean shear, the compressibility effects in isotropic turbulence were weaker than those observed in the free shear layer and perhaps qualitatively different. Given the computational resources available, it was not possible to simulate the spatially-evolving shear layer into a fully-turbulent regime. On the other hand, turbulence forced by uniform shear could be realized at microscale Reynolds number, $Re_\lambda \simeq 75$, sufficient for nonlinearly-evolving strong turbulence (but not with Kolmogorov $-5/3$ inertial range!) using supercomputing resources available in the early 1990s. Thus, compressible uniform shear flow was simulated by Sarkar *et al.* [412], Blaisdell *et al.* [47], Sarkar [409] and Simone, Coleman and Cambon [431]. All of these studies found reduced turbulence levels at high Mach number similar to observations in experimental studies of the free shear layer. However, the earlier studies [412, 47] overemphasized the augmented dissipation due to explicit dilatational terms, namely, the compressible dissipation, $\epsilon_c - (4/3)\nu\overline{d'^2}$ and the pressure dilatation, $\Pi = \overline{p'd'}$, because the turbulence Mach number, $M_t$, and the gradient Mach number, $M_g$ was not representative of values associated with the compressible shear layer; the value of $M_t$ was too large and that of $M_g$ too small. Sarkar [409] changed $M_t$ and $M_g$ independently, and found that, for values of these parameters that are typical of the free shear layer, it was reduced turbulent production, $P_{rod} = R_{12}\partial\widetilde{U}_1/\partial x_2$, and not the dilatational terms, that is responsible for the reduction in the turbulence level, and, furthermore, the gradient Mach number, $M_g$, is the parameter that determines the extent by which turbulent production decreases.

Although simulations of instabilities and transitional states of the spatially evolving shear layer have been performed, there are no simulations into the fully-developed regime. However, the temporally-evolving shear layer in which two streams move in opposite directions with the same speed has been considered. The temporal shear layer, by virtue of a more realistic velocity profile, allows the usual two-dimensional instabilities (Kelvin-Helmholtz billows) as well as secondary instabilities (rib vortices, mushrooms) seen in the experiments, and furthermore, permits entrainment and turbulent transport typical of inhomogeneous turbulent flows. Features such as asymmetric entrainment and modifications to the characteristic convection velocity are not captured by the temporal model. Nevertheless, the temporally-evolving DNS of the inhomogeneous shear layer and uniform shear flow have provided considerable insights into the observed compressibility effects as explained in section 9.5. Recent simulations of the shear layer have been performed by Vreman *et al.*[477], Freund *et al.* [162], and Pantano and Sarkar [354]. The influence of density ratio, separate from that of Mach number, has not been investigated with DNS except for the study of Pantano and Sarkar [354].

The general topic of compressible turbulence modeling, which is addressed in Chapter 10 and 11 of the present monograph, has been reviewed by Wilcox [484] and Speziale and Sarkar [443]. Modeling of the compressible shear layer within the framework of Reynolds stress or two-equation models has been the subject of considerable activity. Without going here through the details, it has been argued that the explicit dilatational terms act as sinks in the turbulence kinetic energy budget and result in reduced turbulence levels. The compressible dissipation, $\epsilon_c$, has been modeled by Zeman [495] based on the the dissipative influence of eddy shocklets and Sarkar *et al.* [416] based on a low-Mach number asymptotic theory for the additional compression/expansion mode in a compressible fluid. Models for the pressure-dilatation have been proposed by Taulbee and VanOsdol [454], Zeman [497], Sarkar [408], Ristorcelli [394], and Hamba [203]. All these models will be detailed in Chapter 10. Although, they are operationally successful in reducing the turbulence levels and thickness growth rate as a function of the Mach number, they are not consistent with the following fundamental observation from DNS of compressible shear flows: the contribution of dilatational terms to the turbulence kinetic energy balance remains small at Mach numbers at which substantial reduction in turbulence levels and thickness growth rate is observed.

## 9.4. A direct simulation of the temporally-evolving shear layer

In sections 9.5 and 9.6 we will discuss observed compressibility effects and provide results from recent direct simulations of the temporally-evolving shear layer for demonstration. These simulations are similar to those reported by Pantano and Sarkar [354] with major differences as follows: the initial thickness of the shear layer is a factor of 4 smaller than that in [354] and the computational domains and grid sizes are somewhat smaller in the simulations discussed here relative to those in [354].

The upper stream has velocity, $U_1$ and density $\rho_1$, while the lower stream has velocity $U_2$ and density $\rho_2$. The lower stream velocity, $U_2$ is positive while the upper stream velocity is $U_1 = -U_2$. Gravitational effects are excluded. The convective Mach number, $M_c = (U_1 - U_2)/(c_1 + c_2)$, and the density ratio, $s_\rho = \rho_2/\rho_1$ are independently varied. The mean pressure is kept constant so that the temperature ratio is inversely proportional to the density ratio. The specific heat ratio and dynamic viscosity are identical between the two free streams. The average density, $(\rho_1 + \rho_2)/2$ is kept constant between simulations.

Three Mach numbers, $M_c = 0.3$, 0.7 and 1.1, and three density ratios, $s_\rho = 1$, 4 and 8 are considered. Since the Navier-Stokes equations are invariant to the transformation, $x_1 \to -x_1$, and $x_2 \to -x_2$, changing a given density ratio of $s_\rho$ to its inverse $1/s_\rho$, does not affect the flow development.

The three-dimensional, unsteady Navier-Stokes equations applicable to a compressible fluid are numerically solved. Up to 6 million grid points are used. The initial streamwise velocity and density profiles have a hyperbolic tangent variation in the cross-stream, $x_2$, direction. Initially isotropic, solenoidal velocity perturbations with a prescribed broadband spectrum are imposed on the mean flow. The perturbation field has a maximum at the centerline and decays to zero in the free streams. The initial thermodynamic fluctuations are designed to limit the compressibility effects induced by initial transients. Thus, the initial pressure fluctuations satisfy the usual Poisson equation while the temperature and density fluctuations are isentropic. Periodic boundary conditions are used in the homogeneous $x_1$ and $x_3$ directions. Non-reflecting boundary conditions along with buffer zones are used in the cross-stream, $x_2$, direction. For the moderate-Reynolds number shear flow without shocks considered here, central schemes of high-order accuracy can be used. Sixth and fourth-order accurate schemes are used for the spatial discretization while a fourth-order, low-storage Runge-Kutta method is used for the time advancement.

The flow develops in time with the statistics being functions of $x_2$ and $t$. Unlike other previous DNS efforts, sufficiently large computational domains and development times are used to allow simulation into a good

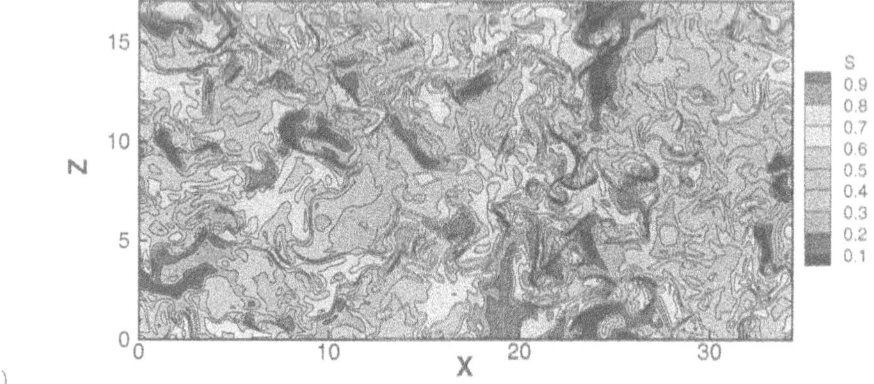


(a)

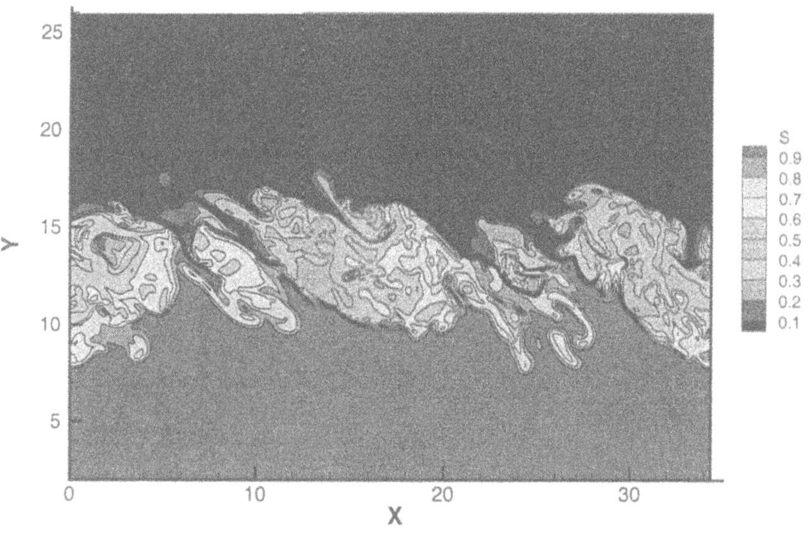


(b)

*Figure 9.1.* Instantaneous scalar visualizations at a late time$\Gamma t\Delta U/\delta_{\omega,0} = 258\Gamma$ wher the shear layer is in a self-similarly evolving state: (a) spanwise cut through the center of the shear layer$\Gamma$ (b) longitudinal section;$M_c = 0.3\Gamma s_\rho = \rho_2/\rho_1 = 1$.

approximation of the final self-similarly evolving state. The final Reynolds number based on the vorticity thickness is as large as $Re_\omega \simeq 10,000$ while the microscale Reynolds number is as large as $Re_\lambda \simeq 140$. Visualizations of a passive scalar at a late time during the self-similarly evolving stage are

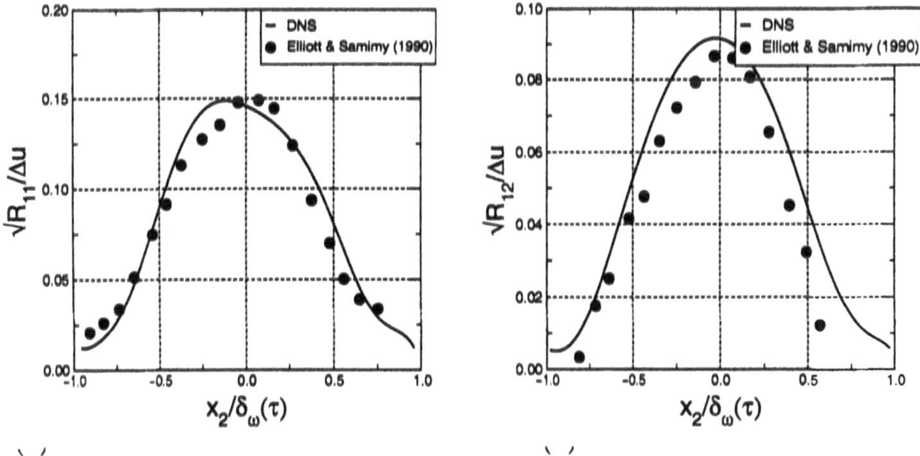


*Figure 9.2.* Comparison of turbulence profiles in DNS simulation at $M_c = 0.7$ with experimental data: (a) streamwise intensity, and (b) shear stress

shown in Fig. 9.1. Unlike the quasi-twodimensional rollers and longitudinal vortices in the early stage, these late-time visualizations indicate strong three-dimensionality and a wide range of scales. Turbulence profiles agree well with laboratory experiments. For example, Fig. 9.2 shows that the profiles of streamwise turbulence level and Reynolds shear stress for the DNS case with $M_c = 0.7$ compares very well with the laboratory data of Elliott and Samimy [146] for a $M_c = 0.64$ shear layer.

## 9.5. The stabilizing effect of Mach number

### 9.5.1. THE THICKNESS GROWTH RATE

It is well-known that, with increasing Mach number, the growth rate of the shear layer thickness decreases. The Langley experimental curve shown in Fig. 9.3 which is a consensus of experimental data from different pre-1972 studies of air/air shear layers shows this stabilizing effect of compressibility.

The later air/air experiments of Elliott and Samimy [146], Debisschop and Bonnet [126], and Chambres *et al.* [76] report results that agree well with the Langley curve. Papamoschou and Roshko [358] and Hall *et al.* [200] who used different combinations of helium, argon and nitrogen, as well as Clemens and Mungal [96] who use both air/air and air/argon systems report even lower growth rates with respect to the Langley curve. In the experiments, the free-stream densities are different, in general, due to both temperature and composition differences. In summary, the experimental

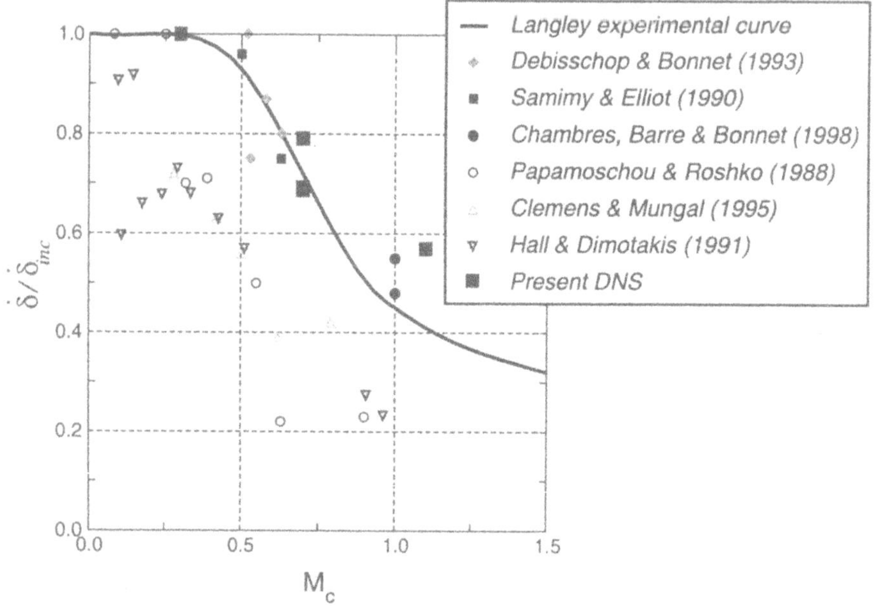


*Figure 9.3.* Dependence of shear layer growth rate on $M_c$.

measurements suggest an influence of density ratio, $s_\rho = \rho_2/\rho_1$, *in addition*, to that of the convective Mach number. The effect of density ratio will be discussed in section 9.6.

### 9.5.2. TURBULENCE INTENSITIES

The peak turbulence level in a shear layer occurs at its centerline where the mean shear has its maximum value. The normalized values of cross-stream and streamwise intensities are plotted in Fig. 9.4.

The cross-stream intensity, $\sqrt{R_{22}}/\Delta U$, decreases as a function of $M_c$ in all data sets. The Reynolds shear stress (not shown here) also decreases with increasing $M_c$ in all data sets. However, there is disagreement about the behavior of the streamwise intensity at high speeds. The experimental data of Goebel and Dutton [186] shows that the streamwise intensity is relatively insensitive to $M_c$ while the other experimental data show a consistent decrease of streamwise turbulent intensity with increasing $M_c$. Our DNS data, plotted in Fig. 9.4, also shows a reduction in $\sqrt{R_{11}}/\Delta U$. However, the DNS data of Vreman *et al* [477] as well as Freund *at al.* [162] do not show such a reduction.

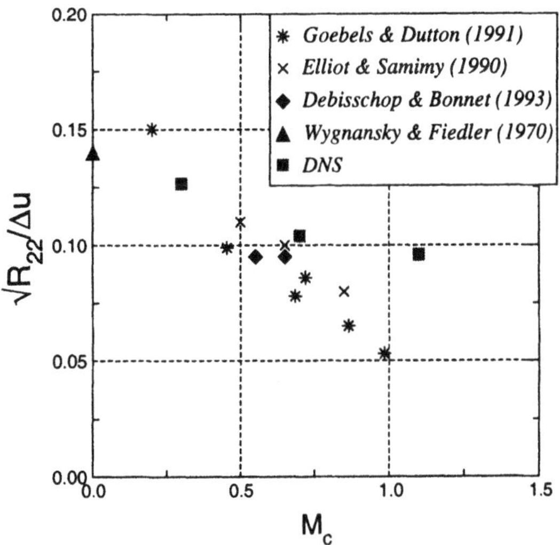


(a)

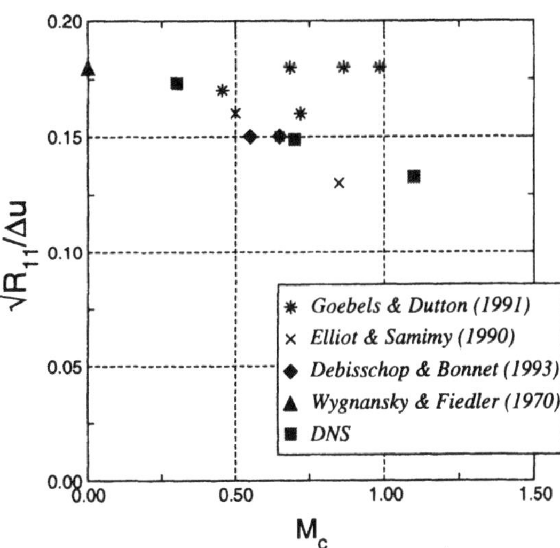


(b)

*Figure 9.4.* Mach number effect on peak turbulence intensities: (a) cross-stream component, and (b) streamwise component.

The anisotropy tensor,

$$b_{ij} = \frac{R_{ij}}{2\widetilde{k}} - \frac{\delta_{ij}}{3}, \tag{9.8}$$

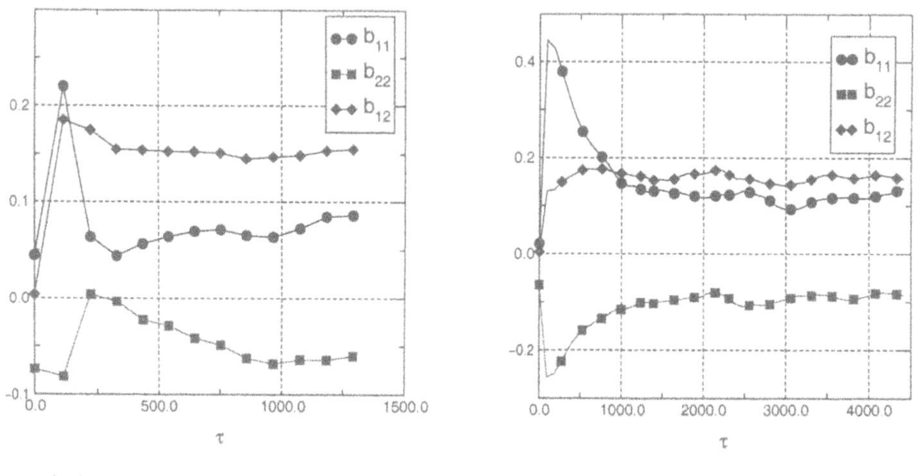


*Figure 9.5.* The evolution of Reynolds stress anisotropy in the DNS: (a) $M_c = 0.3$ case, and (b)$M_c = 1.1$ case.

is an estimate of the large-scale turbulence anisotropy. The data sets that show relatively constant values of streamwise intensity also show substantially increased values of $b_{11}$ because $\widetilde{k}/\Delta U^2$ decreases with increasing $M_c$ in all instances. The evolution of the anisotropy tensor at small and large values of $M_c$ is shown in Fig. 9.5.

At both Mach numbers, there is an initial transient wherein the streamwise anisotropy, $b_{11}$, increases due to direct shear production of streamwise turbulence, followed by a relaxation due to the redistribution of kinetic energy to the other components, and finally the asymptotic approach to an equilibrium level. If the peak values of $b_{11}$ during the early transient are compared, the case with $M_c = 1.1$ has a substantially larger value than that in the $M_c = 0.3$ case. This trend is consistent with the investigations [186, 477, 162] that find streamwise turbulence intensity relatively unaffected by compressibility. However, the later equilibrium values of $b_{11}$ in Fig. 9.5 show relatively little difference between the $M_c = 0.3$ and $M_c = 1.1$ cases. Thus, a possible reason for the controversy regarding the compressibility effect on streamwise turbulence level is that the shear layer in some data sets was in an early stage of evolution while it was at a late, fully-developed stage in other investigations.

The effect of $M_c$ on other components of the anisotropy tensor is of interest. According to our DNS, the effect is small on all components of $b_{ij}$ in the compressible shear layer.

### 9.5.3. AN EXPLANATION OF THE MACH NUMBER EFFECT

The strong reduction of thickness growth rate and turbulence intensities with increasing Mach number is a dramatic phenomenon and, not surprisingly, has prompted many attempts to explain it.

Coherent structures in the shear layer, in particular, the formation and pairing of Kelvin Helmholtz billows, appear to be suppressed at high Mach numbers [358, 405, 200, 96]. However, the observed modification in coherent structures constitute an effect of compressibility and not an explanation. Linear instability theory shows a decrease in the maximal growth rate with increasing $M_c$ and there have been suggestions, for example [340, 406] that linear analysis might explain and perhaps predict the Mach number effect. However a connection of linear stability analysis to turbulent entrainment in a shear layer is difficult to make.

The turbulent energy drain associated with explicit dilatational terms, the compressible dissipation and pressure dilatation, has been identified as a possible cause for the observed compressibility effect. However, DNS data does not support such a mechanism. In the case of uniform shear flow, reduced turbulence levels and normalized turbulence kinetic energy growth rate were shown by Sarkar [409] to be due to reduced turbulence production. In the case of the plane shear layer, Vreman *et al.* [477] derived a relation between the momentum thickness growth rate and the turbulence production that showed that reduction in the former is equivalent to a reduction in the latter. Interestingly, reduction in turbulent production points to a compressibility effect on large-scale dynamics, a feature which is shared with explanations based on coherent structures or linear stability theory.

Unlike the incompressible case, the equations governing flow in the high-speed regime allow for a finite speed of sound which limits upstream and cross-stream influence of pressure perturbations in high-speed flow. A familiar example of this phenomenon is the Mach zone associated with supersonic flight. It should be noted that this effect of limited zone of influence occurs in subsonic conditions too and increases with Mach number. Morkovin [338] conjectured that 'reduced communication' is responsible for the observed reduction of growth rates of instability modes in the high-speed boundary layer and the shear layer. Papamoschou and Lele [356] explored the reduced communication effect in a two-dimensional simulation of the evolution of the pressure field due to an isolated vortex placed in a temporally-evolving free shear layer and found that, with increasing Mach number, the streamwise extent of the pressure field decreases. Breidenthal [56] proposed a sonic-eddy model that assumed that eddies with vertical extent large enough to have supersonic relative velocity do not contribute to

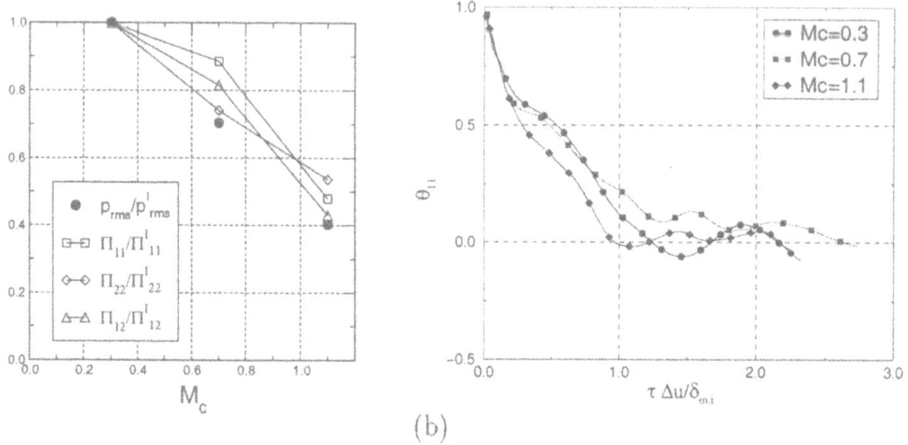


*Figure 9.6.* (a) Mach number effect on the r.m.s. pressure and pressure strain correlation, and (b) The two-time correlation, $\theta_{11}(0, \tau)$.

mixing. However, experimental observations show reduced mixing at Mach numbers where the relative velocity is high but still subsonic.

The detailed behavior of all variables including the pressure field is available in a DNS enabling the measurement of pressure-related quantities. Sarkar [410] investigated the role of pressure and density fluctuations in compressible uniform shear flow. The normalized pressure fluctuations were found to decrease as a function of Mach number and the reduced growth and turbulence levels at high speeds was attributed to be a 'fluctuating pressure' effect associated with an alteration of the pressure field at high speeds. Vreman *et al.* [477] observed reduction in normalized pressure fluctuations with increasing Mach number. They further found that the pressure strain term, $\Pi_{ij}$ decreases to cause the drop in Reynolds shear stress (and turbulent production) with increasing Mach number. This connection between reduced pressure fluctuations and reduced turbulent production has since been verified in other DNS studies, for example, Freund *et al.* [162], and Pantano and Sarkar [354]. Fig. 9.6(a) shows results from our DNS to demonstrate the pressure-related effects of compressibility. A reduction of pressure fluctuations and pressure-strain term with increasing $M_c$ is evident.

The importance of the pressure strain term to sustaining turbulence in shear flows can be seen by specializing eq. (9.6) to the appropriate situation, i.e., a single mean velocity gradient, $\partial\widetilde{U}_1/\partial x_2$. Of the three components of turbulence kinetic energy, only the streamwise component, $R_{11}$, is directly forced by mean shear and that forcing, $P_{11} = -R_{12}\partial\widetilde{U}_1/\partial x_2$, depends on the Reynolds shear stress $R_{12}$. The Reynolds shear stress is produced through

$P_{12}$ by the vertical component $R_{22}$; this can be simply understood by the fact that vertical excursions of a fluid particle that is assumed to conserve streamwise momentum generates Reynolds shear stress if the flow profile has shear. It is clear from eq. (9.6) that the only term that can be a source for $R_{22}$ is the pressure strain component, $\Pi_{22}$. Thus, the pressure strain correlation is critical to sustain turbulence in shear flows.

The key role played by reduced pressure fluctuations and associated reduction in the pressure strain correlation towards causing the observed stabilizing influence of compressibility in the shear layer has been definitively established by the DNS studies [410, 477, 162, 354]. It may be conjectured that the observed 'fluctuating pressure' effect is related to the easily-recognized consequence of compressibility, namely, the 'reduced communication' phenomenon mentioned earlier. Such a mathematical link has been analytically shown by Pantano and Sarkar [354] who explicitly include the effect of finite sound speed by analyzing a *wave equation* for the fluctuating pressure instead of the Poisson equation applicable to incompressible flow. The crucial result of that analysis is that due to the acoustic delay time associated with the finite speed of sound, all components of the pressure-strain term exhibit *monotone* decrease. Furthermore, the appropriate quantity for parameterizing this effect is found to be the gradient Mach number, $M_g = Sl/\bar{c}$.

A critical assumption of the wave equation analysis of Pantano and Sarkar [354] is based on the notion that turbulence has finite memory time. It is assumed that there is temporal decorrelation (an exponential function with characteristic time scale defined by large-scale quantities) associated with the following two-time correlation,

$$\theta_{ij}(\mathbf{r},\tau) = \overline{f'(\mathbf{x}+\mathbf{r},t-\tau)(u'_{i,j}+u'_{j,i})(\mathbf{x},t)}\,. \tag{9.9}$$

Here

$$f' = \frac{\partial}{\partial x_i \partial x_j}\Big(\rho u_i u_j\Big)$$

denotes the nonlinear forcing of the wave equation. Fig. 9.6(b) shows representative plots of the two-time correlation that is required in the analysis. It is clear that there is temporal decorrelation as assumed.

The following physical reason for the reduction in the pressure strain term can be identified. The finite speed of sound in compressible flow causes a time delay, $l/c$, in the travel of pressure signals across a characteristic eddy length $l$ and, thus, causes decorrelation between adjacent points in an "eddy". The ratio of acoustic time delay, $l/c$, to a characteristic flow time ( $1/S$ or $l/\Delta U$) increases with Mach number ($M_g = Sl/c$ or $M_c = \Delta U/2c$) and, as clearly shown by the analysis, the resultant increase in decorrelation inhibits the pressure-strain term.

## 9.6. Variable-density effect in the high-speed shear layer

### 9.6.1. MOMENTUM AND VORTICITY THICKNESS GROWTH RATE

The momentum thickness defined by

$$\delta_\theta = \frac{1}{\rho_o \Delta U^2} \int_{-\infty}^{\infty} \overline{\rho}\left(\frac{\Delta U}{2} - \widetilde{U}_1\right)\left(\frac{\Delta U}{2} + \widetilde{U}_1\right) dx_2 , \tag{9.10}$$

and the vorticity thickness defined by

$$\delta_\omega = \Delta U / (\partial \widetilde{U}_1 / \partial x_2)_{max} , \tag{9.11}$$

are popular measures of the shear layer thickness.
It is important to distinguish between these two thickness measures as will be shown subsequently after analyzing the DNS results.

It is known that variable-density effects influence shear layer growth in the low-speed regime. For example, the results of Brown and Roshko [61] for $s_\rho = 1$, $1/7$ and 7 indicate that the vorticity thickness growth rate increases as a function of $s_\rho$. Approximately a factor of 2 increase in growth rate was observed at a fixed velocity ratio of $\lambda = (U_1 - U_2)/(U_1 + U_2) = 0.8$. The sensitivity to density ratio was observed to decrease with decreasing $\lambda$. Experimental data in the *low-speed* shear layer has been used by Brown [60] and Dimotakis [131] to correlate the effect of non-equal free-stream densities in the shear layer. For example, Brown [60] gives,

$$\frac{d\delta_\omega}{dx} = C_\delta \frac{(1-r)(1+s_\rho^{1/2})}{2(1+rs_\rho^{1/2})} , \tag{9.12}$$

where $r = U_2/U_1$ is the velocity ratio, $s_\rho = \rho_2/\rho_1$ denotes the density ratio, and subscripts 1 and 2 denote the high and low-speed free streams, respectively. Dimotakis [131] improved the correlation by accounting for the asymmetric entrainment of the spatially-developing shear layer. Brown arrived at eq. (9.12) by assuming the following: (a) the *temporal* growth rate in a reference frame moving with the typical convection velocity of the structures is *unchanged*, and (b) the convection velocity shifts to that of the high-density stream.

Eq. 9.12 predicts that the growth rate of a spatially evolving shear layer increases with increasing $r$. Thus, if the low-speed stream has a higher density, the thickness growth rate increases. An alternate procedure based on "equivalent" uniform density flows has been used by Thring and Newby [461] and Ricou and Spalding [393] to describe the *far field* evolution of variable-density jets. Consider a jet of density, $\rho_1$, and diameter, $d_1$, exhausting into a quiescent ambient of density, $\rho_2$. In the far-field where

| $s_\rho$ | $\dot{\delta_\omega}/\Delta U$ | $\dot{\delta_\theta}/\Delta U$ | $\dot{\delta_\theta}/\Delta U$ using eq. (9.14) |
|---|---|---|---|
| 1 | 0.0637 | 0.0107 | 0.0107 |
| 4 | 0.0557 | 0.0082 | 0.0087 |
| 8 | 0.0553 | 0.0057 | 0.0074 |

TABLE 9.1. Dependence of nondimensional growth rates of vorticity and momentum thickness on density ratio at fixed $M_c = 0.7$. For reference, the corresponding value for the incompressible shear layer is $\dot{\delta_\omega}/\Delta U \simeq 0.075$.

the density difference between jet and ambient fluid is small, a notional jet with density equal to that of the ambient fluid can be introduced. Let $U^\star$ and $d^\star$ be the velocity difference and thickness of this notional jet which is "equivalent" in the sense of having the same mass flux, $M_1$, and momentum flux, $J_1$ as the actual jet. The equivalent thickness of the jet is then

$$d^\star = d_1 \frac{\rho_1}{\rho_2} \quad , \quad U^\star = U_1 \ . \tag{9.13}$$

By hypothesis, the self-similar relations giving the downstream evolution of jet velocity and diameter for a uniform-density jet carry over to the variable-density case with $d^\star$ and $U^\star$ replacing $d_1$ and $U_1$, respectively. If the ambient has higher density than the jet fluid, it follows that $d^\star < d$. For a given location, $x$, the value of $x/d^\star$, is larger leading to a *higher* thickness if the jet fluid is *lighter* than that of the ambient. Such an effect is indeed noticed in practice. However, for eq. (9.13) to be exact, the averaged equations should reduce to the constant density form under the transformation, $x_2^\star = (\rho_1/\rho_2)x_2$, which can be seen to be a special case of the Howarth-Dorotnitsyn transformation, $x_2^\star = \int(\rho/\rho_2)dx_2$, used in compressible boundary layer theory. A reduction to constant density form is *not* possible, in general, and, furthermore the mean density profile is required to recover the mean velocity profile. Therefore, eq. (9.13) is useful for only qualitative trends regarding the density effect on the evolution of the jet. In the asymptotic limit, $x/d \to \infty$, the density becomes $\rho/\rho_2 \to 1$, and the concept of an equivalent jet as embodied by eq. (9.13) becomes more accurate.

The DNS results regarding the effect of different density ratios, $s_\rho = 1, 4$ and 8, at fixed $M_c = 0.7$ on the nondimensional growth rates are compared in Table 9.1.

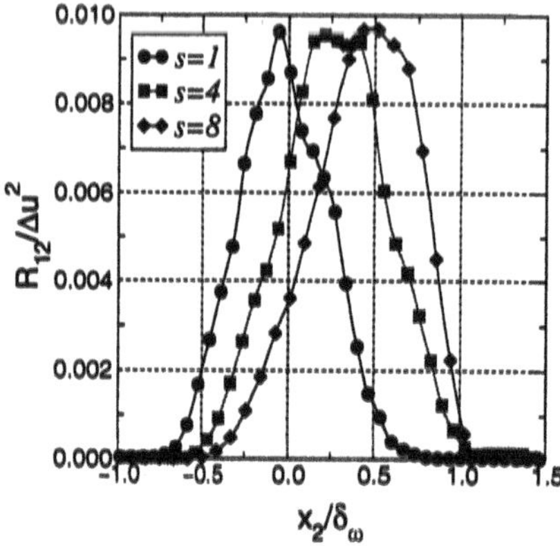


*Figure 9.7.* Reynolds stress $R_{12}$ at $M_c$ = 0.7 for different density ratios.

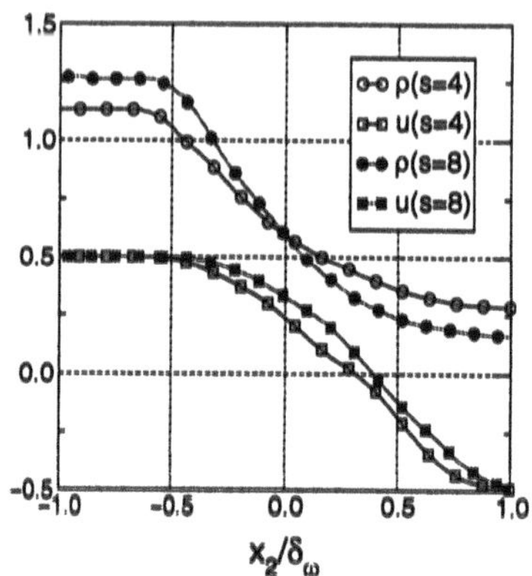


*Figure 9.8.* Mean density and streamwise velocity profiles.

The momentum thickness shows a large systematic decrease with increasing values of $s_\rho$ while the vorticity thickness is less affected by unequal free-stream densities. Thus, the dependence of shear layer thickness on the density ratio depends on its specific definition !

It will be seen later that the 'thickness' of turbulence statistics profiles (Reynolds stresses, r.m.s. thermodynamic fluctuations, Reynolds stress budgets, etc.) does not vary much with density ratio when $\delta_\omega$ is used to nondimensionalize the cross-stream coordinate. From this perspective, the appropriate shear layer thickness is the vorticity thickness and not the momentum thickness.

The turbulent shear stress profiles are plotted in Fig. 9.7, where it is recalled that density-weighted Favre averages are used for defining the turbulent stresses, that is, $R_{ij} = \overline{\rho u_i'' u_j''}/\overline{\rho}$.

The peak value of the shear stress is not affected significantly by the density ratio while its location shifts to the upper, low-density stream. The shear layer is a simple flow where the classical notion that $R_{12}$ is a function of the local value of $\partial\widetilde{U}_1/\partial x_2$ is applicable. Thus, the behavior of peak $R_{12}$ is tied to that of peak $\partial\widetilde{U}_1/\partial x_2$. Since the vorticity thickness is related to the peak $\partial\widetilde{u}_1/\partial x_2$ by eq. (9.11), it is not surprising that the influence of density ratio on $\delta_\omega$ is similar to that on peak $R_{12}$. The dividing streamline, $\widetilde{U} = 0$, also moves to the low-density side approximately coincident with the peak $R_{12}$ location. Fig. 9.8 shows the shift of the dividing streamline.

The momentum thickness depends explicitly on the mean density as shown by eq. (9.10). The influence of density ratio on shear layer momentum thickness which has been analyzed in detail by Pantano and Sarkar [354]

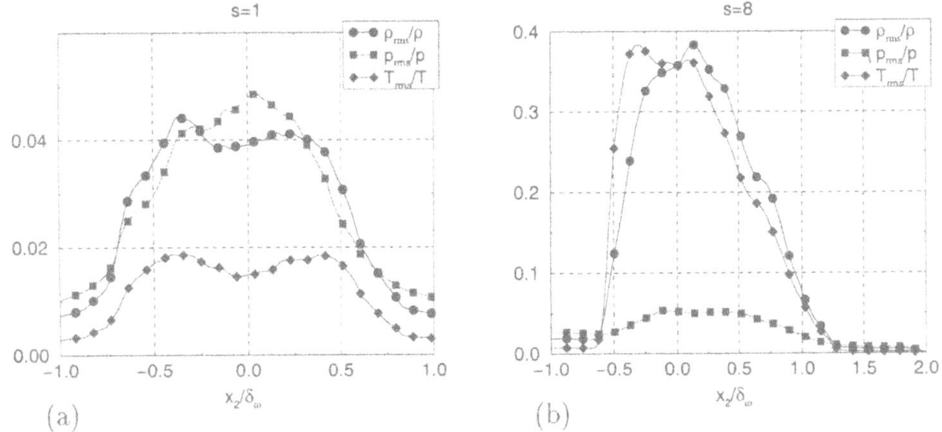


*Figure 9.9.* Profiles of thermodynamic fluctuations: (a) $s_\rho = 1$, and (b) $s_\rho = 8$.

is briefly summarized. By considering the streamwise momentum equation, it is analytically shown that the dividing streamline always shifts to the low-density stream. Using this information and DNS data, model profiles of $\overline{\rho}$ and $R_{12}$ as a function of $\widetilde{U}_1$ are introduced allowing integration of the r.h.s. of eq. (9.10). The general conclusion is that the momentum thickness of the shear layer *must decrease* when $s_\rho \neq 1$. The analysis also leads to the following model specific to the DNS cases,

$$\frac{\dot{\delta}_\theta}{\dot{\delta}_{\theta,1}} = 1 - 0.4\lambda(s_\rho)a(s_\rho) \tag{9.14}$$

where $\lambda(s_\rho) = (s_\rho - 1)/(1 + s_\rho)$, while $a(s_\rho)$ parameterizes the shift of the dividing streamline to the low-density stream. Table 9.1 shows that eq.(9.14) is a good approximation to the observed effect of density ratio on the momentum thickness growth rate.

### 9.6.2. THE THERMODYNAMIC FLUCTUATIONS

Fig. 9.9 shows profiles of $\rho_{\text{rms}}/\overline{\rho}$, $T_{\text{rms}}/\overline{T}$ and $p_{\text{rms}}/\overline{p}$ for the $s_\rho = 1$ and $s_\rho = 8$ cases. The dominance of density and temperature fluctuations over pressure fluctuations in the case with large density ratio, $s_\rho = 8$, is striking. The peak r.m.s. density fluctuations is approximately 35% of the mean density and coincides with the location of maximum mean density gradient.

The acoustic/entropy mode separation of thermodynamic fluctuations is often used in compressible flows. In the classical definition, all the fluctuation in pressure is assumed to be associated with the acoustic mode, and corresponding isentropic density and temperature fluctuations are included

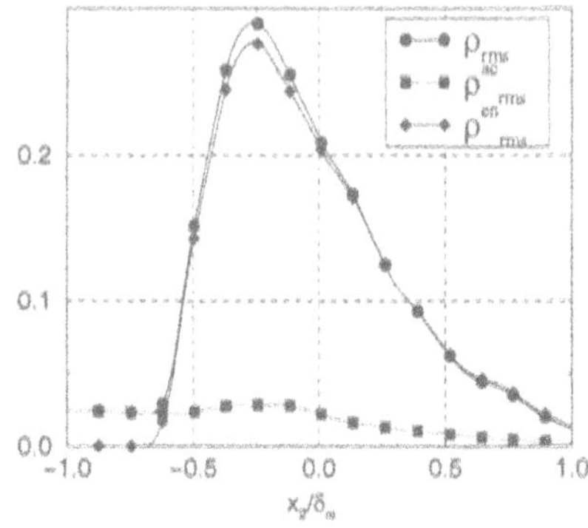


*Figure 9.10.* Acoustic and entropy mode contributions to the density fluctuation when $s_\rho = 8$, $M_c = 0.7$.

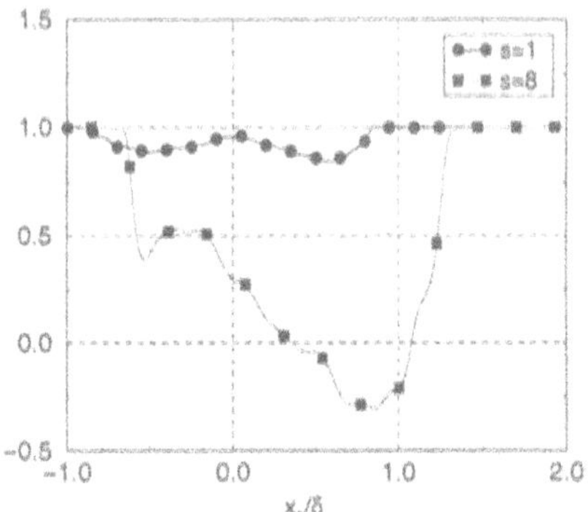


*Figure 9.11.* Correlation between pressure and density fluctuations in the case with $M_c = 0.7, s_\rho = 8$.

in the acoustic mode. Such a definition is not unique (see Chapter 3); for instance, the acoustic pressure may be obtained by subtracting the solution of the Poisson equation for pressure applicable to incompressible flow from the observed pressure fluctuation.
However, we adopt the classical definition so that the acoustic mode is described by,

$$p^{ac'} = p', \tag{9.15}$$

$$\rho^{ac'} = \frac{p^{ac'}}{\overline{c}^2}, \tag{9.16}$$

$$T^{ac'} = \frac{\gamma - 1}{\gamma}\overline{T}\,\frac{p^{ac'}}{\overline{p}}, \tag{9.17}$$

while the entropy mode is given by the remainder,

$$p^{en'} = 0, \tag{9.18}$$

$$\rho^{en'} = \rho' - \rho^{ac'}, \tag{9.19}$$

$$T^{en'} = T' - T^{ac'}. \tag{9.20}$$

Fig. 9.10 shows that almost *all* the density fluctuation in the case with $s = 8$ is associated with the *entropy* mode.

Despite the large density (temperature) fluctuations, the peak r.m.s. pressure fluctuation in Fig. 9.9 is not significantly effected by the density ratio. Evidently, compared to the strong $M_c$ effect that reduces pressure fluctuations, the variable density effect is relatively weak. The profile of the

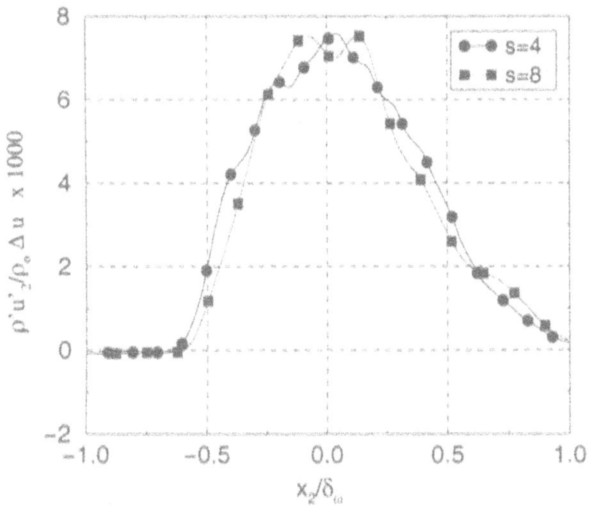


*Figure 9.12.* Profiles of the mass flux for the cases with $s_\rho = 4$ and $s_\rho = 8$.

correlation coefficient, $R(\rho, p)$, defined by

$$R(\rho, p) = \frac{\overline{\rho' p'}}{\rho_{\mathrm{rms}}\, p_{\mathrm{rms}}},$$

is shown in Fig. 9.11. The correlation coefficient is close to unity when $s_\rho = 1$, implying that, in the absence of a mean density contrast, density fluctuations are almost exclusively associated with the acoustic mode. In the case with $s_\rho = 8$, the density fluctuations are not well correlated with pressure in the core of the shear layer indicating that the acoustic mode is no longer dominant. Even so, as shown by the $s_\rho = 8$ curve in Fig. 9.11, $R(\rho, p)$ does not tend to zero in the region with significant density fluctuations. This suggests that there is coupling between the acoustic and entropy modes in this case.

### 9.6.3. THE MASS FLUX

The mass flux, $\overline{\rho' u_2'}$, normalized using the velocity difference, $\Delta U$ and the average layer density, $\rho_0 = 0.5(\rho_1 + \rho_2)$, is shown in Fig. 9.12. It is generally positive and, since the lower stream has the higher density, corresponds to cogradient transport of mass. The turbulence Prandtl number, $Pr_t = \nu_t / D_t$, where $\nu_t$ is the turbulent momentum diffusivity and $D_t$ is the turbulent mass diffusivity has also been calculated. The classical notion of cogradient transport with $Pr_t = O(1)$ applies in this case. $Pr_t$ has a moderate

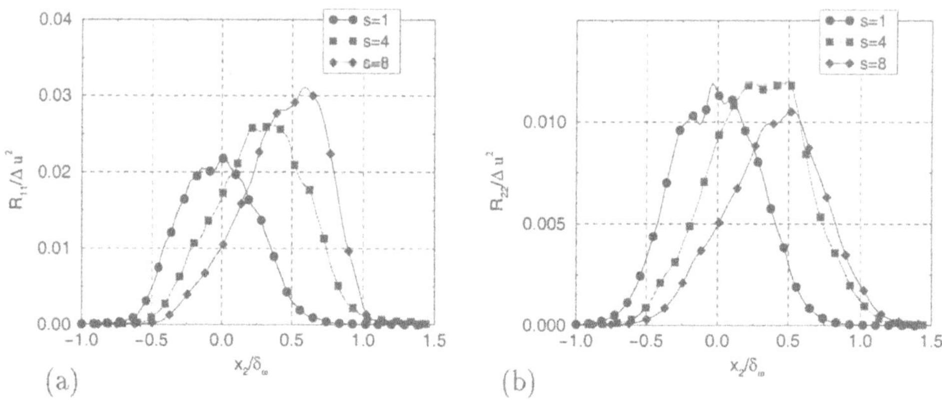


*Figure 9.13.* Normal components of the Reynolds stress tensor: (a) Streamwise, $R_{11}$, and (b) cross-stream, $R_{22}$.

variation across the shearlayer with generally higher values in the upper, low density stream. The lower stream has $Pr_t \simeq 0.7$ in accord with the expectation that, in turbulent flows, scalars have a somewhat higher transport rate compared to the momentum.

### 9.6.4. TURBULENT STRESSES

The profiles of the streamwise component, $R_{11}$, and the cross-stream component, $R_{22}$, are shown in Fig. 9.13. The peak Reynolds stress values shift to the upper, low-density stream where the dividing streamline is located. Although the peak value of $R_{11}$ shows a systematic increase with density ratio, the peak value of $R_{22}$ is not significantly affected. Recall that, as shown by Fig. 9.7, the peak value of $R_{12}$ is not significantly affected by the density ratio.

## 9.7. Concluding remarks

Experimental and DNS investigations have allowed a detailed picture of compressibility effects in the high-speed shear layer. Viewing compressibility effects on turbulence as a consequence of 'acoustic' and 'entropy' modes is helpful. The remarkable decrease in the growth rate of the shear layer at high Mach number is recognized now to be an 'acoustic' mode effect ie; due to the reduction in normalized pressure fluctuations and consequently reduced influence of the pressure gradient term in the momentum balance. A subtle chain connects the reduced growth rate to ultimately the limitation on the space-time influence of pressure perturbations imposed by the finite speed of sound in compressible flow. The temporally-evolving DNS has been key to develop our current level of understanding and now further DNS studies of the spatially-evolving turbulent shear layer are necessary to complete the picture.

In the case of a temporally-evolving shear layer with unequal free stream densities the effect of the density ratio, $s_\rho$, on the vorticity thickness growth rate is relatively weak compared to that on the momentum thickness growth rate. The spatially-evolving case has been considered by a few experimental studies; however, more detailed DNS investigations are now required. With increasing values of $s_\rho$, the entropy mode dominates the acoustic mode. Unlike the acoustic mode, the effect of the entropy mode on the turbulence is relatively weak.

The Mach number effect in the shear layer arises from the coupling of the 'acoustic' mode to turbulence. The reacting high-speed shear layer is complicated by the additional couplings possible due to the entropy mode associated with heat release. Further work is necessary to have a complete description of compressibility effects in the high-speed reacting shear layer.

Our understanding of the compressible free shear layer needs to be placed in the context of other compressible turbulent flows. For instance, the gradient Mach number, $M_g = Sl/c$, appears to be the parameter that directly determines the strong compressibility effects in the free shear layer related to the fluctuating pressure i.e., acoustic mode. In the wall boundary layer, the value of $M_g$ is much smaller that that in the free shear layer at comparable values of mean Mach number suggesting that acoustic mode effects on turbulence would be less important in the boundary layer. The role of the gradient Mach number in the boundary layer as well as other flows deserves investigation.

Large eddy simulation (LES) is a promising tool for prediction of turbulent flows. Since compressibility effects in the free shear layer are, to leading order, related to changes in the large-eddy dynamics, LES is potentially of value. However, further work is necessary to develop compressible formulations of subgrid models. Also, in reacting flows as well as shock/turbulence interaction problems, small-scale compressibility effects are, in principle, important and new subgrid models need to be developed.

# CHAPTER 10

# FIRST-ORDER MODELING

*This chapter opens the last part of the monograph, which is more specifically dedicated to an engineering audience. It begins with a general presentation and discussion of prediction methods for turbulent flows, based on statistical — or Reynolds — averaged Navier-Stokes equations (RANS). Then, the incidence of density changes and the incorporation of variable-density and compressibility effects in first-order closure models are analyzed with respect to (i) "modifications" to incompressible schemes and (ii) introduction of additional "specific contributions" to non-constant density flows. At last, some zero-, one-, two- and three-equation models are reviewed.*

## 10.1. Introduction

The ultimate objective of this third part of the monograph is to discuss tractable models that have been developed to calculate quantities of interest and practical relevance in turbulent flows of variable density fluid. Turbulence is considered here as a moderate-to-high Reynolds number flow, produced by unsteady, non-linearly interacting instabilities generating a large dynamic range of 3-D velocity and vorticity fluctuations in space and time. It is assumed that the motion is governed by the Navier-Stokes equations, along with the continuity equation, the first law of thermodynamics and the equation of state, as detailed in Chapter 4. For practical applications, several predictive methods[1] are available at the present time:

- Large Eddy Simulations (LES), in which the motion equations are solved for a *filtered* velocity field, which is representative of the large scale turbulent motion. Hence, a model is required for the smaller-scale motions which are not represented;
- Probability Density Function methods (PDF), in which a model transport equation is solved for a probability density function of the fluctuating velocity and other single or joined scalar fluctuations;
- Single point modeling of Reynolds Averaged Navier-Stokes equations, the so called RANS approach, in which closure schemes are introduced to provide a set of statistical equations governing one point moments.

Only the last approach will be addressed hereafter.

[1] At present, Direct Numerical Simulation (DNS) is not considered as a predictive method for industrial applications.

The conventional hierarchy of such turbulence closure approaches is well known. It is mainly based upon a simple division between *first*, *second* and other higher order closure levels. In first-order models, turbulence or Reynolds stresses are directly coupled with the mean flow, using an eddy-viscosity concept or more generally linear or non-linear constitutive schemes. Now, since turbulence is not a fluid property but depends on the motion itself, the eddy-viscosity is to be prescribed as a function of some flow characteristics. This can be achieved algebraically, yielding zero-equation models, or by solving one or more additional transport equations, producing one-equation, two-equation... turbulence models.
In second-order models[2], transport equations are solved for the Reynolds stress tensor and all other second-order moments, yielding Reynolds Stress (transport) Equations models (RSE).

## 10.2. Synopsis of one-point turbulence modeling status

### 10.2.1. CONSTANT DENSITY FLOWS

Historically, since the late 60's, two groups have mainly contributed to the development of one-point modeling under the leadership of B.E. Launder, now at UMIST and J.L. Lumley, now at Cornell University. Since then, a wide amount of literature has been published on the topic by many other authors. Up to date reviews of first and second-order closure schemes for *incompressible* turbulent flows can be found in Schiestel [421], Chen & Jaw [89], Piquet [366], Pope [370], and Chassaing [84].

It is commonly agreed that eddy-viscosity models:

- perform quite satisfactorily in quasi-parallel, equilibrium, 2-D, wall-attached and free flows;
- are relatively numerically 'robust'.

On the other hand, the generic problems associated with linear eddy-viscosity models have been known for many years and include the inability to:

- track rapid, inviscidly induced mean flow alterations;
- predict counter-gradient fluxes;
- predict negative production zones;
- generate turbulence induced secondary flows.

With RSE modeling, such flaws are not present and significant improvements have been obtained, concerning the representation of:

[2]Algebraic stress models (ASM) should also be mentioned. They implicitly determine the local Reynolds stresses as a function of the turbulence kinetic energy, its dissipation rate, and mean velocity gradients. Due to the approximations they involved, they are simpler but less general and accurate than RSE. They are not considered here.

- pressure-strain correlations;
- low-Reynolds number effects;
- anisotropy in the near-wall region.

Nevertheless, some problems are still present with this type of models. They concern:

- the dissipation (or length-scale) equation(s);
- the gradient diffusion (turbulent transport) assumptions.

In constant density flows, some of the dominant physical processes of the turbulent regime can be captured with statistical, single point, closure models. However, such a modeling approach is neither sufficiently developed nor intrinsically adapted to account for all turbulence features, such as coherent structures and free flow intermittency, or energy cascade and small-scale intermittency.
In simple free shear layers, for instance, many observations reveal the presence of large (coherent) structures which are specific to each type of flow (plane or round jet, wakes, mixing layers). Such evidence, in turn, suggests that the energy-containing lower wave number portion of the turbulence spectrum should be flow-dependent. Hence the problem of turbulence modeling cannot simply be reduced to how reflecting modifications to a common basic state, mainly according to Kolmogorov's ideas. As suggested by Bushnell [65], "this is one of the root causes of the *variable* constants required thus far for all modeling approaches".
Another incidence of such large scale motions on the physics of turbulence is the irregular and intermittent behavior at the edge of free flows, with direct consequences on the mixing process, entrainment and expansion. Many proposals have been made to account for free flow intermittency, but at the present time, none of them is actually involved in practical predicting tools for engineering applications.

### 10.2.2. VARIABLE DENSITY FLOWS

An increasing number of what is considered as variable density effects has been now identified from experimental investigations, theoretical analysis such as stability analysis and Rapid Distortion Theory, or direct numerical simulations. As briefly presented in Chapter 2, the influence of variable density and compressibility on a turbulent flow is manifold and may results in:

- mean temperature/concentration effects;
- mean flow bulk dilatation/compression effects;
- additional baroclinic torque generation/destruction;
- additional instability mechanisms;

- alterations of turbulent eddies interactions and energy cascade;
- shock-induced modifications;
- dilatational contributions to turbulence;
- variation of the physical coefficients and bulk viscosity effects.

During 1975-85 decade, an important effort was dedicated by Ha Minh-Chassaing and co-workers, [80], [87], [193], [195], [85], [194], [473], [197], [332], [199] to the extension of constant density first and second-order closure schemes to variable density turbulent flows, in low and high speed fluid motions. The contributions due to Launder's and Lumley's groups in low-speed modeling for scalar fields were also extended to heterogeneous flows, considering the equations governing concentration or temperature correlations.
At that time and using density weighted averages, the main question, as pointed out by Janicka and Lumley [232] in 1980, was to know to what extent "model assumptions which are developed for constant density flows can be adopted for closure of density-weighted moments for variable density flows."

Later on, a serious attempt to extend RSE closure to super/hypersonic flows was promoted by Speziale-Sarkar *et al.* [443] at NASA Langley. Since then, there have been many proposals to model *new* compressibility effects in high-speed flows, by (i) adding "corrections" to the model "constants" via terms depending on a turbulence Mach number, (ii) introducing "extra-compressibility terms" accounting for dilatational effects (compressible or dilatational dissipation, pressure dilatation correlation). Both procedures are reviewed in the present chapter, as far as they deal with first-order closure schemes.

A second point emerged, giving rise to a slightly different discussion in variable and constant density situations. It directly addresses the choice of the closure level. For "theoretical" reasons, which will become clear in the next chapter, second-order level appears to be particularly suitable in modeling variable density flows. Thus, when reviewing first-order closure schemes, it is worth keeping in mind the following statement by Fulachier *et al.* [173] in 1989 "The improvements obtained by a second-order modeling are so evident, and the ability of this type of modeling to predict the effects of density differences in complex flows so clear, that it is to be recommended that such model be used even for industrial purposes."

## 10.3. The first-order modeling issue

When deriving closure schemes for the mean motion equations of variable density fluids, one is faced with a three-fold challenge:

- selecting the intrinsic mechanisms which are actually responsible for the dominant variable density effects that are observed in a given flow configuration;
- identifying, in the open set of equations, the corresponding terms that can be handled at a given closure level;
- deriving proper closure schemes to such terms, capturing the correct underlying physics.

Although present in the modeling issue for constant density fluid flows, the challenge is even more crucial in variable density situations. Let us now give some examples.

***Identification of intrinsic mechanisms.*** The search for the explanation of the reduction in the mixing layer growth rate through compressibility can be considered as an illustrative example of the difficulty in identifying intrinsic mechanisms of turbulence in modeling variable density fluid motions.

After the discovery by Passot and Pouquet [359] in 1987, from numerical simulations, of the distinct possibility of random shock-like structures to appear in the flow domain, even when the turbulence or r.m.s. Mach number $M_t$ is subsonic, the presence of "shocklets" was taken as one of a plausible explanation of the phenomenon. However, by the same time, a new controversy about the actual mechanism responsible for the Mach number stabilizing effect to be introduced in turbulence modeling was opened.

- In 1990, Zeman [495] proposed the dilatation dissipation concept to account for such an effect.

- In 1989, to improve the modeling of shock wave/boundary layer interaction in flows over compression ramps, Grasso & Speziale [187] argued that the pressure-dilatation correlation[3] $\overline{p'\partial u'_j/\partial x_j}$ and the turbulent mass flux coupled with the mean pressure gradient $(\overline{\rho' u'_i}/\overline{\rho})(\partial\overline{P}/\partial x_i)$ play an important role in this situation.

- In 1990, Nichols [346], addressing the same flow configuration, suggested that a turbulent "velocity-density" dissipation term $\overline{U_i}\,\overline{\rho' u'_j}(\partial\overline{U_i}/\partial x_j + \partial\overline{U_j}/\partial x_i)$ has to be introduced in the modeling of compressibility effects in that case.

A similar lack of consensus can also observed in the choice of the additional functions which are used as arguments in the various closure schemes accounting for density/compressibility effects. The example of the $(k - \epsilon)$ model is particularly illustrative in this respect. Several proposals have been made to extend the incompressible version of this model to compressible situations, generally adopting the turbulence Mach number $M_t = \sqrt{2\overline{k}}/\overline{c}$,

[3] $f'$ denotes a Reynolds fluctuation.

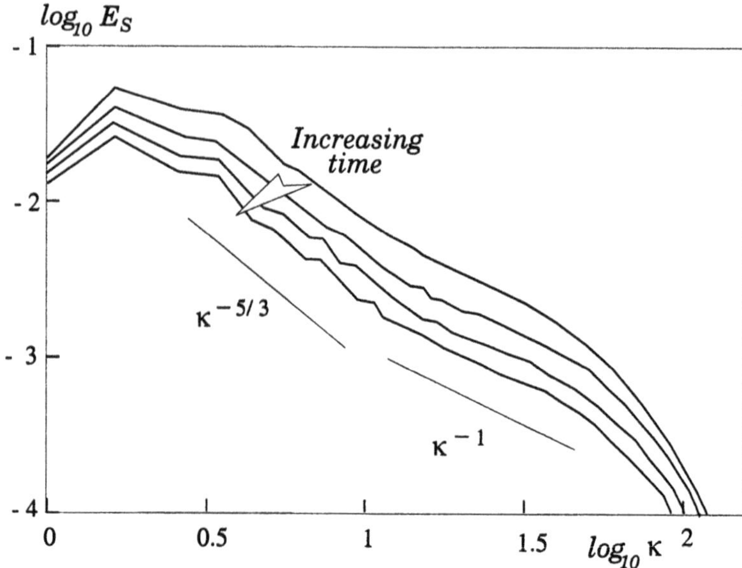


*Figure 10.1.* Spectrum of the solenoidal velocity fluctuations in a decaying homogeneous isotropic supersonic turbulence, adapted from Pouquet [371].

where $\bar{c}$ is the mean speed of sound, as the parameter accounting for compressibility effects. Some of them are reviewed in the present chapter. However such straightforward extensions have revealed to be not entirely satisfactory, and the need for additional functions has been investigated. Various candidates have been suggested and will be examined in §10.12.

***The underlying physics.*** In modeling constant density fluid turbulence, single point closure of mean transport equations in fully developed, high Reynolds number flows may be addressed first. Far from the walls or in free turbulent flows, this leads to consider that turbulence can reach some equilibrium state, depending on the boundary conditions, but with only a limited memory of its past state. Accordingly, the energetic process is statistically depicted as obeying the Kolmogorov cascade, and large scale organization effects, if present, are not explicitly taken into consideration. To transpose "incompressible" closure schemes to variable density fluid turbulence, it is (*generally implicitly*) assumed that compressibility effects do not radically change the physics, so that most of the incompressible closure schemes can be adapted to compressible turbulence.

Although not always justified, this procedure will be adopted in the following presentation, in which, from a "practical" modeling point of view, compressibility effects are considered as (i) modifying existing incompressible schemes and (ii) introducing new additional terms into the incompres-

sible closure formulation.
However, it is worth recalling that this attitude corresponds to a reduction in the physics. Let us give an illustration of this point. Direct *3-D* numerical simulations of supersonic homogeneous and isotropic turbulence by Pouquet [371] (see Fig.10.1) reveal a double scaling of the spectrum $E_S$ of the *solenoidal* part of the velocity fluctuations: (i) as $\kappa^{-5/3}$ for the energetic wave number range $\kappa < \kappa_t$, where $\kappa_t$ is the Taylor wave number, and (ii) as $\kappa^{-1}$, for the small scale structures $\kappa_t < \kappa < \kappa_\eta$, where $\kappa_\eta$ is the Kolmogorov wave number. To the author's knowledge, such a feature has not yet been included in turbulence modeling.

## 10.4. The open set of equations and the closure issue

Regarding the instantaneous equations governing various types of turbulent motion of a variable density fluid, several approximate models to the full variable-density Navier-Stokes equations can be derived. This question has been addressed in Chapter 3, where such approximate models have been presented, depending on various physical assumptions concerning the dominant mechanisms driving density variations.
With respect to the general objective of the monograph, closure schemes[4] that have been developed for such simplified situations will not be considered here. Hence, due to the wide variety of variable density turbulent flows, the full set of Navier-Stokes equations is adopted (see Chapter 4).

The non-linearities present in these equations generate unknown moments in any *finite* set of *averaged* equations deduced from the instantaneous ones. When deriving "modeled equations", one basically aims at producing a closed set of equations by recovering only that part of information, lost from the averaging procedure, which is required at a given level of statistical description. As far as *first*-order closure is concerned, this level of description simply refers to the "*mean motion equations*". These equations have been detailed in Chapter 5, adopting either conventional or density

[4]Such models are largely restricted to the situation for which they have been derived. For example, Boussinesq's approximations, or more generally weak compressible assumptions, yield simplification of the pressure role and make the modeling issue relatively easily tractable as an extension from the isovolume/incompressible regime. As a direct consequence, the density variance can be readily deduced from the temperature or concentration variance, as solution to a modeled transport equation for a *quasi-passive* contaminant.
On the other hand, in compressible turbulent flows, even within the limit of linear Kovasznay's modes decomposition (see Chapter 3), density fluctuations are included in both acoustic ($p' \neq 0$, $\rho' \neq 0$, and $s' = \omega' = 0$) and entropy ($s' \neq 0$, $\rho' \neq 0$, and $\omega' = p' = 0$) modes. Therefore, in compressible boundary layers and channel flows, for instance, the density variance is not a good parameter to describe compressibility effects accurately.

weighted averages viz., respectively:

$$F = \overline{F} + f' = \widetilde{F} + f'', \qquad \text{where} \qquad \widetilde{F} = \overline{\rho F}/\overline{\rho}$$

and noticing that: $\overline{f'} = 0\,, \quad \widetilde{f''} = 0\,, \quad \widetilde{f'} \neq 0\,, \quad \overline{f''} \neq 0\,,$

and $\qquad \overline{F} - \widetilde{F} \equiv f'' - f' = -\overline{\rho' f''}/\overline{\rho} = -\overline{\rho' f'}/\overline{\rho} = -\overline{\rho f'}/\overline{\rho}\,.$

Adopting density-weighted or Favre's averages[5], and considering a binary mixture of perfect gases (see Chapter 5), the equations governing the mean motion are:

$$\textit{Mean continuity}: \quad \frac{\partial \overline{\rho}}{\partial t} + \frac{\partial(\overline{\rho}\widetilde{U}_k)}{\partial x_k} = 0\,, \tag{10.1}$$

$$\textit{Mean momentum}: \quad \frac{\partial(\overline{\rho}\widetilde{U}_i)}{\partial t} + \frac{\partial(\overline{\rho}\widetilde{U}_i\widetilde{U}_j)}{\partial x_j} = \overline{\rho}\widetilde{F}_i - \frac{\partial \overline{P}}{\partial x_i} - \frac{\partial(\overline{\rho}\,\widetilde{u''_i u''_j})}{\partial x_j} + \frac{\partial \overline{\tau}_{ij}}{\partial x_j}\,, \tag{10.2}$$

$$\textit{Mean internal energy}: \quad \frac{\partial(\overline{\rho}\,\widetilde{e})}{\partial t} + \frac{\partial(\overline{\rho}\,\widetilde{e}\,\widetilde{U}_j)}{\partial x_j} = \overline{P}\frac{\partial \widetilde{U}_i}{\partial x_i} - \overline{P\frac{\partial u''_i}{\partial x_i}} + \overline{\tau}_{ij}\frac{\partial \widetilde{U}_i}{\partial x_j} + \overline{\tau_{ij}\frac{\partial u''_i}{\partial x_i}} + \frac{\partial \overline{q}_i}{\partial x_i} - \frac{\partial(\overline{\rho}\,\widetilde{e'' u''_j})}{\partial x_j}\,, \tag{10.3}$$

$$\textit{Mean mass fraction}: \quad \frac{\partial(\overline{\rho}\widetilde{C})}{\partial t} + \frac{\partial(\overline{\rho}\widetilde{C}\widetilde{U}_j)}{\partial x_j} = -\frac{\partial(\overline{\rho}\,\widetilde{\gamma'' u''_j})}{\partial x_j} + \frac{\partial \overline{q}_{m_j}}{\partial x_j}\,. \tag{10.4}$$

In the previous equations, $\tau_{ij}$ stands for the viscous stress tensor, $q_j$ and $q_{m_j}$ for the molecular heat and mass fluxes respectively.

With respect to the previous equations, the closure issue is concerned with two types of terms:

[5]The equations governing the *mean* properties of variable density fluid motions have been derived in Chapter 5. As compared with the incompressible, constant density situation, new non-linearities due to density variations introduce additional correlations. These correlations can be handled in different ways. Two of them have been more specifically discussed, with reference to binary regrouping (density weighted or Favre's averaging) and ternary regrouping (conventional or Reynolds averaging). The two formulations differ by the way the extra correlation terms due to density variation, the so called d.f.c. (density fluctuation correlations), are handled. This formal distinction may have a direct incidence on the expression of the closure schemes.
Favre averaged equations bear great *formal* resemblance to those governing incompressible flows, since most d.f.c.'s are explicitly "eliminated" from the formulation, but are implicitly present in mean mass-weighted operators.

– (i) turbulent diffusion (or transport) terms, viz. $\overline{\rho}\,\widetilde{\alpha'' u_j''} \equiv \overline{\rho \alpha'' u_j''}$;
– (ii) correlations with the Favrian velocity fluctuation $u_i''$.

The modeling of the former can be sought from formal extensions of the corresponding constant density fluxes, where, for instance, the Reynolds stress tensor $\rho_0 \overline{u_i' u_j'}$ is simply changed to $\overline{\rho u_i' u_j'}$ or $\overline{\rho u_i'' u_j''}$.
The latter introduce new *dilatational* or *compressible* contributions that are specific to variable density fluid motions, as we shall see now.

10.4.1. PRESSURE-DILATATION CORRELATION

In eq.(10.3), the coupling term between the instantaneous pressure and the Favrian dilatation fluctuation reads:

$$\overline{P \frac{\partial u_i''}{\partial x_i}} = \overline{P}\frac{\partial \overline{u_i''}}{\partial x_i} + \overline{p' \frac{\partial u_i''}{\partial x_i}} \, .$$

The last term in the right-hand-side introduces the pressure-dilatation correlation $\Pi_d$, which can be equally written as (see Chapters 5 and 6):

$$\Pi_d \equiv \overline{p'\vartheta'} = \overline{p' \frac{\partial u_i''}{\partial x_i}} = \overline{p' \frac{\partial u_i'}{\partial x_i}} \qquad (10.5)$$

This term is specific to non-isovolume *fluctuating* motions.

10.4.2. DILATATION DISSIPATION

Similarly, the coupling term between the viscous stress and the Favrian dilatation fluctuation in eq.(10.3) reads:

$$\overline{\tau_{ij} \frac{\partial u_i''}{\partial x_i}} = \overline{\tau}_{ij} \frac{\partial \overline{u_i''}}{\partial x_i} + \overline{\tau_{ij}' \frac{\partial u_i''}{\partial x_i}} \, ,$$

where $\tau_{ij}'$ denotes the fluctuation *centered* on the mean value $\overline{\tau}_{ij}$.
Now, as shown in Chapter 6-§6.4.4, the last term in the right-hand-side expression can be rewritten as (neglecting viscosity fluctuations):

$$\overline{\rho}\,\overline{\epsilon} \equiv \overline{\tau_{ij}' \frac{\partial u_i''}{\partial x_i}} = \overline{\rho}\,\overline{\epsilon}_s + \overline{\rho}\,\overline{\epsilon}_d + \overline{\rho}\,\overline{\epsilon}_{nh} \, . \qquad (10.6)$$

Here, $\overline{\epsilon}_s$, $\overline{\epsilon}_d$ and $\overline{\epsilon}_{nh}$ denote respectively the solenoidal, dilatational and non-homogeneous contributions to the dissipation rate:

$$\overline{\epsilon}_s = 2\frac{\overline{\mu}}{\overline{\rho}}\overline{\omega_{ij}'\omega_{ij}'} \, , \quad \overline{\epsilon}_d = \frac{4}{3}\frac{\overline{\mu}}{\overline{\rho}}\overline{\vartheta'^2} \, , \quad \overline{\epsilon}_{nh} = 2\frac{\overline{\mu}}{\overline{\rho}}\left(\frac{\partial^2(\overline{u_i' u_j'})}{\partial x_i \partial x_j} - 2\frac{\partial(\overline{\vartheta' u_j'})}{\partial x_j}\right) \, , \qquad (10.7)$$

where $\omega'_{ij} = (\partial u'_i/\partial x_j - \partial u'_j/\partial x_i)/2$.

### 10.4.3. THE CLOSURE PROBLEM

To summarize, the closure problem of the open set of equations governing the mean motion of a variable density fluid requires to derive closure schemes for different types of terms which:

- are formally analogous to those considered in constant density flows, e.g., the "diffusion" terms, say $\overline{\rho}\,\widetilde{\alpha'' u''_j}$;
- directly originate from density fluctuations, such as the turbulent mass flux $\overline{u''_i}$;
- are specific to density fluctuations, namely $\Pi_d$ and $\overline{\epsilon}_d$.

Some closure schemes for each type of terms are separately reviewed in the following sections.

## 10.5. Turbulent momentum transport modeling

### 10.5.1. EDDY-VISCOSITY REPRESENTATION OF REYNOLDS STRESSES

A widely used representation of the Reynolds stresses in incompressible flows is based on the so-called Boussinesq scheme, introducing an eddy or turbulent viscosity $\mu_t$:

$$-\rho_0\overline{u'_i u'_j} + \frac{2}{3}\rho_0\overline{k}\delta_{ij} = 2\mu_t\overline{S}_{ij} \equiv \mu_t\left(\frac{\partial \overline{U}_i}{\partial x_j} + \frac{\partial \overline{U}_j}{\partial x_i}\right), \tag{10.8}$$

or

$$-\rho_0\, b_{ij} = 2\mu_t\overline{k}\,\overline{S}_{ij}\,, \tag{10.9}$$

where $b_{ij} = (\overline{u'_i u'_j}/\overline{k}) - \frac{2}{3}\delta_{ij}$ is the Reynolds stress anisotropy tensor. $\rho_0$ denotes the *constant* density of the *incompressible* fluid in an *isovolume* motion and $\overline{k} \equiv \frac{1}{2}\overline{u'_i u'_i}$ is the turbulence kinetic energy.

Apart from the general validity of such a linear relation (see §10.5.4), transposition of eq.(10.8) in variable density flows raises at least three specific questions:

- What accounts for the "flux", i.e., the compressible equivalent to the incompressible Reynolds stress tensor?
- What accounts for the "forces" driving the flux, i.e., the mean strain rate tensor in isovolume/incompressible flows?
- Are the "incompressible" eddy-viscosity expressions acceptable in variable density fluid situations?

Some answers to these questions will be now examined.

*Direct transpositions*
A straightforward transposition of eq.(10.8) to variable density flows yields the following expressions, based on Reynolds and Favre decompositions respectively:

$$-\overline{\rho}\,\overline{u_i'u_j'} + \frac{2}{3}\,\overline{\rho}\,\overline{k}\delta_{ij} = 2\mu_t(\overline{S}_{ij} - \frac{1}{3}\frac{\partial \overline{U}_l}{\partial x_l}\delta_{ij})\,, \tag{10.10}$$

and

$$-\overline{\rho}\,\widetilde{u_i''u_j''} + \frac{2}{3}\,\overline{\rho}\,\widetilde{k}\delta_{ij} = 2\mu_t(\widetilde{S}_{ij} - \frac{1}{3}\frac{\partial \widetilde{U}_l}{\partial x_l}\delta_{ij})\,, \tag{10.11}$$

where $\overline{k} = \frac{1}{2}\overline{u_l'u_l'}$ and $\widetilde{k} = \frac{1}{2}\widetilde{u_l''u_l''}$ in eqs.(10.10) and (10.11), respectively. The previous relations are not equivalent, as it can be seen by converting Favre averages into Reynolds averages, according to the simple analytical relationship given in Chapter 5 (see Table 5.2). When Favre averages are used, eq.(10.11) is the most widely adopted expression, beginning with Jones and Launder [238], Jones [237], in 1979 and Ha Minh *et al.* [195] in 1981.
Nevertheless, the following alternate formulations have been proposed.

**Wilcox-Rubesin (1980).** A more complex relation was introduced in 1980 by Wilcox and Rubesin [485] as part of a ($k$-$\omega^2$) model, where $\omega = \epsilon/k$ is the dissipation rate per unit kinetic energy. It reads

$$-\overline{\rho}\,\widetilde{u_i''u_j''} + \frac{2}{3}\,\overline{\rho}\,\widetilde{k}\delta_{ij} = 2\mu_T(\widetilde{S}_{ij} - \frac{1}{3}\widetilde{S}\delta_{ij}) + \frac{8}{9}\widetilde{k}\frac{\widetilde{S}_{im}\widetilde{R}_{mj} + \widetilde{S}_{jm}\widetilde{R}_{mi}}{\beta^*\omega^2 + 2\widetilde{S}_{nm}\widetilde{S}_{nm}}\,, \tag{10.12}$$

where $\beta^* = 0.09$ is a model coefficient.

**Dussauge-Quine (1988).** Dussauge and Quine [141] used the following eddy-viscosity in a ($k$ - $\epsilon$) model for the turbulent shear stress

$$-\overline{\rho}\widetilde{u''v''} = \frac{C_\mu}{[1 - \frac{3}{2}\beta C(2)\, M^2]^2}\,\overline{\rho}\frac{\widetilde{k}^2}{\overline{\epsilon}}\,,$$

with $C_\mu = 0.09$, $C(2) = 1.5$. $M$ denotes the local Mach number, and

$$\beta = \frac{\alpha(\gamma - 1)}{C_1 - 1 + P_{rod}/\epsilon}\,,$$

where $\gamma = C_p/C_v = 1.4$ for air, and $C_1 = 1.5$. $P_{rod}$ denotes the production rate in the turbulence kinetic energy equation. Several values of the last model coefficient have been tested, ranging between $-1.35$ and $-0.8$.
Applied to the calculation of a supersonic shear layer, the model gives

encouraging results, since it predicts a decrease of the spreading rate as the Mach number increases, in contrast with the usual ($k$ - $\epsilon$) model which appears practically insensitive to such an effect.

**Taulbee-VanOsdol (1991).** Extra terms due to turbulent fluxes are included in the model adopted by Taulbee and VanOsdol [454] since

$$-\overline{\rho}\,\widetilde{u_i''u_j''} + \frac{2}{3}\,\overline{\rho}\,\widetilde{k}\delta_{ij} = \mu_t\left(\frac{\partial \widetilde{U}_i}{\partial x_j} + \frac{\partial \widetilde{U}_j}{\partial x_i} - \frac{2}{3}\frac{\partial \widetilde{U}_l}{\partial x_l}\delta_{ij}\right) - \overline{\rho}\,\overline{u_i''}\,\overline{u_j''} + \frac{1}{3}\overline{\rho}\,\overline{u_l''}\,\overline{u_l''}\delta_{ij}\,. \tag{10.13}$$

This scheme was introduced in a (k-$\epsilon$-$\overline{\rho'^2}$-$\overline{u''}_i$) model to predict the adiabatic-wall boundary layer and the compressible free shear layer. As noticed by the authors, "the extra terms appearing in the equation are probably not too important" in the first situation. Indeed, according to Kistler's data, $\overline{u''}/\overline{U}$ in this case is about $-0.0011$ and $-0.0028$ for $M_\infty = 1.7$ and $4.7$, respectively.

### 10.5.2. EDDY-VISCOSITY EXPRESSIONS

In contrast with the molecular viscosity which can be considered as a macroscopic physical property of the *fluid*, the eddy-viscosity is tightly linked with the properties of the *flow* field. Hence, the next step in the modeling procedure based on such a concept is to prescribe the eddy-viscosity, and more generally, all turbulence transport coefficients, such as the thermal and mass diffusivities. This can be achieved from different ways, associated with:

- *zero-equation* models, based on algebraic expressions of the eddy-viscosity;
- *one-equation* models, in which the eddy-viscosity is directly derived from a modeled transport equation, or deduced from another transportable quantity;
- *two-equation* models, where two distinct characteristics are introduced as transportable functions governing eddy-viscosity variations.

***Mixing-length models.*** When restricted to two-dimensional boundary layers flows, the eddy-viscosity is obtained from a *mixing length*, as given by:

$$\overline{\mu}_t = \overline{\rho}\,\ell_m^2 \left|\frac{\partial \overline{U}}{\partial y}\right| \,. \tag{10.14}$$

The *mixing length* $\ell_m$ is to be prescribed algebraically as a local function of position in the flow field. Some expressions of $\ell_m$ are recalled in section 10.9.

***Turbulence-kinetic-energy models.*** From simple dimensional considerations, the eddy-viscosity can be written as:

$$\overline{\mu}_t = \overline{\rho}\, u_{ref} \times l\,, \tag{10.15}$$

where $u_{ref}$ — resp. $l$ —, is a characteristic velocity scale — resp. length scale —, of the fluctuating turbulent motion.
Adopting $u_{ref} \propto \sqrt{\overline{k}}$ and $l \propto \ell_m$, we reduce eq.(10.15) to

$$\overline{\mu}_t = C\overline{\rho}\,\sqrt{\overline{k}}\,\ell_m\,, \tag{10.16}$$

where $C$ stands for a nondimensional constant.
To be adopted as a closure scheme, eq.(10.16) requires that the local amount of turbulence kinetic energy is known. This can be achieved by solving a model transport equation for this function, introducing the one-equation ($\overline{k}$) turbulence model. Closure schemes involved in the modeling of this equation in variable density situation are discussed in section 10.10.2.

***Two-equation models.*** The next step in prescribing the eddy-viscosity from eq.(10.15) consists in solving a second modeled transport equation for the characteristic length scale $l$, producing a two-equation ($k$- $l$) turbulence model. In actual fact, this idea can be generalized to obtain the eddy-viscosity from any dimensionnally correct group of two independent functions, based on any combination of suitable powers of characteristic velocity and length scales. Formally, this can be written as:

$$\overline{\mu}_t \propto \overline{\rho}\,\overline{k}^m \zeta^n\,. \tag{10.17}$$

Some examples of the additional transportable function $\zeta$, which has been used in predicting variable density fluid motions are given in table 10.1. Additional references can be found in Wilcox [484], pages 459 & 460.

Since 1975, the constant density version of the two-equation ($k$ - $\epsilon$) model has become one of the most popular in this category and is widely adopted in predicting turbulent flows for industrial applications. It has also been extended to predicting variable density and compressible fluid motions and will be detailed in section 10.11.1.

### 10.5.3. COMPRESSIBILITY EFFECTS ON THE EDDY-VISCOSITY CONCEPT

In most proposals to extending eq.(10.15) to variable density and/or compressible situations, Batchelor's scaling $\overline{\epsilon} \propto \overline{k}^{3/2}/\ell$, is adopted for the dissipation rate. Thus, it is assumed that only *one* characteristic length scale

TABLE 10.1. Some examples of two-equation ($k - \zeta$) models

| Author - Ref. | Year | Additional function $\zeta$ |
|---|---|---|
| Kolmogorov [256] | 1942 | |
| Wilcox [483] | 1988 | $\overline{k}^{1/2} / l (\equiv \omega)$ |
| Saffman [403] | 1970 | |
| Spalding [441] | 1972 | $\overline{k}/l^2 \ (\equiv \omega^2)$ |
| Wilcox & Traci [486] | 1976 | |
| Davidov [123] | 1961 | |
| Harlow & Nakayama [209] | 1967 | $\overline{k}^{3/2}/l \ (\equiv \overline{\epsilon})$ |
| Jones & Launder [238] | 1972 | |
| Cousteix, Saint-Martin, Messing, Bézard, Aupoix [108] | 1997 | $\overline{\epsilon}/\overline{k}^{1/2} \ (\equiv \varphi)$ |

$l \propto \ell$ accounts for *both* turbulent transport, say $l$, and energy transfer by fluctuating motions, say $\ell$. In high turbulence Reynolds number flows, it results, for instance, that:

$$\nu_t = C_\mu \frac{\overline{k}^2}{\overline{\epsilon}} \,. \tag{10.18}$$

Directly adopting eq.(10.18) in variable density and/or compressible fluid motions is questionable for, at least, the following reasons:

a) A possible consequence of compressibility effects is that the model coefficient $C_\mu$ (0.09, in incompressible flows) could change with the turbulence Mach number $M_t = \sqrt{\overline{k}}/c$. Such modifications have been actually proposed, as reviewed by Nichols [346].

b) As recalled by Ristorcelli [394], even in constant density, high Reynolds number flows, the ratio $l/\ell$ slightly depends on the type of flow, so that the coefficient in eq.(10.18), should be flow dependent.

c) In the context of compressible turbulent flows and even restricting Batchelor's scaling to the solenoidal dissipation, the choice of a *single* characteristic length scale is even more questionable. This can be seen from different results obtained, for instance, in compressed turbulence (Zeman & Coleman [499]), in shock-turbulence interaction (Jamme [231]) or in a compressible mixing layer (Freund *et al.* [163]).

For the latter situation, the characteristic length scale of the energy containing eddies $\ell$ exhibits a dependence on the convective Mach number which is different from the transverse large-eddy length scale $l$. It seems then plausible that compressible mean flow interacts directly with the large eddies which govern the production of turbulence kinetic energy and the expansion rate. On the other hand, turbulent dissipation occurs mainly at high frequency, i.e., involves small-scale fluctuations that are less influenced by the mean flow and associated compressibility effects. Thus, a possible way to handle this specific action of compressibility could consist in adopting a multiple-scale model, as proposed by Liou *et al.* [300] in 1995.

In 1997, using such a two-scale method, Yoshizawa et al. [492] reached the conclusion that compressibility effects on Boussinesq's scheme were two-fold. One consequence should be a modified expression of the eddy-viscosity, the other comes from the deviation of the Reynolds stress from a turbulent viscosity representation, which is written using the *Lagrange* derivative of the mean velocity and the *spatial* derivatives of the mean density and internal energy. The Reynolds stress representation proposed by these authors reads:

$$\overline{u_i'u_j'} - \frac{2}{3}\overline{k}\delta_{ij} = -2\nu_{t_C}(\overline{S}_{ij} - \frac{1}{3}S_{nn}\delta_{ij}) \ + \ \text{non linear correction} \\ + \ \text{mean density gradient contribution} \ + \ \text{mean internal energy contribution}\,, \tag{10.19}$$

where $\overline{S}_{ij} = (\partial\overline{U}_j/\partial x_i + \partial\overline{U}_i/\partial x_j)/2$.
The contributions due to mean density and internal-energy gradients are supposed to increase near a shock, where these gradients become large. The non-linear correction includes $\overline{S}_{ij}$-$\overline{S}_{ij}$ and $\overline{S}_{ij}$-$\overline{R}_{ij}$ parts (see next section). The compressible correction to the eddy-viscosity $\nu_{t_C}$ is written as follows:

$$\nu_{t_C} = \frac{1}{[1 + A(I_\rho/M_t)]^{3/2}}\,\nu_t\,, \tag{10.20}$$

where $\nu_t = C_\mu\overline{k}^2/\overline{\epsilon}$ is the usual expression for the incompressible eddy-viscosity, with $C_\mu = 0.09$. Hence, from eq.(10.20), compressibility effects result from the influence of the turbulence density intensity $I_\rho = \sqrt{\overline{\rho'^2}}/\overline{\rho}$ and the turbulence Mach number $M_t$. Two values of the model coefficient have been investigated ($A = 5$ and $A = 10$). The differences observed with these values on the prediction of the spreading rate of a compressible shear layer cannot be significantly corroborated, due to the scattering of the experimental data.

### 10.5.4. NON-LINEAR CONSTITUTIVE RELATIONSHIP

Even in incompressible flows, a linear expression such as eq.(10.8) has no firm basis, neither mathematically nor physically. In particular, it results from eq.(10.9) that the principal axes of the Reynolds stresses are systematically aligned with those of the mean rate of strain tensor, a point which is in disagreement with experimental evidence in homogeneous shear flows, for instance. From a theoretical point of view, such linear expressions can be considered as crude approximations to a more general constitutive expression of the turbulent stresses, see, for instance, Lumley [304].

As a first step toward non-linear eddy-viscosity models in incompressible fluid motions, the following second-order expansion, Lumley (1970) [304], Saffman (1976) [404] can be derived

$$-\overline{u_i u}_j + \frac{2}{3}\overline{k}\,\delta_{ij} = 2\nu_t \overline{S}_{ij} - D\frac{\overline{k}^3}{\overline{\epsilon}^2}\left(\overline{S}_{ik}\overline{R}_{kj} + \overline{S}_{jk}\overline{R}_{ki}\right)$$

where
$$\overline{R}_{kj} = \frac{1}{2}\left(\frac{\partial \overline{U}_k}{\partial x_j} - \frac{\partial \overline{U}_j}{\partial x_k}\right)$$

and one additional constant $D$ is introduced.
This relation was adapted in 1980 by Wilcox and Rubesin [485] as part of a $(k - \omega^2)$ model for compressible flows, yielding:

$$-\overline{\rho}\,\widetilde{u_i'' u_j''} + \frac{2}{3}\,\overline{\rho}\,\widetilde{k}\delta_{ij} = 2\mu_t(\widetilde{S}_{ij} - \frac{1}{3}\frac{\partial \widetilde{U}_l}{\partial x_l}\delta_{ij}) + \frac{8}{9}\widetilde{k}\frac{\widetilde{S}_{im}\widetilde{R}_{mj} + \widetilde{S}_{jm}\widetilde{R}_{mi}}{\beta^*\omega^2 + 2\widetilde{S}_{nm}\widetilde{S}_{nm}},$$

where $\beta^* = 0.09$ is a model coefficient.
With this scheme, the model improves the prediction of the anisotropy of the normal stresses in the logarithmic region of a compressible boundary layer, with a fair restitution of pressure gradient effects.

The promising idea behind the derivation of more general non-linear eddy-viscosity models in incompressible fluid motions is the expected capability to capture turbulence effects that are directly linked with Reynolds stress anisotropy in complex situations, such as secondary flows in ducts or two-component behaviour of turbulence near a solid surface. It is hoped that predictions based on such types of closure could be as good as those obtained with second-order models and cheaper, in terms of computing resources. Several non-linear constitutive relations have been proposed. The first one, by Pope [369] in 1975, has been extended to three dimensional flows and non-inertial reference frames by Gatski & Speziale [175], in 1993. Various other formulations have been proposed by Shih *et al.* [426] in 1993, Craft *et al.* [111] in 1996, Wang [478] in 1997, in addition to proposals derived from the renormalization group theory, see, for instance, Rubinstein & Barton [400]. They are briefly reviewed in Chassaing [84].

A comparative study of various linear and non linear eddy-viscosity schemes in two- and three-equation models was published in 2000 by Barakos and Drikakis [29]. These authors aim at assessing the capability and accuracy of the models in predicting transonic flows featuring shock/boundary-layer interaction and separation. The general conclusion is balanced: as compared with those obtained from *linear* eddy-viscosity models, the predictions are improved when using *non-linear* models. However the computed results are still far from the experimental data. For the flow over a bump for example (ONERA Case C, as provided by Délery at the 1980-81 Stanford Conference), the maximum of the turbulent shear stress profile is not predicted at the same location as that given by the measurements, and its value is underpredicted by about 40% to 50%.

Since it has not yet been demonstrated that compressible anisotropy effects can be accounted for by formal transposition of non-linear incompressible expressions, and owing to the fact that, even in incompressible flows, the implementation of some non-linear eddy-viscosity formulations may become cumbersome, this type of closure will not be discussed in more detail here.

## 10.6. Turbulent heat/mass transport modeling

### 10.6.1. GRADIENT DIFFUSION ASSUMPTION

For heat and mass turbulent fluxes, the counterpart of the gradient diffusion assumption for turbulent momentum transport are:

$$-\overline{\rho}\widetilde{f''u''_j} = \overline{\rho}\,\Gamma_t \frac{\partial \widetilde{F}}{\partial x_j} \quad \text{or} \quad -\overline{\rho f'u'_j} = \overline{\rho}\,\Gamma_t \frac{\partial \overline{F}}{\partial x_j}\,, \tag{10.21}$$

where $f'$ (resp. $f''$) denotes temperature or mass-fraction fluctuations with respect to conventional (resp. density-weighted) averages.
Eq.(10.21), can be viewed as a direct transposition of Fourier's and Fick's laws for *molecular* heat and mass *transfer* to *turbulent transport*, where $\Gamma_t(L^2T^{-1})$ stands for a heat or mass *turbulence* diffusivity. Hence, the ratio of the thermal ($\mathrm{a}_t$) and mass ($\mathcal{D}_t$) turbulence diffusivity to the turbulence kinematic eddy-viscosity $\nu_t$ results in turbulence Prandtl and Schmidt numbers:

$$\sigma_t = \frac{\nu_t}{\mathrm{a}_t} = \frac{\nu_t}{\mathcal{D}_t}\,. \tag{10.22}$$

In addition to the validity of the gradient diffusion assumption, eq.(10.22) raises the question of the value of the turbulence Prandtl-Schmidt numbers in compressible and variable density flows. As reported by Cousteix & Aupoix [107], no systematic effects of Mach number in boundary layers flows have been observed on the turbulence Prandtl number, which is rather flow-dependent: 0.8∼0.9 (flat plate boundary layer), 0.7 (jet) and 0.5 (wake).

### 10.6.2. ALTERNATIVE TO GRADIENT DIFFUSION SCHEMES

Even in incompressible fluid motions, there exists several classes of turbulent flows where simple eddy-viscosity and gradient diffusion schemes are not appropriate. The reason can be found from at least two arguments, in terms of (i) time scale ratio and (ii) level of anisotropy, Pope [370].

- (i) As pointed out by Corrsin [103], the gradient diffusion scheme requires that the time behaviour of the diffusing motions should be locally and instantaneously dominated by the time scale imposed by the mean velocity gradient, say $(\partial \overline{U}/\partial y)^{-1}$ in a thin shear layer. This can be presumably achieved by the small-scale turbulent eddies but not by the large scale turbulent motions;
- (ii) A linear relationship between stress and strain tensors can be mathematically derived when assuming isotropic conditions. Hence, the deviatoric part of the Reynolds stress tensor cannot be determined from the local mean strain rates in flows where high anisotropy levels are developed, due to curvature, secondary flows or swirl, for instance.

Examples of local and non-local formulations, accounting for curvature and large scale structure effects (intermittency, convective transport) are reviewed in Chassaing [84].

Some of the previous flaws can be obviously transposed to variable density flows. In low Mach number flows, a severe limitation of such simple gradient-type approximations, due to large scale motions, can be expected in fire flows where very large density gradients are present.
In 1999, Shimomura [428] derived a new model for thermally driven turbulent flow, by applying a two-scale direct-interaction approximation to the Rehm-Baum [379] equations governing a low Mach number flow (see Chapter 3). The model is theoretically deduced by considering the fluctuating fields without perturbing the buoyant term and assuming an isobaric evolution for the thermal pressure ($\rho T = C^{te}$), both in space and time. The result, to be used as a turbulence closure in a first-order model, is as follows:

$$-\overline{u_i' u_j'} = \nu_t (\frac{\partial \overline{U}_i}{\partial x_j} + \frac{\partial \overline{U}_j}{\partial x_i}) - \frac{2}{3}(\overline{k} + \nu_t \frac{\partial \overline{U}_l}{\partial x_l})\, \delta_{ij}\,, \tag{10.23}$$

$$-\overline{\theta' u_j'} = \mathrm{a_t}(\frac{\partial \overline{\mathrm{T}}}{\partial \mathrm{x_j}}) + \eta_0 \frac{\partial \overline{\mathrm{P}}}{\partial \mathrm{x_j}}\,, \tag{10.24}$$

with:

$$\nu_t = 0.12 \frac{\overline{k}^2}{\overline{\epsilon}}, \ \mathrm{a_t} = 0.14 \frac{\overline{\mathrm{k}}^2}{\overline{\epsilon}}(1 + 0.27 \frac{\overline{\mathrm{k}}}{\overline{\epsilon}} \frac{\mathrm{Q_s}}{\rho_0 \mathrm{C_P T_0}}) \text{ and } \eta_0 = 0.32 \frac{1}{\mathrm{T_0}} \frac{\overline{\theta'^2}}{\overline{\epsilon}_\theta}\,.$$

Here, $Q_s$ is the heat source and $\overline{\epsilon}_\theta = 2\lambda \overline{(\partial \theta'/\partial x_i)^2}$ the turbulent dissipation rate of the thermal variance $\overline{\theta'^2}$.

As compared with the usual gradient type expressions — eqs.(10.10) and (10.11) —, it appears that the main modification concerns the turbulent heat flux, due to the presence in eq.(10.24) of the mean pressure gradient term. Also noticeable is the change in the turbulence thermal diffusivity $a_t$, as a function of the heat source when $Q_s \neq 0$. Without external heat release, the equivalent turbulence Prandtl number is about 0.86.
These closure schemes were applied to the prediction of the natural turbulent convection along a heated vertical plate. The model prediction of the longitudinal heat flux profile $\overline{u'\theta'}$ is in better agreement with the experimental data.

## 10.7. Modeling d.f.c terms

According to the classification introduced in §10.4.3, we turn now to the second point to be addressed in first-order closure of variable density turbulent flows, viz. the modeling of first-order d.f.c. terms $\overline{\rho' f'} \equiv \overline{\rho f'} = -\overline{\rho}\,\overline{f''}$. Particular attention is given to the turbulent mass flux $\overline{u_i''}$ in relation with Favre's averaged formulation of the mean motion and turbulence kinetic energy equations.

*Linearized approximations*
Approximate expressions of the turbulent mass flux $\overline{\rho u_i'}$ can be easily obtained when considering *separately* one of the following three situations:

(a) perfect compressible gas (homogeneous composition);
(b) isobaric evolution (homogeneous composition);
(c) isothermal, ideal mixing of two non-reactive species.

The corresponding equations of state are respectively:

$$(a)\ \frac{1}{\rho} = R\frac{T}{P}, \qquad (b)\ P = C^t \times \rho^n, \qquad (c)\ \frac{1}{\rho} = \frac{C}{\rho_1} + \frac{1-C}{\rho_2}. \qquad (10.25)$$

In eq.10.25($b$), $n$ is the polytropic coefficient ($n = 0$: isobaric evolution, $n = 1$: isothermal evolution, $n = C_p/C_v$: isentropic evolution).
In eq.10.25($c$), $C$ denotes the mass fraction of one of the two species and $\rho_1$ and $\rho_2$ are the densities of each pure species.
First-order approximate expressions of density fluctuations can be deduced from the previous expressions, assuming small fluctuations, viz., $\rho'/\overline{\rho} \ll 1$, $p'/\overline{P} \ll 1$, $\theta'/\overline{\theta} \ll 1$ and $\gamma'/\overline{\gamma} \ll 1$, where $\rho'$, $p'$, $\theta'$ and $\gamma'$ are the fluctuations of density, pressure, temperature, and mass-fraction corresponding to the Reynolds averages $\overline{\rho}$, $\overline{P}$, $\overline{T}$ and $\overline{C}$, respectively. They read:

$$(a)\ \frac{\rho'}{\overline{\rho}} \simeq \frac{p'}{\overline{P}} - \frac{\theta'}{\overline{T}} \qquad (b)\ n\frac{\rho'}{\overline{\rho}} \simeq \frac{p'}{\overline{P}} \qquad (c)\ \frac{\rho'}{\overline{\rho}^2} \simeq \frac{\rho_1 - \rho_2}{\rho_1\rho_2}\gamma' \quad (10.26)$$

Thus the turbulent mass fluxes are linearly linked with pressure-velocity, temperature-velocity and mass-fraction-velocity correlations:

$$(a)\,\frac{\overline{\rho' u_i'}}{\overline{\rho}} = \frac{\overline{p' u_i'}}{\overline{P}} - \frac{\overline{\theta' u_i'}}{\overline{T}}, \quad (b)\, n\frac{\overline{\rho' u_i'}}{\overline{\rho}} = \frac{\overline{p' u_i'}}{\overline{P}}, \quad (c)\,\frac{\overline{\rho' u_i'}}{\overline{\rho}} = \frac{\rho_1 - \rho_2}{\rho_1 \rho_2}\overline{\rho}\,\overline{\gamma' u_i'}\,. \quad (10.27)$$

Hence, for a non isothermal evolution ($n \neq 1$) of a perfect gas, it can be deduced that:

$$\frac{\overline{\rho' u_i'}}{\overline{\rho}} = \frac{1}{n-1}\frac{\overline{\theta' u_i'}}{\overline{T}}\,. \qquad (10.28)$$

The approximate expression 10.27($c$) was used in 1987 by Shih *et al.* [427] in second-order modeling of a variable density mixing layer.
In compressible flows, and with the aid of additional assumptions, eq.(10.28) can be used to get the expected scheme, as first proposed by Rubesin.

**Rubesin (1976).** The simplification introduced by Rubesin [397] in 1976, consists in assuming that the constant total-temperature condition ($C_pT + U_jU_j/2 = C^t$) also applies to the turbulent field. Hence, to first-order, one can assume that $C_p\theta' + \overline{U}_i u_i' \simeq 0$, from which, in *2-D* thin shear layer flows ($\overline{U}_1 (\equiv \overline{U}) >> \overline{U}_2$), it results that

$$\frac{\overline{\rho' u'}}{\overline{\rho}} \simeq -\frac{\overline{U}}{(n-1)C_p\overline{T}}\overline{u'^2} \simeq -\frac{\overline{U}}{(n-1)C_p\overline{T}}\overline{k}\,. \qquad (10.29)$$

This expression, with with $n = 1.2$ as recommended by Rubesin, leads to a systematically negative mass flux. This is not the case with eq.(10.28).

**Galmes-Dussauge-Dekeyser (1983).** In a modified version of the Jones-Launder ($k$ - $\epsilon$) model, Galmes *et al.* [174] suggested to account for compressibility effects in the prediction of supersonic boundary layers with non-zero pressure gradient, by including the mean pressure term $\overline{u''}(\partial\overline{P}/\partial x)$. The closure scheme adopted for the turbulent mass flux $\overline{\rho' u'}/\overline{\rho}$ is based on the strong Reynolds analogy[6]. For a *2-D* boundary layer, it reads:

$$\frac{\overline{\rho' u'}}{\overline{\rho}} = R_{\rho u}(\gamma - 1)M^2\overline{\rho}\frac{\widetilde{u''^2}}{\widetilde{U}},$$

where $M$ is the Mach number. The value adopted for the correlation coefficient between density and velocity fluctuations $R_{\rho u}$ is 0.8.
It can be noticed that the previous expression is not galilean invariant. Finally, it is to be added that in [174], the eddy-viscosity is also modified by

[6]The model is only valid in boundary layers over adiabatic walls, not too far from equilibrium.

the previous compressibility term, as mentioned in §10.5.3. The expression is:

$$\mu_t = C_\mu \frac{\overline{\rho}\widetilde{k}^2}{\overline{\epsilon}(1 - A_\rho/\overline{\epsilon})} \, .$$

**Vandromme-Ha Minh (1985).** A direct transposition of eq.(10.29) using Favre's averages was extensively adopted by Ha Minh and Vandromme [332], [471] to predict various types of compressible flows, including shock-boundary layer interaction. It reads:

$$\overline{u_i''} = \frac{\widetilde{U}_j}{(n-1)C_p\widetilde{T}} \widetilde{u_i''u_j''} \, . \tag{10.30}$$

**Dussauge-Quine (1988).** A slightly different expression was used by Dussauge and Quine [141] in the prediction of *2-D* supersonic mixing layers:

$$\overline{\rho' u_i'} = C(i)(\gamma - 1)M^2 \overline{\rho} \frac{\widetilde{u_1''u_i''}}{\widetilde{U}} \, ,$$

where $M$ denotes the local Mach number, and subscript 1 refers to the direction of mean advection. The value of the model constant $C(i)$ is adjusted to the flux component: $C(1) = 0.8$ and $C(2) = 1.5$.

Basically, most of the previous schemes are directly derived from the utilization of the strong Reynolds analogy of Morkovin (see Chapter 4, section 4.4.2), yielding the generic form $\overline{u''} \propto \overline{u''^2}/\widetilde{U}$. According to the DNS data of Zeman [496], such an expression is of wrong sign for the response of an initially isotropic turbulence to a normal shock. It results in gross errors in the turbulence kinetic energy balance (gain, instead of loss of energy), which yields Zeman to "strongly suggest that the application of SRA be avoided in modeling turbulent boundary layers in the presence of shocks."

*Gradient diffusion hypothesis*
As reviewed by Purwanto [376] in 1994, various proposals suggested to model d.f.c. terms $\overline{\rho' u_i'}$, $\overline{\rho'\theta'}$,..., using a generalized gradient diffusion expression of the form:

$$-\overline{\rho' u_i'} = D_t \frac{\partial \overline{\rho}}{\partial x_i} \, , \tag{10.31}$$

where $D_t$ denotes a turbulence diffusivity.
A closure scheme of this type was adopted by Milinazzo & Saffman [331] in 1976 to model the dominant turbulent mass flux $(\overline{\rho' v'})$ in the mean

continuity equation of an inhomogeneous, two-dimensional mixing layer. It reads

$$-\overline{\rho' v'} = \frac{\mu_t}{\sigma_\rho} \frac{1}{\overline{\rho}} \frac{\partial \overline{\rho}}{\partial y} .$$

The value of turbulence Prandtl-Schmidt mass diffusion number is ranging between 0.25 and 1.

Such gradient diffusion schemes were also introduced very early in combustion applications[7] (see Kent and Bilger [245] in 1977 and Janicka and Kollmann [233] in 1979, for instance). This explains why the validity of such gradient diffusion schemes was first questioned in premixed flames. In this case, measurements showed that the local heat release drives the turbulent heat flux in a direction opposite to that predicted by eq.(10.31). In 1982, the measurements in a nonpremixed flame by Driscoll *et al.* [134] exhibit clear evidence of counter-gradient diffusion of density for the axial turbulent flux component, so that the gradient diffusion assumption is seldom obeyed for $\overline{\rho' u'}$. On the contrary, the radial flux $\overline{\rho' v'}$ was found to follow a gradient diffusion relation throughout the flow field.

Some years later, the gradient diffusion relation for turbulent mass flux was also questioned in non-reactive, binary gas jets. In 1989, the results of Zhu *et al.* [503] validated a gradient diffusion relation

$$\overline{\rho' u'} = -D_t \frac{\partial \overline{\rho}}{\partial x} , \qquad (10.32)$$

only for a downstream distance $x/D_0 \geq 6$ in a premixed helium/air round jet, with 50% of helium and $D_0 = 9.5\,\text{mm}$. In the near field of the jet, eq.(10.32) is not valid over a substantial part of the flow, except near the core of the jet. Now considering the radial flux component, the corresponding gradient relation

$$\overline{\rho' v'} = -D_t \frac{\partial \overline{\rho}}{\partial y} , \qquad (10.33)$$

only applies for $x/D_0 \geq 5.1$. But, unlike the axial flux, eq.(10.33) is not valid inside the potential core of the jet, but is true over the rest of the flow field.

[7] In 1975, Lockwood and Naguib [302] proposed a ($k$ - $\epsilon$) model, based on conventional averaged equations, to predict free turbulent diffusion flames in a round jet. Although not *explicitly* associated with the turbulent mass flux, the modeled equation of the turbulence kinetic energy $\overline{k}$ incorporates a density gradient term, to account for "the generation of $\overline{k}$ due to the density fluctuations". In a *2-D*-thin-shear-layer situation, it reads

$$C_{omp} = C_\rho \overline{\gamma'^2} \frac{\mu_t}{\sigma_\rho} \frac{\partial \overline{\rho}}{\partial x} ,$$

where $\gamma'$ is the fluctuation of the mixture fraction, $C_\rho$ a model constant and $\sigma_\rho$ the turbulence Schmidt number.

In 1990, So *et al.* [436] extended the measurements to the self-preserving region of a helium/air, isothermal round jet with the same exit diameter $D_0$. The exit Reynolds number was 4,300 and the jet-to-ambient fluid density ratio 0.64. In this case, mean quantities, turbulent mass fluxes and second-order turbulent correlations achieve self-preservation at a downstream location of about 24 $D_0$. For $13.8 \leq x/D_0 \leq 24.46$, the axial turbulent mass flux $\overline{\rho' u'}$ and the mean density gradients have opposite signs only in the central region of the jet ($r \leq 0.8\delta$, where $\delta$ is the half-velocity width of the jet). Thus, according to the authors, a gradient diffusion relation does not apply in the outer region of the jet, as a consequence of the physics of the mixing which is mainly controlled by large-scale motions in this region. Finally, as shown in Chapter 6, limitations to gradient diffusion formulations can be pointed out analytically in free, isobaric jets.

Gradient diffusion expressions for the turbulent mass flux were also introduced in various first and second-order models to predict compressible flows, Taulbee & VanOsdol [454] in 1991, Sarkar & Lakshmanan [413] in 1991, for instance. In the latter reference, similar gradient diffusion schemes are introduced for both mass and heat turbulent fluxes:

$$\overline{\rho' u'_i} = -\frac{\nu_t}{\sigma_\rho}\frac{\partial \overline{\rho}}{\partial x_i} \quad \text{and} \quad \overline{\theta' u'_i} = -\frac{\nu_t}{\sigma_T}\frac{\partial \overline{T}}{\partial x_i},$$

where $\nu_t = 0.09\overline{k}^2/\overline{\epsilon}$. The values of the turbulent Schmidt and Prandtl numbers are $\sigma_\rho = \sigma_T = 0.7$.
An original contribution was proposed by Rubesin [398] in 1990. Within the *2-D* thin shear layer approximations it reads:

$$\overline{u''} = \frac{\gamma - 1}{n - 1} c_e \frac{\widetilde{k}}{\widetilde{\epsilon}} \frac{1}{\widetilde{a}^2} \frac{\overline{\rho u'' v''}}{\overline{\rho}} \frac{\partial \widetilde{h}}{\partial y}. \tag{10.34}$$

In eq.(10.34), $\gamma$ is the isentropic coefficient, $\widetilde{h}$ and $\widetilde{a}$ are the enthalpy and speed of sound of the mean flow ($\widetilde{a}^2 = (\gamma - 1)\widetilde{h}$), $\widetilde{k}$ and $\widetilde{\epsilon}$ the turbulence kinetic energy and its dissipation rate in a Favrian formulation. The value of the model constant is $c_e$=0.35.
This scheme was developed to improve a previous proposal by Rubesin in 1976 —see eq.(10.29) —, in the prediction of super and hypersonic boundary layers with non-adiabatic conditions. In particular, it can be noticed that in a flow where the Reynolds stress has a given sign, the sign of $\overline{u''}$ is different below or above the point where $\widetilde{h}$ is extremum.

*Transport equation of the turbulent mass flux*
An open transport equation for the turbulent mass flux $\overline{\rho' u'_i} \equiv -\overline{\rho}\,\overline{u''_i}$ can be obtained by manipulation of the equations governing velocity and density

fluctuations (see Chapter 5-§5). Further simplification to the results given in that chapter can be obtained, by assuming, for instance, a linearized form of $1/\rho$ (Liou & Shih [299]):

$$\frac{1}{\rho} = \frac{1}{\overline{\rho}}(1+\frac{\rho'}{\overline{\rho}})^{-1} = \frac{1}{\overline{\rho}}[1-\frac{\rho'}{\overline{\rho}}+(\frac{\rho'}{\overline{\rho}})^2+...] \simeq \frac{1}{\overline{\rho}} - \frac{\rho'}{\overline{\rho}^2} + \mathcal{O}(\frac{\rho'^2}{\overline{\rho}^3}) .$$

**Jones (1979).** Jones [237] was probably the first author to propose a modeled transport equation for the turbulent mass, based on the following general form in high turbulence-Reynolds-number flows:

$$\overline{\rho}(\frac{\partial \overline{u_i''}}{\partial t} + \widetilde{U}_j \frac{\partial \overline{u_i''}}{\partial x_j}) = \text{Production} \ + \ \text{Diffusion} \ - \ \text{Dissipation} . \qquad (10.35)$$

**Zeman (1991).** Dealing with the response of an initially isotropic turbulence to a normal shock, Zeman [496] suggested the following transport equation:

$$\frac{D\overline{\rho' u'_i}}{Dt} \approx -\frac{\overline{\rho' u'_i}}{\tau_a} - \overline{u_i'' u_j''}\frac{\partial \overline{p}}{\partial x_j} - \overline{\rho' u'_j}\frac{\partial \widetilde{U}_i}{\partial x_j} .$$

The first term in the right-hand-side drives the turbulent mass flux to relax to zero after the shock, on the fast acoustic time scale $\tau_a = 0.4 M_t \tau_t$, where $\tau_t = \widetilde{k}/\overline{\epsilon}_s$ is the vortical turbulence time scale based on the solenoidal dissipation (see Chapter 6).
Within the shock, this equation yields a negative normal component of the turbulent mass flux ($\overline{\rho' u'_1} < 0$), so that the mean pressure coupling term ($d$) in eq.(10.52) damps turbulence, as expected from the Rayleigh-Taylor analogy (Chapter 2).

**Taulbee and VanOsdol (1991).** Taulbee and VanOsdol [454] proposed the following closure schemes to the modeling of eq.(10.35):

$$\text{Production} = -\overline{\rho}\,\overline{u_j''}\frac{\partial \widetilde{U}_i}{\partial x_j} + \frac{1}{\overline{\rho}}(\overline{\rho u_i'' u_j''} - \overline{\rho' u_i'' u_j''})\frac{\partial \overline{p}}{\partial x_j} , \qquad (10.36)$$

$$\text{Diffusion} = \frac{\partial}{\partial x_j}[\mu_t(\frac{\partial \overline{u_i''}}{\partial x_j} + \frac{\partial \overline{u_j''}}{\partial x_i} - \frac{2}{3}\frac{\partial \overline{u_m''}}{\partial x_m}\delta_{ij}) + \frac{1}{3}\overline{\rho' u_m'^2}\delta_{ij}] , \qquad (10.37)$$

$$\text{Dissipation} = C_{u_2}\,\overline{\rho}\,\frac{\overline{\epsilon}}{\widetilde{k}}\,\overline{u_i''} . \qquad (10.38)$$

The production term in eq.(10.36) includes two contributions involving mean velocity and density gradients. They are both "exact", i.e., directly

deduced from transportable quantities, when adopting the eddy-viscosity formulation of the authors (see eq.(10.13). Due to the various assumptions introduced during its derivation, the dissipation scheme (10.38) is restricted to low Mach number situations where $\sqrt{\overline{\rho'^2}}/\overline{\rho} \ll 1$ and the density pressure-gradient correlations are negligible. The model constant is $C_{\underline{u_2}} = 5.3$.
The cross-stream averaged mass fluctuating velocity profiles $(\overline{v''})$ calculated with this model in a flat plate boundary layer can be compared with those obtained from a simple gradient hypothesis:

$$\overline{\rho}\,\overline{v''} = \nu_t \frac{\partial \overline{\rho}}{\partial y} \quad \text{with} \quad \nu_t = C_\mu \frac{\overline{k}^2}{\overline{\epsilon}} \, .$$

The same shape of profiles results from both formulations, the profiles deduced from a simple gradient hypothesis being systematically lower than those obtained with the previous model. The departure between the two computed profiles is all the more important than the free stream Mach number is high.

## 10.8. Modeling density effects

We consider now the final question introduced in §10.4.3. It is concerned with the modeling of the terms that are specific to variable density fluid turbulence, namely the *pressure-dilatation correlation* and the *dilatational* or *compressible dissipation*. Some of the various schemes, that have been derived so as to be used in first-order closure models, are reviewed in this section, beginning with the pressure-dilatation correlation:

$$\Pi_d = \overline{p'\vartheta'} = \overline{p' \frac{\partial u_i''}{\partial x_i}} \equiv \overline{p' \frac{\partial u_i'}{\partial x_i}} \, .$$

### 10.8.1. PRESSURE-DILATATION CORRELATION

*Review of some closure schemes*

**Viegas-Horstman (1978).** Several expressions were developed by Rubesin and co-workers at NASA-Ames for the compressibility contribution in a turbulence kinetic energy modeled equation. As an example, we just mention here the proposal by Viegas and Horstman [476] in 1978:

$$\overline{p'\vartheta'} = \xi \overline{\rho} \frac{\widetilde{k}}{\gamma} M^2 \frac{\partial \widetilde{U}_i}{\partial x_i} \, , \tag{10.39}$$

where $M$ is the local Mach number and $\xi = 0.73$ a model constant.
The derivation of this scheme is based on the same kind of assumptions as

those discussed in section 10.7 (Rubesin's scheme, 1976).

**Horstman (1987).** Various inclusions of compressibility effects into the Jones-Launder model were examined by Horstman [218]. In the $\widetilde{k}$ equation, the following scheme for the pressure-dilatation correlation was used

$$\overline{p'\vartheta'} = C\frac{n}{\gamma}(\frac{\gamma-1}{n-1})^2\overline{\rho}\widetilde{k}\,M^2\frac{\partial\widetilde{U}_i}{\partial x_i},$$

where $M$ is the local Mach number. The model constants are $n = 1.2$ and $C = 0.12$.
A similar modification (with a factor $0.3\,\widetilde{k}/\overline{\epsilon}$) was introduced into the dissipation equation.

**Zeman (1991).** In homogeneous, shocklets free, decaying turbulence or shear driven turbulence, Zeman [497] suggested that the pressure variance equation reduces to

$$\overline{p'\vartheta'} = -\frac{1}{2\overline{\rho}\,a^2}\frac{D\overline{p'^2}}{Dt},$$

where $a$ stands for the local speed of sound and $D/DT$ is the material derivative with respect to the mean motion. In other words, according to this relation, the pressure-dilatation correlation is proportional to the rate of change of potential energy $\overline{p'^2}$ due to compression work.
Hence, the closure issue is focused on the pressure variance equation, as discussed in a following section, see eq.(10.57).

**Aupoix** *et al.* **(1990).** Using direct numerical simulations for homogeneous, compressible turbulence, Aupoix *et al.* [24] proposed to model the evolution of the pressure-dilatation term from the following transport equation:

$$\frac{\mathrm{d}\overline{p'\vartheta'}}{\mathrm{dt}} = -C_1\overline{\rho}\frac{1}{\tau_a}M_t^2\frac{\mathrm{d}\overline{k}}{\mathrm{dt}} - C_2\frac{1}{\tau_a}\overline{p'\vartheta'},$$

with,

$$M_t = \frac{\sqrt{2\overline{k}}}{a},\ \tau_a = \frac{\overline{k}^{3/2}}{\overline{\epsilon}\,a},\ C_1 = 0.25,\ C_2 = 0.2\,.$$

This model was only validated in homogeneous sheared compressible flows.

**Durbin-Zeman (1992).** Applying rapid distortion theory to the analysis of compressed turbulence, Durbin and Zeman [138] suggested that the pressure-dilatation term makes a rapid contribution which is proportional to $\overline{p'^2}\times\partial\widetilde{U}_i/\partial x_i$. Hence, a new expression has been proposed, in which a rapid part, accounting for the rapid compression contribution, is added to the slow relaxation term in eq.(10.57):

$$2\overline{\rho}\, a^2\, \overline{p'\vartheta'} = \frac{\overline{p'^2} - p_e^2}{\tau_a} + C_d\, \overline{p'}^2\, \frac{\partial \widetilde{U}_i}{\partial x_i} \,. \tag{10.40}$$

Here, $p_e^2$ is the equilibrium pressure variance and $\tau_a$ an acoustic relaxation time scale (see §10.12.2). The model constant is $C_d = (5-3\gamma)/6$ as determined from RDT, so that $C_d = 0$ for mono-atomic gases and 0.066 for air.

**Zeman-Coleman (1991).** As pointed out by Zeman and Coleman [499], the isotropic model — eq.(10.40) — is entirely inadequate for 1D compression, since it does not distinguish between spheric (*3-D*) and anisotropic (directional) compression. These authors suggested to add a new contribution, solely due to compression anisotropy:

$$\overline{p'\vartheta'} = \frac{1}{2\overline{\rho}a^2}\left(\frac{\overline{p'^2} - p_e^2}{\tau_a} + C_d\overline{p'^2}\frac{\partial \widetilde{U}_i}{\partial x_i}\right) - C_A\frac{\sqrt{\overline{p'^2}}}{\overline{P}M_t^2}\widetilde{k}\tau S_{ij}^* S_{ij}^* \,, \tag{10.41}$$

where $S_{ij}^* = \frac{1}{2}(\partial\widetilde{U}_i/\partial x_j + \partial\widetilde{U}_j/\partial x_i - \frac{2}{3}\partial\widetilde{U}_l/\partial x_l\delta_{ij})$ is the trace-free deformation tensor. The *vortical* turbulence time scale $\tau = 2\widetilde{k}/\overline{\epsilon}_s$ is based on the *solenoidal* dissipation rate $\overline{\epsilon}_s$.

**Sarkar (1992).** The analysis of Sarkar [408] is concerned with the evolution of compressible flow statistics for time intervals larger than the acoustic time scale. Hence, the compressible part of the pressure-dilatation correlation $\overline{p'^C\vartheta'}$ can be neglected, as compared with the incompressible one, $\overline{p'^I\vartheta'}$. Here, $p'^I$ is the *incompressible pressure* associated with the solenoidal part of the velocity vector. In such situations, the pressure-dilatation correlation can be formally written as

$$\overline{p'\vartheta'} \simeq \overline{p'^I\vartheta'} = \overline{p'^S\vartheta'} + \overline{p'^R\vartheta'} \,.$$

As inferred from the previous relation, the closure scheme of $\overline{p'^I\vartheta'}$ is made of two additive terms, as it is usual in pressure-strain modeling for constant density flows. Thus the incompressible pressure contribution is split into the so-called *slow* and *rapid* parts.

– The slow part is modeled as

$$\overline{p'^S\vartheta'} = \alpha_3\overline{\rho}\epsilon_s M_t^2 \,,$$

where $\epsilon_s$ is the solenoidal dissipation associated with isovolume velocity fluctuations, and $M_t = \sqrt{2\widetilde{k}}/\overline{c}$ is the turbulence Mach number.

– The rapid part — associated with $p'^R$ — that reacts instantaneously to a change in the mean velocity gradient, can be approximated by the algebraic relation

$$\overline{p'^R\vartheta'} = 2\,\overline{\rho}\frac{\partial \overline{U}_m}{\partial x_n}A_{mn} + 2\,\overline{\rho}\frac{\partial \overline{U}_m}{\partial x_m}A_{nn} ,$$

provided that an equilibrium scaling exists in the flow. Hence, additional compressible correlations do not dominate the incompressible terms. As shown by Sarkar, $A_{mm} = q_C^2 \equiv \overline{u_i'^C u_i'^C}$, where $u_i'^C$ is the dilatational part of the fluctuating velocity. Introducing

$$A_{mn} = \frac{1}{3}q_C^2\,\delta_{mn} + B_{mn} ,$$

the deviatoric tensor $B_{mn}$, modeled by the leading-order term in a Taylor-series expansion around the isotropic state, is taken as proportional to the anisotropy tensor $a_{ij} = \overline{u_i'u_j'} - (2\overline{k}\delta_{ij}/3)$. Assuming that, for $M_t < 0.5$, the anisotropic part of $A_{mn}$ varies as $M_t q^2$, the final scheme for the rapid part is:

$$\overline{p'^R\vartheta'} = \frac{8}{3}\alpha_4 M_t^2\overline{\rho}\frac{\partial \overline{U}_m}{\partial x_m}\overline{k} + \alpha_2 M_t\overline{\rho}\frac{\partial \overline{U}_m}{\partial x_n}a_{mn}\overline{k} .$$

The final form of the pressure-dilatation model is

$$\overline{p'\vartheta'} = \alpha_2 M_t\overline{\rho}\frac{\partial \overline{U}_m}{\partial x_n}a_{mn}\overline{k} + \alpha_3 M_t^2\overline{\rho}\epsilon_s + \frac{8}{3}\alpha_4 M_t^2\overline{\rho}\frac{\partial \overline{U}_m}{\partial x_m}\overline{k} . \qquad (10.42)$$

**Ristorcelli (1997).** The analysis of Ristorcelli [394] treats turbulence in which compressible effects are generated by the turbulent motions, under several hypothesis and, in particular,

- (i) a compact-source assumption: a turbulent eddy is small with respect to the length scale of its acoustic radiation;
- (ii) a compact flow assumption: the size of the turbulent field, $D$ is small or in the order of the acoustic scale; $D/\lambda \leq 1$, where $\lambda$ is the characteristic length scale of the propagation of pressure and density fluctuations.

In this regard, homogeneous compressible DNS, as treated by Sarkar *et al.* [416], for example, is not a compact flow, since it is concerned with turbulence of scale $\ell$ irradiated by an *infinite* external acoustic field generated by turbulence whose statistics are the same as those of the local turbulent region.

According to Ristorcelli's assumptions, the pressure-dilatation is found to be a non-equilibrium phenomena. It is modeled as the sum of a slow and rapid part and includes the substantial derivative following a mean fluid particle of a relative time scale based on the non-dimensional strain and rotation rates of the mean motion. Restricted to low-$M_t^2$ situations, it scales as

$$\overline{p'\vartheta'} \propto M_S^2[\frac{P_{rod}}{\overline{\epsilon}_s} - 1] \,. \tag{10.43}$$

Here $P_{rod}$ stands for the production rate of the turbulence kinetic energy, $\overline{\epsilon}_s$ is the *solenoidal* dissipation (as modeled in next section) and $S = (S_{ij}S_{ij})^2$ the trace of the square of the mean strain rate matrix $S_{ij} = (\partial\overline{U}_i/\partial x_j + \partial\overline{U}_j/\partial x_i)/2$.
Hence, from eq.(10.43), conditions for the pressure-dilatation correlation to be important, are, that the square of the strain (gradient) Mach number $M_S = M_t S\overline{k}/\overline{\epsilon}_s$ is large and turbulence far from the energetic equilibrium $P_{rod} \neq \overline{\epsilon}_s$. Now, depending upon the departure from energetic equilibrium, eq.(10.43) shows that the pressure-dilatation can be either positive or negative, and therefore the gradient Mach number can equally have a stabilizing or destabilizing effect.

**Hamba (1999).** The following scheme, due to Hamba [203], is part of the modeling of the pressure variance equation (see §10.12.2). It reads

$$\overline{p'\vartheta'} = -(1 - C_{pd3}\chi_p) \times [C_{pd1}M_t^2\frac{D(\overline{\rho}k)}{Dt} + C_{pd2}\gamma\overline{\rho}M_t^2\overline{k}\,\frac{\partial\overline{U}_i}{\partial x_i}] \,. \tag{10.44}$$

In this expression, $D/Dt$ stands for the material derivative following the mean motion and $\chi_p$ is a non-dimensional pressure variance corresponding to the ratio of potential to kinetic energy for weak fluctuations

$$\chi_p = \frac{\overline{p'^2}}{2\overline{\rho}^2\overline{c}^2\overline{k}} \,. \tag{10.45}$$

To achieve good agreement between model predictions and direct numerical simulations, the constants are set to $C_{pd1} = 1.2$ and $C_{pd3} = 6$. The value of $C_{pd2}$ needs not to be prescribed in homogeneous shear flows where the mean velocity divergence is zero.

*Discussion*

Dealing with compressible homogeneous turbulence at a moderate Mach number, Sarkar *et al.* [416] suggested in 1991 that, for the purpose of turbulence modeling, the effect of the pressure-dilatation correlation can be absorbed in the model of the compressible dissipation (see the next section). This statement is based on DNS of isotropic turbulence which indicates that the average of $\overline{p'\vartheta'}$ over its oscillations is significantly smaller than the compressible dissipation.

Also shown by direct numerical simulation results, the pressure-dilatation contribution to the evolution of turbulence kinetic energy is more important in homogeneous shear compressible flows than in decaying turbulence. In this case, the major contributor to the pressure-dilatation comes from the *incompressible pressure* $p'^I$ associated with the solenoidal velocity.

To some extent, this explains why the model derived by Sarkar [408] in 1992 — with $\alpha_2 = 0.15$ and $\alpha_3 = 0.2$ in eq.(10.42) —, can be considered as a reasonable approximation for weakly inhomogeneous flows, such as compressible shear layer and flat plate boundary layer, far from the wall.

Although the pressure-dilatation term $\overline{p'\vartheta'}$ may be negligible in the $\overline{k}$ equation for homogeneous compressible shear flow, it cannot be inferred that there is no need for an appropriate closure scheme, for at least two reasons:

- The model derived by Zeman and Coleman [499], with $C_A = 0.0008$ for the closure parameter in eq.(10.41), is capable of replicating the important feature of kinetic to pressure energy transfer, specific to 1D compression and the resulting different amplification rates of $\overline{p'^2}$ for low and high Mach number situations;
- The pressure-dilatation correlation plays an important role in the pressure variance equation (see §10.12.2), so that the modeling of this term is to be addressed when deriving a closure expression to this equation. In compressible homogeneous shear flow, for instance, the DNS results of Hamba [203] in 1999 show that the pressure-dilatation correlation is the dominant term in the pressure variance equation.
  As far as Hamba's scheme is concerned — eq.(10.44) —, it can be observed that
  - if $\chi_p$ is neglected and approximating $D(\overline{\rho}\overline{k})/Dt$ by $\rho(P_{rod} - \epsilon)$, then eq.(10.44) is almost the same as the model derived by Sarkar [408] in 1992;
  - compressibility effects are represented by the factor $(1 - C_{pd3}\chi_p)$, which amounts to 0.88 and 0.58 for $M_t$ equal to 0.1 and 0.3, respectively, thus reducing the pressure-dilatation growth rate when increasing the turbulent Mach number.

Finally, in both Ristorcelli's and Hamba's schemes, the pressure-dilatation correlation is related to the non-equilibrium or unsteady properties of the turbulent field. This is particularly apparent with the material derivative of the turbulence kinetic energy in Hamba's scheme (eq.(10.44)) and a bit more intricate in the *almost* algebraic expression of Ristorcelli (see eq. (87) in [394]). As pointed out by the latter, the model produces predictions which are consistent with the study by Simone *et al.* [431] where the observed behavior is related to the anisotropy component $b_{12}$ and leads to negligible contributions in a quasi-equilibrium situation, as it is the case in channel flows and boundary layers without strong pressure gradients.

### 10.8.2. DILATATION DISSIPATION

*Review of some models*
We turn now to the modeling of the last term which is specific to variable density turbulent flows, viz. the dilatation or compressible dissipation $\overline{\epsilon}_d$, eq.(10.7).

**Zeman (1990).** According to Zeman [495], the dilatation dissipation is considered as proportional to the solenoidal dissipation

$$\overline{\epsilon}_d = c_d\, F(M_t, K)\, \overline{\epsilon}_s\ . \tag{10.46}$$

The adjustable model constant is $c_d = 0.75$. The scalar function $F$ depends on the rms or turbulence Mach number $M_t = \sqrt{2\widetilde{k}}/\overline{c}$, where $\overline{c} = \sqrt{\gamma R\widetilde{T}}$, and on the probability distribution function of the turbulence, through the kurtosis of the p.d.f. of the velocity fluctuation, $K = \overline{u''^4}/(\overline{u''^2})^2$ (ranging from 4 to 20). The function $F(M_t, K)$, of order unity, is found to increase rapidly for $0 < M_t < 2$ and reaches a quasi constant level for $M_t > 2$ (close to unity, for $K = 6$).
This closure was introduced in a second-order model to predict compressible mixing layers. It appeared able to predict the reduction of layer growth rates as a function of the convective Mach number in agreement with the experimental data.

**Sarkar-Erlebacher-Hussaini-Kreiss (1991).** Sarkar *et al.* [416] come to a very similar expression, but postulate a fundamentally different basis. The model is simply

$$\overline{\epsilon}_d = \alpha_1 M_t^2\, \overline{\epsilon}_s\ , \tag{10.47}$$

where the model constant is $\alpha_1 = 1$, based on direct numerical simulations of the decay of isotropic compressible turbulence at a Reynolds number based on the Taylor micro-scale $R_\lambda = 15$. The turbulence Mach number is $M_t = \sqrt{2\overline{k}}/\overline{c}$, where $\overline{c}$ is the local mean speed of sound.

**Wilcox (1994).** As shown in Wilcox [484], page 185, the dilatation-dissipation concept can be introduced in a ($k$ - $\omega$) model ($\omega$ is the Kolmogorov frequency of the energy bearing eddies $\omega = \sqrt{\overline{k}}/l \equiv \overline{\epsilon}/(C_\mu \overline{k})$ ). It merely consists in letting the closure coefficients $\beta$ and $\beta^*$ of the model vary with the turbulence Mach number, thus departing from their incompressible values $\beta_0$ and $\beta_0^*$ as

$$\beta^* = \beta_0^*[1+\xi^* F(M_t)] \qquad \text{and} \qquad \beta = \beta_0 - \beta_0^*\xi^* F(M_t)\,,$$

with $\quad F(M_t) = (M_t^2 - M_{t_0}^2)\mathcal{H}(M_t - M_{t_0})\,, \quad \xi^* = 3/2\,, \quad M_{t_0} = 1/4\,,$

where $\mathcal{H}(x)$ is the Heaviside step function.

**Ristorcelli (1997).** Within the same frame of analysis as that previously mentioned in deriving the pressure-dilatation closure scheme, Ristorcelli [394] succeeded in producing a representation for the effects of the compressible (or dilatational) dissipation $\overline{\epsilon}_d$ as a sum of slow ($\overline{\epsilon}_d^s$) and rapid ($\overline{\epsilon}_d^r$) portions.

The ratio $\overline{\epsilon}_d/\overline{\epsilon}_s$ is found to be a function of the turbulence Reynolds number $R_t = 4\overline{k}^2/9\nu\overline{\epsilon}$, scaling as

$$\frac{\overline{\epsilon}_d^s}{\overline{\epsilon}_s} \propto \frac{M_t^4}{R_t} \qquad \text{and} \qquad \frac{\overline{\epsilon}_d^r}{\overline{\epsilon}_s} \propto \frac{M_t^2 M_S^2}{R_t}\,, \tag{10.48}$$

for high-$R_t$ and low-$M_t^2$ non-equilibrium flows, respectively. Here, $M_t$ and $M_S$ are the turbulence and strain (gradient) Mach numbers, respectively. An important feature of the previous model is the dependence of the dilatation dissipation on the viscosity: for fixed $M_t$, $\overline{\epsilon}_d$ vanishes at a sufficiently high turbulence Reynolds number.

*Discussion*

Prediction of the reduction in the spreading rate of a compressible mixing layer can be taken as one of the most challenging issues to turbulence modeling of compressible effects in such flows. Indeed, most "classical" extensions of incompressible models, performed without incorporating dilatational terms, fail to predict the reduction growth rate when the convective Mach number increases. This can be observed in Figure 10.2 for first-order ($k$ - $\omega$) and second-order ($R_{ij}$) closure models as well. The ($k$ - $\epsilon$) model generates predictions similar to those of ($k$ - $\omega$), if not worse.

In this figure, the spreading rate of the compressible mixing layer $C_\delta$ is defined as

$$C_\delta = \frac{d\delta}{dx}\left(\frac{U_1+U_2}{U_1-U_2}\right).$$

It is normalized by its incompressible value $C_{\delta_0}$. The shear layer thickness $\delta(x)$ is the distance between the points where the mean velocity is, respectively, $U_2 + 0.1(U_1 - U_2)$ and $U_2 + 0.9(U_1 - U_2)$. The ratio $C_\delta/C_{\delta_0}$ is plotted against the convective Mach number

$$M_c = \frac{U_1 - U_2}{c_1 + c_2}\,,$$

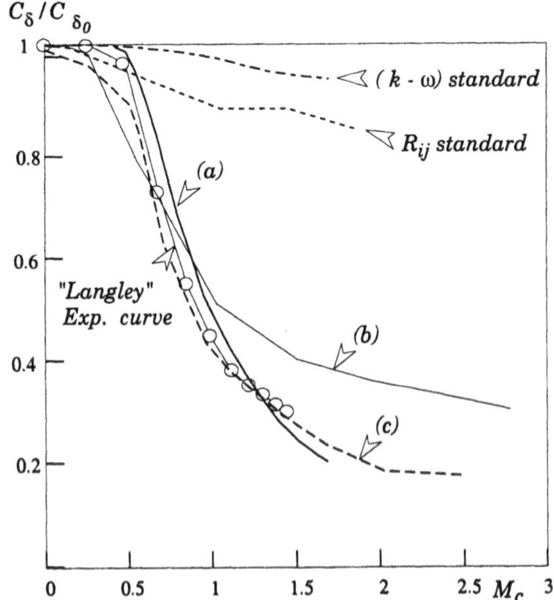


*Figure 10.2.* Prediction of the spreading rate of a compressible mixing layer according to (a) Wilcox [484], (b) Sarkar *et al.* [416] and (c) Zeman [495].

where $c_1$ and $c_2$ stand for the speed of sound in the two incident free streams.

As shown Fig.10.2, a better agreement with experimental data is obtained when incorporating compressible or dilatation effects in the closure schemes. Except the first-order model of Wilcox [484], all other closure schemes, as derived by Zeman [495], eq.(10.46) and Sarkar *et al.* [416], eq.(10.47), have been introduced in second-order models to predict the reduction of the mixing layer growth rate as a function of the convective Mach number (see, for instance, Sarkar and Lakshmanan [413] in 1991 for the latter).

These examples can be considered as representative of a "second generation"[8] of compressible models [495], [415], [416], [484], in which it was presumed that *explicit* dilatational terms could account for compressibility effects, namely the pressure-dilatation correlation and the dilatation dissipation.
Later literature on the topic, Sarkar [409] in 1995, Vreman *et al.* [477] in 1996, Hamba [203] in 1999, *inter alia*, pointed out important *implicit* effects of compressibility, associated with structural changes of the turbulent field.

[8]The "first generation" only involves simple adjustments (turbulence Mach number parameterization) and formal extensions of "incompressible" closure schemes.

Moreover, even though the importance of dilatational terms is difficult to assess, *a priori* from theoretical analysis, or check, *a posteriori* from experimental data, further developments — Ristorcelli [394] and Simone *et al.* [431] for example —, support the conclusion that compressible dissipation is negligible in most compressible flows at a moderate turbulence Mach number (see the $M_t^2$ and $M_t^4$ scalings in eqs.(10.48), for instance).

Hence, one can reasonably question the capability of this early generation of compressible models (based on specific modeling of explicit dilatational terms) in capturing the actual mechanism responsible for predicted compressibility effects and suspect whether an effect attributed to compressible dissipation could not actually represent an alteration in production, as recently suggested by Guézengar et al. [191].

## 10.9. Examples of zero-equation models

After having analyzed *separate* closure schemes accounting for variable density or compressibility effects, this section introduces a second part of the chapter which is devoted to the presentation of some *complete* models that have been used to predict such types of turbulent flows. The review does not aims at being exhaustive, but simply illustrative of some salient contributions.

Owing to the great number of algebraic expressions of eddy-viscosity, the presentation is limited here to those which have been used in predicting compressible turbulent flows for practical applications. For a more detailed review, the reader is referred to Vandromme [469] or Cousteix & Aupoix [107], for instance.

**Michel-Quémard-Durant (1969).** In the mixing length scheme proposed by Michel *et al.* [327], the eddy-viscosity in a turbulent boundary layer is given by:

$$\mu_t = \overline{\rho} F^2 \, \ell_m^2 \, | \frac{\partial \overline{U}}{\partial y} | \, .$$

The mixing length $\ell_m$ is taken as a function of the distance to the wall $y$

$$\frac{\ell_m}{\delta} = 0.085 \, tanh( \frac{\chi}{0.085} \frac{y}{\delta} ) \, , \quad \text{with} \quad \chi = 0.41 \, .$$

Here $\delta$ is the conventional boundary layer thickness. $F$ is a modified version of the Van Driest damping function:

$$F = 1 - exp[ - \frac{\ell_m \sqrt{\overline{\rho} \, \tau}}{26 \, \chi \, \mu} ] \, ,$$

where $\tau = \mu\frac{\partial\overline{U}}{\partial y} - \overline{\rho u'v'}$ is the total shear stress. The counterpart of this scheme for the turbulent heat flux is:

$$-\overline{\rho v'\theta'} = -\frac{1}{\sigma_t}\overline{\rho}F^2\,\ell_m^2\,\frac{\partial\overline{U}}{\partial y}\frac{\partial\overline{T}}{\partial y},$$

where $\sigma_t = 0,89$ is the turbulence Prandtl number.
This model is able to reproduce quite satisfactorily compressibility effects in supersonic turbulent boundary layers not too far from equilibrium. It was also included in a integral method by Michel *et al.* [326] to predict satisfactorily adiabatic boundary layers.

**Cebeci-Smith-Mosinskis (1970).** The Cebeci-Smith-Mosinskis (C.S.M.) scheme [74] is an extension to compressible flows of Cebeci's model. Several adaptations of this kind have been proposed. For a rather complete inventory, the reader is referred to the book by Cebeci and Smith [73].
In the C.S.M. model, the eddy-viscosity is derived from a two-layer formulation:

$$Inner\ region: \quad 0 \le y \le y_c \quad \mu_{ti} = \overline{\rho}\,\kappa^2\,y^2(1-e^{-y/A})^2\,\frac{\partial\overline{U}}{\partial y},$$

$$Outer\ region: \quad y_c \le y \le \delta \quad \mu_{to} = \alpha\,\overline{\rho}\,U_E\,\delta_1/\Gamma,$$

where $y_c$ is the distance from the wall where the two expressions are matched ($\mu_{to} = \mu_{ti}$).
In the first expression, $\kappa = 0.41$, $\alpha = 0.0168$, $\delta_1 = \int_0^\infty(1-\overline{U}/U_E)dy$. $\Gamma$ is the intermittency correction factor:

$$\Gamma = [1+5.5(\frac{y}{\delta})^6]^{-1}\,.$$

In the inner region, the damping coefficient $A$ is defined as

$$A = A^+\frac{\nu}{N}(\frac{\tau_w}{\rho_w})^{1/2}(\frac{\overline{\rho}}{\rho_w})^{-1/2},$$

where $A^+ = 26$ and $N$ is given by the following expression, accounting for pressure gradient and wall blowing (transpiration velocity $V_w$):

$$N = \frac{\mu}{\mu_e}(\frac{\rho_e}{\rho_w})^2\frac{P^+}{V_w^+}[1-exp(11.8\frac{\mu_w}{\mu}V_w^+)] + exp(11.8\frac{\mu_w}{\mu}V_w^+)\,,$$

with

$$P^+ = (\frac{\nu_e U_e}{u_\tau^3}\frac{\partial U_e}{\partial x}) \qquad V_w^+ = \frac{V_w}{u_\tau} \quad u_\tau^2 = \frac{\tau_w}{\rho_w}\,.$$

In these formulae, subscripts $w$ and $e$ refer to wall and external flow conditions respectively.

**Baldwin-Lomax (1978).** One of the major drawbacks of the C.S.M model is revealed in separated flows, where $\delta_1$ is not well defined. This is not the case with the Baldwin-Lomax (B.L.) scheme, where an implicit matching is adopted between two eddy-viscosity values $\nu_{ti}$ and $\nu_{to}$. The local eddy-viscosity is taken as the lowest value between the two previous ones, respectively defined by:

$$Inner\,region: \ \nu_{ti} = l_m^2 \Omega \quad where \ l_m = \chi y \left(1 - e^{y^+/A^+}\right) \quad \text{and} \ \Omega = \left\|\vec{\Omega}\right\|,$$

$$Outer\,region: \ \nu_{to} = \alpha C_{cp} F \widetilde{I}_K (x, y),$$

$$\text{where} \quad F = \min\left[y_{max} f_{max}; C_w y_{max} U_{dif}^2 / f_{max}\right], \ f_{max} = \frac{1}{\chi} \sup_y \left[l_m \Omega\right]$$
$$\text{and } \widetilde{I}_K = \left[1 + 5,5 \left(C_K y / y_{max}\right)^6\right]^{-1}.$$

$\vec{\Omega}$ is the mean vorticity vector and $y_{max}$ the distance from the wall where $[l_m\Omega](y)$ is maximum. $U_{dif}$ stands for the mean velocity difference over $y_{max}$.
The model constants and parameters are given in table 10.2.

TABLE 10.2. Model constants of the Baldwin-Lomax eddy-viscosity model.

| Constant | $\chi$ | $\alpha$ | $A^+$ | $C_{cp}$ | $C_\omega$ | $C_K$ |
|---|---|---|---|---|---|---|
| Value | 0.41 | 0.0168 | 26 | 1.6 | 0.25 | 0.3 |

*Discussion*
Algebraic eddy-viscosity models were primarily derived for closing the averaged Navier-Stokes equations under thin shear layer approximations and were mainly dedicated to wall bounded flows. They are easy to use, cheap, and produce rather accurate results in boundary layers that are not too far from equilibrium, even at high Mach numbers, Cousteix & Aupoix [107]. Mean flow predictions of attached supersonic boundary layers with C.B.S. and B.L. models are generally found in reasonably good agreement with experimental data.

## 10.10. One-equation models

The following table give some examples of one-equation models that have been proposed for predicting incompressible flows. As far as applications to variable density and compressible flows are concerned, only two of them will be detailed here, based on modeled transport equations for (i) the eddy-viscosity and (ii) the turbulence kinetic energy.

TABLE 10.3. Examples of one-equation incompressible turbulence models

| Transportable function | Author | Ref. | Year |
|---|---|---|---|
| Turbulence kinetic energy | Prandtl | [372] | 1945 |
| | Emmons | [148] | 1954 |
| Turbulence shear stress | Bradshaw & Ferriss | [53] | 1972 |
| Eddy viscosity | Nee & Kovasznay | [345] | 1968 |
| | Secundov | [424] | 1971 |
| | Baldwin & Barth | [28] | 1990 |
| | Spalart & Allmaras | [440] | 1992 |
| | Gulyaev, Kozlov & Secundov | [192] | 1993 |
| | Durbin, Mansour & Yang | [139] | 1994 |

### 10.10.1. EDDY-VISCOSITY TRANSPORT EQUATION

The first model based on an eddy-viscosity transport equation was proposed by Nee & Kovasznay [345] in 1968. After an eclipse of more than twenty years, this type of closure is again in use for practical applications, thanks to the proposals of Baldwin & Barth [28], and more recently, Spalart & Allmaras [440]. In the latter models, the eddy-viscosity is taken as:

$$\nu_t = \alpha \hat{\nu} \,, \tag{10.49}$$

where $\alpha$ is a model coefficient and $\hat{\nu}$ an effective eddy-viscosity, which is governed by a model transport equation of the general form[9]:

$$\frac{\partial \hat{\nu}}{\partial t} + \overline{U}_j \frac{\partial \hat{\nu}}{\partial x_j} = D_{iff} + \text{Source} - \text{Sink} \,. \tag{10.50}$$

[9]The Spalart-Allmaras model can predict a laminar solution. The complete version of this model includes a "trip" term to initiate transition in a smooth manner. For sake of shortness, the detailed expressions of this term are not given in Table 10.4.

TABLE 10.4. Details of Baldwin-Barth and Spalart-Allmaras eddy-viscosity transport equation models, in reference to eqs.(10.49) and (10.50).

| Term | Baldwin-Barth | Spalart-Allmaras |
|---|---|---|
| Diffusion | $(\nu + \frac{\nu_t}{\sigma})\frac{\partial^2 \hat{\nu}}{\partial x_j \partial x_j}$ | $\frac{1}{\sigma}\frac{\partial}{\partial x_j}[(\nu+\hat{\nu})\frac{\partial \hat{\nu}}{\partial x_j}]$<br>$+$<br>$\frac{C_{b2}}{\sigma}\frac{\partial \hat{\nu}}{\partial x_j}\frac{\partial \hat{\nu}}{\partial x_j}$ |
| Source | $(C_{\epsilon 2} f_2 - C_{\epsilon 1})\sqrt{\hat{\nu} P}$, where<br>$f_2 = \frac{C_{\epsilon 1}}{C_{\epsilon 2}} + (1 - \frac{C_{\epsilon 1}}{C_{\epsilon 2}})(\frac{1}{\chi y^+} + D_1 D_2) \times$<br>$[\sqrt{D_1 D_2} + \frac{y^+}{\sqrt{D_1 D_2}}(\frac{1}{A_0^+} e^{-\frac{y^+}{A_0^+} D_2} + \frac{1}{A_2^+} e^{-\frac{y^+}{A_2^+} D_1})]$<br>and<br>$P = \nu_t[(\frac{\partial \overline{U}_i}{\partial x_j} + \frac{\partial \overline{U}_j}{\partial x_i})\frac{\partial \overline{U}_i}{\partial x_j} - \frac{2}{3}\frac{\partial \overline{U}_k}{\partial x_k}\frac{\partial \overline{U}_l}{\partial x_l}]$ | $C_{b1}\hat{S}\hat{\nu}$, where<br>$\hat{S} = (2\overline{\Omega}_{ij}\overline{\Omega}_{ij})^{1/2} + \gamma \frac{\hat{\nu}}{\chi^2 y^2}$<br>$\gamma = 1 - \frac{\zeta}{1+\alpha\zeta}$<br>and<br>$\zeta = \hat{\nu}/\nu$ |
| Sink | $\frac{1}{\sigma}\frac{\partial \nu_t}{\partial x_j}\frac{\partial \hat{\nu}}{\partial x_j}$ | $C_{w1} f_w (\frac{\hat{\nu}}{y})^2$, where<br>$f_w = g\,(\frac{1+C_{w3}^6}{g^6 + C_{w3}^6})^{1/6}$<br>$g = r + C_{w2}(r^6 - r)$<br>$r = \hat{\nu}/\hat{S}\chi^2 y^2)$ |
| $\alpha$ | $C_\mu D_1 D_2$, where<br>$D_1 = 1 - e^{-y^+/A_0^+}$, *and* $D_2 = 1 - e^{-y^+/A_2^+}$ | $\frac{\zeta^3}{\zeta^3 + C_{v1}^3}$ |

TABLE 10.5. Model constants.

| Model | Constants |
|---|---|
| Baldwin & Barth | $C_\mu = 0.09$ $C_{\epsilon 1} = 1.2$ $C_{\epsilon 2} = 2$ $A_0^+ = 26$ $A_2^+ = 10$<br>$\sigma = 0.7$ $\chi = 0.41$ |
| Spalart & Allmaras | $C_{b1} = 0.1355$ $C_{b2} = 0.622$ $C_{v1} = 7.1$ $C_{w1} = 3.2391$<br>$C_{w2} = 0.3$ $C_{w3} = 2$ $\sigma = 2/3$ $\chi = 0.41$ |

The one-equation model of Spalart & Allmaras has been used by Wong [487] to predict supersonic separated flows in an axisymmetric configuration that were not validated by the authors of the model [440]. According to Wong, the model is able to predict relaminarization but cannot be trusted to predict accurately the onset and duration of transition. Nevertheless, this corresponds to an improved situation as compared with the results obtained from an algebraic eddy-viscosity model (Baldwin-Lomax) which cannot predict relaminarization without *ad hoc* tuning of the constants.

## 10.10.2. TURBULENCE KINETIC ENERGY MODEL EQUATION

A widely adopted expression of the eddy-viscosity in incompressible turbulence reads:

$$\nu_t \propto \overline{\rho}\hat{u}\, l\,, \tag{10.51}$$

where $\hat{u}$ and $l$ are characteristic velocity and length scales of the turbulent motion, respectively.

Assuming that $l$ can be prescribed algebraically, the characteristic turbulence velocity is generally taken as $\hat{u} \propto \sqrt{\widetilde{k}}$ or $\sqrt{\overline{k}}$ (it is recalled that $\widetilde{k} \equiv \widetilde{u''_i u''_i}/2$ — resp. $\overline{k} \equiv \overline{u'_i u'_i}/2$ — denotes the kinematic turbulence kinetic energy, according to density weighted or classical averages). This function is governed by an exact transport equation which reads, when Favre's decomposition is adopted (see Chapter 5):

$$\underbrace{\overline{\rho}\left(\frac{\partial \widetilde{k}}{\partial t} + \widetilde{U}_j \frac{\partial \widetilde{k}}{\partial x_j}\right)}_{(a)} = \underbrace{-\overline{\rho}\widetilde{u''_i u''_j}\frac{\partial \widetilde{U}_i}{\partial x_j}}_{(b)} \underbrace{- \frac{\partial}{\partial x_j}\left[\frac{1}{2}\overline{(\rho u''_i u''_i u''_j)} + \overline{p' u''_j}\right]}_{(c)} \underbrace{- \overline{u''_i}\frac{\partial \overline{P}}{\partial x_i}}_{(d)} \underbrace{+ \overline{p' \frac{\partial u''_i}{\partial x_i}}}_{(e)}$$

$$+ \underbrace{\frac{\partial \overline{(\tau_{ij} u''_i)}}{\partial x_j}}_{(g)} \underbrace{- \overline{\tau_{ij}\frac{\partial u''_i}{\partial x_j}}}_{(h)} . \tag{10.52}$$

In eq.(10.52):

($a$) is the mean material variation of the turbulence kinetic energy, based on a *transport* formulation. Making use of the mean continuity equation, it can be easily transformed into a *conservative* expression:

$$\frac{\partial \overline{\rho}\widetilde{k}}{\partial t} + \frac{\partial(\overline{\rho}\widetilde{k}\widetilde{U}_j)}{\partial x_j} \equiv \frac{\partial(\overline{\rho}\widetilde{k})}{\partial t} + \widetilde{U}_j \frac{\partial(\overline{\rho}\widetilde{k})}{\partial x_j} + \overline{\rho}\widetilde{k}\,\frac{\partial \widetilde{U}_j}{\partial x_j}\,;$$

($b$) is the mean shear production;

($c$) is the turbulent transport or diffusion of turbulence kinetic energy, including pressure effects;

($d$) corresponds to an energy transfer by coupling of the turbulent mass flux with the mean pressure field. It will be briefly called the mean pressure work, and is specific to variable density flows *and* Favre formulation;

($e$) is the pressure-dilatation correlation, which can also be written as $\overline{p'(\partial u'_i/\partial x_i)}$. It is specific to non-solenoidal velocity fluctuations.

The last two terms in eq.(10.52) are accounting for molecular viscosity effects, since $\tau_{ij}$ is the *instantaneous* viscous stress tensor, viz. for a Newton-Stokes behavior:

$$\tau_{ij} = 2\mu S_{ij} - \frac{2}{3}\mu \frac{\partial U_l}{\partial x_l}\delta_{ij} .$$

Thus:
($g$) is the work done by external viscous forces in the fluctuating motion;
($h$) is the Favre-averaged dissipation rate.

Now, as far as the *modeled* expression is concerned, the generic form of the turbulence kinetic energy equation can be taken as:

$$\frac{\mathrm{D}}{\mathrm{D}t}(\overline{\rho}\widetilde{k}) = P_{rod}^{(k)} + D_{iff}^{(k)} - \overline{\epsilon}_s + A_\rho , \tag{10.53}$$

where $\mathrm{D}/\mathrm{D}t$ denotes the material derivative with reference to the mean flow.
In eq.(10.53), $P_{rod}^{(k)}$, $D_{iff}^{(k)}$, $\overline{\epsilon}_s$ are the production, diffusion and dissipation terms, the closure schemes of which are 'similar' to those developed for incompressible or constant density flows (direct extension from the incompressible situation). Hence, $A_\rho$ accounts for all compressible or variable density contributions. Its modeling requires specific closure schemes being derived for:
- the turbulent mass flux;
- the pressure-dilatation correlation[10];
- the dilatation dissipation.

Some of the corresponding closure schemes, dedicated to the modeling of $A_\rho$, have been already discussed in the previous sections. Therefore, we are simply concerned here with closure schemes for the production, diffusion and solenoidal dissipation terms.

***Production.*** When direct transpositions of the linear eddy-viscosity concept are adopted in variable density fluid motions, — eqs.(10.10) and (10.11) —, the production term is obtained explicitly:

$$P_{rod}^{(k)} = -\overline{\rho}\,\widetilde{u_i''u_j''}\frac{\partial \widetilde{U}_i}{\partial x_j} = 2\mu_t\widetilde{S}_{ij}\frac{\partial \widetilde{U}_i}{\partial x_j} - \frac{2}{3}\mu_t\left(\frac{\partial \widetilde{U}_l}{\partial x_l}\right)^2 - \frac{2}{3}\overline{\rho}\widetilde{k}\frac{\partial \widetilde{U}_l}{\partial x_l} .$$

[10] By regrouping all pressure contributions in eq.(10.52), only one pressure term is obtained, $u_i''\frac{\partial P}{\partial x_i}$. In Aupoix *et al.* [24], an exact transport equation is derived for this term, assuming a perfect gas equation of state. However, as pointed out by the authors, a term-by-term closure of this equation cannot be applied, since some of them go to infinity and only the difference remains finite. Thus the splitting of the pressure terms, as given in eq.(10.52) is generally the one adopted in the modeling process.

***Diffusion.*** A widely adopted closure scheme for the diffusion term in eq.(10.52) is based on a generalized gradient-type expression:

$$D_{iff}^{(k)} = \frac{\partial}{\partial x_j}\left(\frac{\mu_t}{\sigma_k}\frac{\partial \widetilde{k}}{\partial x_j}\right),$$

where the turbulent transport coefficient (or diffusivity) of $\widetilde{k}$ is a model constant, taken as a turbulence Prandtl-Schmidt number $\sigma_k$.
This expression was introduced, for instance, in 1972 by Jones and Launder [238], in a $(k - \epsilon)$ model, with $\sigma_k = 1$.

***Solenoidal dissipation.*** In the modeling of the energy transport equation, the solenoidal or incompressible contribution to the dissipation $(\overline{\epsilon}_s)$ is simply deduced from $\widetilde{k}$ and $l$. Assuming that Batchelor's incompressible scaling still applies in variable density flows, the expression reads:

$$\overline{\epsilon}_s \propto \frac{\widetilde{k}^{3/2}}{l},$$

introducing only one model constant.

## 10.11. Two-equation models

In two-equation models for incompressible fluid motions, the characteristic length scale $l$ in eq.(10.51) is obtained from any product of a given power of the turbulence kinetic energy with another transportable quantity of suitable dimension.
Some examples of such available combinations that have been used in variable density flows are reported in Table 10.1. Only three of them will be discussed here.

### 10.11.1. THE $(\kappa - \epsilon)$ MODEL

In 1979, Jones [237] suggested that the eddy-viscosity can be obtained from the turbulence kinetic energy and the dissipation rate, according to a direct extension of the usual incompressible relation (Jones & Launder [238], 1972). Using Favre averages, this scheme reads

$$\mu_t = C_\mu \overline{\rho}\frac{\widetilde{k}^2}{\overline{\epsilon}}, \qquad (10.54)$$

where $C_\mu = 0.09$ is a model constant, and $\overline{\epsilon}$ the dissipation rate, obtained from a model transport equation.

As reviewed by Vandromme and Ha Minh [472], the common form of the transport equation for the dissipation rate in high Reynolds number compressible flows can be written as (see also Rubesin [398]):

$$\frac{\partial(\overline{\rho}\overline{\epsilon})}{\partial t}+\frac{\partial(\overline{\rho}\overline{\epsilon}\widetilde{U}_j)}{\partial x_j}=-C_{\epsilon 1}\frac{\overline{\epsilon}}{\overline{\overline{k}}}\overline{\rho}\widetilde{u''_i u''_j}\frac{\partial \widetilde{U}_i}{\partial x_j}+D_\epsilon-C_{\epsilon 2}\overline{\rho}\frac{\overline{\epsilon}^2}{\overline{\overline{k}}}$$

$$+\underbrace{C_{\epsilon 3}\frac{\overline{\epsilon}}{\overline{\overline{k}}}\overline{p'\frac{\partial u''_i}{\partial x_i}}}_{(d)}-\underbrace{C_{\epsilon 4}\frac{\overline{\epsilon}}{\overline{\overline{k}}}\overline{u''}_i\frac{\partial \overline{P}}{\partial x_i}}_{(e)}-\underbrace{C_{\epsilon 5}\overline{\rho}\,\overline{\epsilon}\frac{\partial \widetilde{U}_i}{\partial x_i}}_{(f)}\ . \qquad (10.55)$$

In the constant density situation, eq.(10.55) reduces to the first line, where the right-hand-side terms are the shear production, the diffusion and the dissipation. The second line is thus accounting for compressible effects. It is made of three additional terms: $(d)$ and $(e)$ are simply the counterparts for the pressure-dilatation and mean pressure contributions, already present in the turbulence kinetic energy equation. The last one $(f)$ is introduced to accounting for the dependence of turbulence length scale on passing through a shock wave. The constant model is $C_{\epsilon 5}=1/3$ in isotropic turbulence and unity otherwise.

When dilatation (compressible) dissipation schemes are introduced, this equation is adopted with adapted values of the constants to calculate the solenoidal part of the dissipation rate ($\overline{\epsilon}_s$). We shall now just give two examples.

**Ha Minh *et al.* (1981-91).** A simplified version of eq.(10.55) was intensively used by Ha Minh and co-workers [195], [199], [280], [198], [196], [471], [470], for the conventional dissipation rate, as part of ($k$ - $\epsilon$) model and second-order closure models to predict wall bounded flows, including shock-boundary layer interaction.

In this case, the low Reynolds number corrections of Jones-Launder [239] are adopted, in addition to the following assumptions, [195]:

- The gradient diffusion is simply $D_\epsilon=\frac{\partial}{\partial x_j}(\frac{\mu_t}{\sigma_\epsilon}\frac{\partial\overline{\epsilon}}{\partial x_j})$ with $\sigma_\epsilon=1.3$;
- When present[11], compressibility correction is limited to the mean pressure gradient term, adopting a Rubesin-derived expression for the turbulent mass flux — eq.(10.30) —, with $C_{\epsilon 4}=1$.

When applied to the shock-induced boundary layer separation, one of the most striking results found by Vandromme and Ha Minh (see [471], for instance), is the great sensitivity of the prediction to the value of $C_{\epsilon 1}$. In the transonic flow over a bump (Délery & Le Diuzet [130]), the "lambda"

[11] In [280], no compressibility terms are added.

shape of the shock wave is not observed with $C_{\epsilon 1} = 1.28$, while it is with $C_{\epsilon 1} = 1.57$. However, even in this case, the model is not able to produce the correct level of turbulence kinetic energy.

**Taulbee-VanOsdol (1991).** In Taulbee and VanOsdol [454], the usual (incompressible) form of eq.(10.55) is used to calculate the *solenoidal* dissipation, in addition to a closure model for the dilatation-dissipation. Coupled transport equations of the density variance and turbulent mass flux ($\overline{u_i''}$) are also included in the model.
For *2-D* thin shear layers, the modeled dissipation equation reads

$$\bar{\rho}\widetilde{U}\frac{\partial\bar{\epsilon}}{\partial x} + \bar{\rho}\widetilde{V}\frac{\partial\bar{\epsilon}}{\partial y} = \frac{\partial}{\partial y}(\frac{\mu_t}{\sigma_\epsilon}\frac{\partial\bar{\epsilon}}{\partial y}) + C_{\epsilon 1}\frac{\bar{\epsilon}}{\widetilde{k}}\mu_t(\frac{\partial\widetilde{U}}{\partial y})^2 - C_{\epsilon 2}\bar{\rho}\frac{\bar{\epsilon}^2}{\widetilde{k}} .$$

The "standard" values of the "incompressible" constants are used ($C_{\epsilon 1} = 1.44$, $C_{\epsilon 2} = 1.92$, $\sigma_\epsilon = 1.3$).
With this model, the prediction of the growth rate of a compressible mixing layer is in reasonable agreement with the experimental data.

***Discussion.*** Most proposals extending incompressible versions of the ($k$ - $\epsilon$) model to variable density and/or compressible situations, consist in adding compressibility corrections to the baseline incompressible formulation of the model. Some of them have been reviewed in section 10.8. With such ($k$ - $\epsilon$) closure types, it is worth noticing that, from computational grounds, specific schemes accounting for variable density effects in the dissipation equation cannot be introduced without a similar counterpart into the turbulence kinetic energy equation and vice versa.
Although not discussed here, the same kind of variable-density or compressible extension can also be applied to other types of two-equation models, such as the ($k$ - $\omega$) model proposed by Wilcox & Traci [486], or the ($k$ - $\phi$) model by Cousteix et al. [108] However, in all cases, Batchelor's scaling $\bar{\epsilon} \propto \widetilde{k}^{3/2}/\ell$, is adopted, assuming that only *one* characteristic length scale $l \propto \ell$ is accounting for *both* turbulent transport and energy transfer by fluctuating motions.
As discussed in §10.5.3, such an assumption could be at least questionable in variable density and/or compressible turbulent flows.

## 10.12. Three-equation models

As it has been seen in the previous section, different modifications of the incompressible versions of two-equation models have been proposed, (i) including changes in the incompressible schemes, and (ii) introducing new additional terms, specific to density variations.

With the aim of modeling latter contributions, one can presume that it could be more suitable to account for variable density and compressibility effects by adding, at least, one more additional transportable function, yielding a three-equation model.

Several proposals have been made for such extra transport equations. In 1979, Chen and Chen [91] derived a modeled transport equation for the temperature variance in order to predict vertical buoyant jets. The pressure variance was introduced by Zeman [497] in 1991 for predicting homogeneous compressible shear flows, and revisited by Hamba [203] in 1999. Three-equation models ($k$ - $\epsilon$ - $\overline{\rho'^2}$) using a density variance equation, have been developed in order to predict compressible mixing layers by Lejeune *et al.* [287], [286] in 1996 and 1997, Yoshizawa *et al.* [492] in 1997, and separated flows by Duranti & Pittaluga [137] in 2000. In Taulbee and VanOsdoll [454], the density variance is also obtained from a modeled transport equation, the closure of which is achieved with a model transport equation for the turbulent mass fluxes.

### 10.12.1. DENSITY VARIANCE EQUATION

By direct manipulation of the open set of equations, an exact transport equation can be derived for the density variance $\overline{\rho'^2}$ (see Chapter 4, eq.(4.32) and also Taulbee and VanOsdoll [454], for instance). Some models that have been derived for closing this equation are now presented.

**Taulbee-VanOsdol (1991).** The modeled form of the density variance transport equation proposed by Taulbee and VanOsdol [454] in *2-D* thin shear layers is

$$\widetilde{U}\frac{\partial\overline{\rho'^2}}{\partial x}+\widetilde{V}\frac{\partial\overline{\rho'^2}}{\partial y}=-2\overline{\rho'^2}\,\frac{\partial\widetilde{U}_j}{\partial x_j}+2\overline{\rho}\,\overline{v''}\,\frac{\partial\overline{\rho}}{\partial y}+\frac{\partial}{\partial y}\left(\frac{\mu_t}{\sigma_\rho}\frac{\partial}{\partial y}\left(\frac{\overline{\rho'^2}}{\overline{\rho}}\right)\right)-C_\rho\frac{\overline{\epsilon}}{\widetilde{k}}\frac{\overline{\rho'^2}}{\rho}, \tag{10.56}$$

where the values of the constants $\sigma_\rho = 1$ and $C_\rho = 5.3$ are chosen by trial and error to get reasonable agreement with experimental data.
In reference [454], the compressible flat plate boundary layer over an adiabatic wall is calculated with the low-Reynolds version of the ($k$ - $\epsilon$) model of Chien [94], *without compressibility corrections.* The density fluctuations variance is obtained from eq.(10.56), where the spanwise component $\overline{v''}$ of the turbulent mass fluxis calculated from a modeled transport equation, based on closure schemes (10.35), (10.36) and (10.37). Thus $k$ and $\epsilon$ equations can be solved independently from $\overline{\rho'^2}$, $\overline{u''}$ and $\overline{v''}$ equations.

The turbulence density intensity $(\overline{\rho'^2})^{1/2}/\overline{\rho}$ and the longitudinal turbulent mass flux normalized by the local mean velocity $\overline{u''}/\overline{U}$ are given in table 10.6. The predicted values are compared with the experimental data of

Kistler and the results obtained from Morkovin's strong Reynolds analogy (see Chapter 4, section 4.4.2):

$$I_\rho \equiv \frac{\sqrt{\overline{\rho'^2}}}{\overline{\rho}} = (\gamma - 1)M^2 \frac{\sqrt{\overline{u'^2}}}{\overline{U}} \quad \text{and} \quad \frac{\overline{u''}}{\overline{U}} = -(\gamma - 1)M^2 \frac{\overline{u'^2}}{\overline{U}^2} .$$

It is seen that the overall agreement between predictions and experiment is reasonable for the two values of the Mach number.
Finally, for this flow configuration, it should be added that according to the authors, the dominant production in the model equation (10.37), comes from the mean density gradient, so that a good modeling of the turbulent normal mass flux $\overline{v''}$ is crucial, and cannot be achieved with gradient type diffusion schemes.

TABLE 10.6. Comparison of turbulence density intensity and longitudinal turbulent mass flux in a flat plate boundary layer over an adiabatic wall, at $y/\delta = 0.5$, from Taulbee and VanOsdol [454].

| | $I_\rho$ | | $\overline{u''}/\overline{U}$ | |
|---|---|---|---|---|
| | $M_\infty = 1.7$ | $M_\infty = 4.7$ | $M_\infty = 1.7$ | $M_\infty = 4.7$ |
| Morkovin's Analogy | 0.035 | 0.131 | - 0.0016 | - 0.0124 |
| Model [454] | 0.028 | 0.120 | - 0.0008 | - 0.0028 |
| Kistler's Exp. | 0.027 | 0.115 | - 0.0011 | - 0.0028 |

**Lejeune-Kourta-Chassaing (1996).** To predict high-speed turbulent mixing layers, up to a convective Mach number of 0.8, the following model transport equation for the density variance was derived by Lejeune *et al.* [287]:

$$\frac{D\overline{\rho'^2}}{Dt} = -\overline{\rho'^2}\frac{\partial \widetilde{U}_j}{\partial x_j} + 2\frac{\nu_t}{\sigma_\rho}\left(\frac{\partial \overline{\rho}}{\partial x_j}\right)^2 + \frac{\partial}{\partial x_j}\left(\mu_t \frac{\partial}{\partial x_j}\left(\frac{\overline{\rho'^2}}{\overline{\rho}}\right)\right) - 2\frac{\overline{\rho}^2}{\gamma \overline{P}}\overline{p'\frac{\partial u'_j}{\partial x_j}} ,$$

where $\gamma = C_p/C_v$.
The scheme for the pressure-dilatation term is

$$\overline{p'\frac{\partial u'_j}{\partial x_j}} = \alpha \frac{\overline{\rho'^2}}{\overline{\rho}^2}\frac{\gamma \overline{P}}{M_t}\frac{\overline{\epsilon}}{\overline{k}} ,$$

where $M_t$ is the turbulence Mach number.
With $\alpha = -0.05$, $\sigma_\rho = 0.7$, the model predicts turbulence kinetic energy

profiles which are in good agreement with the measurements, but exhibits a too loose dependence of the mixing layer growth rate on the convective Mach number.

### 10.12.2. PRESSURE VARIANCE EQUATION

**Zeman (1991).** The phenomenological closure of the pressure variance equation proposed by Zeman [497] in homogeneous compressible shear flows is based on the heuristic argument that, in the absence of forcing, the pressure fluctuations (i) tend to relax to an equilibrium value depending on the turbulence Mach number, $p_e(M_t)$, and (ii) at a rate set by an acoustic timescale $\tau_a \propto L/a$ over an eddy of size $L$. Thus:

$$-2\overline{\rho}\, a^2\, \overline{p'\vartheta'} = \frac{D\overline{p'^2}}{Dt} = -\frac{\overline{p'^2} - p_e^2}{\tau_a}\,. \tag{10.57}$$

The acoustic time scale is taken as $\tau_a = 0.2\,\tau M_t$, where the turbulence time scale $\tau$ is related to vortical turbulence since $\tau = 2\widetilde{k}/\epsilon_s$, where $\epsilon_s$ denotes the solenoidal dissipation. Finally it is inferred from DNS data and theory that the equilibrium pressure variance can be taken as

$$p_e^2 = 2(\frac{M_t^2 + M_t^4}{1 + M_t^2 + M_t^4})\,\overline{\rho}^2\,\widetilde{k}\,a^2\,.$$

As shown by Zeman and Coleman [499], the model mechanism of relaxation to equilibrium is supported by direct numerical simulation data of homogeneous shear turbulence at moderate shear rates.

**Hamba (1999).** Restricting the analysis to homogeneous flows, the pressure variance equation reduces to, Hamba [203]:

$$\frac{D\overline{p'^2}}{Dt} = -2\gamma\overline{P}\,\overline{p'\vartheta'} - \epsilon_p + Res\,, \tag{10.58}$$

where $\vartheta' = \partial u_i'/\partial x_i$, $\epsilon_p$ is the pressure-variance dissipation and $Res$ stands for residual terms which can be considered as negligible.
At sufficiently high turbulence Mach number, the dissipation $\epsilon_p$ does not depend on the viscosity $\nu$, and can be modeled as

$$\epsilon_p = C_{\epsilon p1}(\gamma - 1)\frac{\overline{p'^2}}{\overline{k}P_{rod}}\epsilon\,, \tag{10.59}$$

where $P_{rod} = -\overline{u_i'u_j'}\partial\overline{U}_i/\partial x_j$ is the production rate of turbulence kinetic energy. The value of the model parameter $C_{\epsilon p1}$, which depends on the

turbulence Mach number $M_t$, is $C_{\epsilon p1} = 1.3$ for $M_t = 0.3$.
Although the pressure-dilatation term $\overline{p'\vartheta'}$ may be negligible in the $\overline{k}$ equation, it plays an important (actually dominant) role in the $\overline{p'^2}$ equation. The scheme for the pressure-dilatation correlation is as given in eq.(10.44).

## 10.13. Closure models for buoyant flows

Many problems involving buoyant modification to turbulence occur in natural environment where, to the exception of fire flows situations, the density change is never more than a few percent of the mean value. The present part of the monograph is not dedicated to the modeling of such types of flow configurations. Readers with direct interest in buoyancy affected turbulent flows may refer to Launder [274] or Craft [109], *inter alia.*

The following example is merely given to illustrate the way body-force influence was included in early modeling of transport ($k$ - $\epsilon$) equations and scalar variance equations (see e.g. Gibson & Launder [181] in 1976, Chen & Chen [91] in 1979).

**Chen and Chen (1979).** These authors [91] adapted the standard incompressible version of the ($k$ - $\epsilon$) model to predict vertical buoyant jets. They derived a three-equation ($k$ - $\epsilon$- $\overline{\theta'^2}$) model, adding a transport equation for the temperature variance $\overline{\theta'^2}$. With the thin shear layer approximations, the main modifications to the standard ($k$ - $\epsilon$) closure concern the expressions of the significant Reynolds stresses and heat flux components, which become:

$$\text{Reynolds stresses :}\quad -\overline{u'v'} = \frac{1-c_0}{c_1}\frac{\overline{v^2}}{\overline{k}}\left(1+\frac{\overline{k}\,g\,\partial\overline{T}/\partial y}{c_h\,\overline{\epsilon}\,T_a\,\partial\overline{U}/\partial y}\right)\frac{\overline{k}^2}{\overline{\epsilon}}\frac{\partial\overline{U}}{\partial y},$$

$$\overline{v'^2} = c_2\,\overline{k}\,,$$

$$\text{Turb. heat fluxes :}\quad -\overline{u'\theta'} = \frac{\overline{k}}{c_h\,\overline{\epsilon}}\left[\overline{u'v'}\frac{\partial\overline{T}}{\partial y} + \overline{u'\theta'}(1-c_{h1})\frac{\partial\overline{U}}{\partial y} + g(1-c_{h1})\frac{\overline{\theta'^2}}{T_a}\right],$$

$$-\overline{v'\theta'} = \frac{1}{c_h}\frac{\overline{v^2}}{\overline{k}}\frac{\overline{k}^2}{\overline{\epsilon}}\frac{\partial\overline{T}}{\partial y}\,.$$

The previous expressions are obtained from the exact transport equations of the second-order correlations $\overline{u'_iu'_j}$ and $\overline{u'_i\theta'}$ when neglecting convection and diffusion terms. $T_a$ is the ambient temperature, $g$, the gravitational constant, and the model constants are $c_0 = 0.55$, $c_1 = 2.2$, $c_2 = 0.53$, $c_h = 3.2$ and $c_{h1} = 0.5$. The incompressible closure of $\overline{k}$ and $\overline{\epsilon}$ equations is unchanged, except for the production terms, in which the additional buoyant contributions are respectively added:

$$P^k_{rod} = -\overline{u'v'}\frac{\partial\overline{U}}{\partial y} + g\,\frac{\overline{u'\theta'}}{T_a} \qquad \text{and} \qquad P^\epsilon_{rod} = C_{\epsilon 1}\frac{\overline{\epsilon}}{\overline{k}}P^k_{rod}\,.$$

## 10.14. Final discussion and concluding remarks

Before coming to scientific conclusions, it is worth keeping in mind the following two general points:

- Turbulence modeling of variable density flows addresses a wide range of situations, due to the various origins of density variations. Scientific faithfulness should require the review to be exhaustive, which is not the case of the present one. Progress in turbulence modeling has developed in a way far different from a logical, gradual, systematic approach. Historical faithfulness results in an impressionistic picture, which could make turbulence modeling look like a black art for non-specialists, as could probably appear from the present review;
- This chapter is restricted to reviewing some of the aspects of the modeling of density variations and compressible effects in low and high speed flows in first-order closure models. This choice is not motivated on scientifically argued grounds and simply results from pedagogical considerations: second-order modeling is addressed in next chapter. Therefore, it would be unwise to conclude from this review that first-order is an appropriate closure level for capturing the compressible effects that are now documented.

Within the previous frame, the following scientific conclusions can be drawn:

• In low-speed, non-reactive fluid motions, density variations arise from changes in temperature and/or composition, which can produce high density-intensity levels ($\sqrt{\overline{\rho'^2}}/\overline{\rho}$) in turbulent flows. As far as the modeling of statistical averaged equations is concerned, one is faced with the specific closure of turbulent mass flux and, more generally, correlations with density fluctuation (d.f.c.). Such terms, which have no equivalence in constant density flows, are present even when using density-weighted averages. Free shear flows can be predicted by using direct extensions of closure schemes derived for incompressible and buoyant flows, provided a suitable closure for such d.f.c. is adopted. In free jets, it has been demonstrated that a gradient-type diffusion closure for d.f.c. is not necessarily appropriate, depending upon the signs of mean velocity and density gradients.

• In non-separated high-speed boundary layers, up to supersonic free stream Mach numbers $M_\infty < 3$ to 5, and channel flows, the effects of density and pressure fluctuations under adiabatic conditions at the wall are small. Hence, compressibility effects are mostly due to *mean* density and temperature variations, and mean velocity profiles can be recovered from that in incompressible turbulence, using the Van Driest transformation [220]. To some extent, this explains why direct extensions of incompressible closure schemes to that of density-weighted moments for variable density flows ("first" generation), do not yield irrelevant predictions, as demonstrated by Ha Minh and co-workers [195], [199], [198], [196].

• In eddy-viscosity models, structural changes of the turbulence field associated with modifications in the Reynolds stress anisotropy tensor cannot be introduced explicitly into closure schemes, as it is the case with second-order modeling (see next chapter). Nevertheless, some consequences of such dominant effects can be indirectly and subtly accounted for in first-order closure models. This is basically the reason for improving the non-equilibrium schemes derived for the pressure dilatation correlation, Ristorcelli [394], Hamba [203], as part of the closure of the pressure variance transport equation.

• Even within the limit of Kovasznay's modes decomposition, density fluctuations are included in both acoustic ($p' \neq 0$, $\rho' \neq 0$, and $s' = \omega' = 0$) and entropy ($s' \neq 0$, $\rho' \neq 0$, and $\omega' = p' = 0$) modes. With nonadiabatic walls, density fluctuations can exist with little compressible turbulence effects [221]. Therefore, for boundary layers and channel flows, the density variance is not a suitable choice of an independent function to accurately account for compressibility effects in a ($k$ - $\epsilon$) model, for instance.

• In high-speed free shear flows, direct extensions of incompressible models, similar to those adopted in predicting wall bounded flows, fail to predict the strong compressibility effects observed in such flows. In a "second" generation of new compressible models, specific dilatational terms (pressure-dilatation correlation, compressible or dilatation dissipation) were presumed to account for such effects.

• This statement was not confirmed by later literature on the topic. In 1995, Sarkar [409] showed that the reduced growth rate of turbulence kinetic energy in homogeneous shear flow was primarily due to the reduced level of turbulence production, as a consequence of a change in the anisotropy of the Reynolds stress due to compressibility. In 1996, Vreman *et al.* [477] showed that reduced pressure fluctuations are responsible for the reduction in the growth rate of mixing layers via the pressure-strain term. In 1999, Hamba [203] confirmed the conclusion that, in compressible homogeneous shear flow, the anisotropy of the Reynolds stress, which reduces the turbulence production, is primarily due to the decrease in the pressure-strain term $\Pi_{12}(=\overline{p'\partial u_1'/\partial x_2})$ which, in turn, results from the reduced level of pressure fluctuations.

Since dilatational corrections for pressure and dissipation terms in this second generation of compressible models were introduced in second-order closure models, Zeman [495], Sarkar and Lakshmanan [413], the observed improvements are hardly separable from those resulting from other modifications.

• The observations quoted in the previous item are based on DNS studies. They are not yet entirely confirmed by experimental evidence, due to technical difficulty in the measurements. For example, there is no clear indication of changes in the Reynolds stress anisotropy in [146] or [77], as opposed to [186].

*CHAPTER 11*

# SECOND-ORDER MODELING

*In addition to first-order closure models reviewed in the previous chapter, second-order closure schemes are discussed in the present one to complete the review on single point modeling. Accordingly, this chapter is intended as providing some insights on where second-order turbulence models have reached in accounting for several distinct effects due to density variation in low speed motions and compressibility in high Mach number flows.*

## 11.1. Introduction

In incompressible fluid motion, second-order turbulence modeling is concerned with the closure of transport equations for second-order moments: $\rho_0\overline{u_i'u_j'}$, $\rho_0\overline{u_i'f'}$ and $\rho_0\overline{f'^2}$, where $f'$ stands for any turbulent fluctuation of a passive scalar contaminant and $\rho_0$ is the (constant) density of the fluid.

In variable density fluid motions, a similar approach can be adopted, based on equivalent correlations which are now including density variations, such as $\overline{\rho u_i'u_j'}$, with a ternary regrouping (centered fluctuations), or $\overline{\rho u_i''u_j''}$, with a binary regrouping using Favrian fluctuations (see Chapter 5).

From a general point of view, second-order modeling in variable density turbulent flows addresses transport equations for the following moments $\overline{\rho u_i''u_j''}$, $\overline{\rho\,\theta''u_j''}$, $\overline{\rho\gamma''u_j''}$, $\overline{\rho\,\theta''^2}$ and $\overline{\rho\gamma''^2}$, where $\theta''$ and $\gamma''$ denote the Favrian fluctuations of temperature ($T$) and mass fraction ($C$) in a binary mixture respectively.
To avoid solving an increasing number of additional transport equations in compressible flows, scalar correlations are often derived from a generalized gradient diffusion assumption. In this case, second-order modeling is mostly concerned with deriving closure schemes for the Reynolds stress transport equation. This question will be thus discussed first.

## 11.2. The modeling issue of RST equation

### 11.2.1. THE OPEN TRANSPORT EQUATION

The Reynolds stress transport equation in a compressible fluid motion has been derived in Chapter 5. By simple rearrangement of the right-hand-side terms of eq.(5.36) and using Favre's averaging $(\widetilde{u''_i u''_j} \equiv \overline{\rho u''_i u''_j}/\overline{\rho})$, it reads (see also Chapter 9):

$$\frac{\partial(\bar{\rho}\widetilde{u''_i u''_j})}{\partial t} + \frac{\partial(\bar{\rho}\widetilde{u''_i u''_j}\tilde{U}_k)}{\partial x_k} = \bar{\rho}P_{ij} - \frac{\partial(T_{ijk})}{\partial x_k} + \overline{\rho}\,\Pi^*_{ij} + \Sigma_{ij} - \bar{\varepsilon}_{ij}\,, \qquad (11.1)$$

where the right-hand-side contributions are:

Turbulent production: $P_{ij} = -\left(\widetilde{u''_i u''_k}\dfrac{\partial\tilde{U}_j}{\partial x_k} + \widetilde{u''_j u''_k}\dfrac{\partial\tilde{U}_i}{\partial x_k}\right)$,

Pressure strain: $\Pi^*_{ij} = \dfrac{1}{\bar{\rho}}\left[\overline{p'(\dfrac{\partial u''_i}{\partial x_j} + \dfrac{\partial u''_j}{\partial x_i})}\right]$,

Transport: $T_{ijk} = \overline{\rho u''_i u''_j u''_k} + \overline{p' u'_i}\delta_{jk} + \overline{p' u'_j}\delta_{ik}$
$\qquad -(\overline{\tau'_{jk}u''_i} + \overline{\tau'_{ik}u''_j})$,

Mass flux coupling: $\Sigma_{ij} = \overline{u''_i}\left(\dfrac{\partial\bar{\tau}_{jk}}{\partial x_k} - \dfrac{\partial\bar{P}}{\partial x_j}\right) + \overline{u''_j}\left(\dfrac{\partial\bar{\tau}_{ik}}{\partial x_k} - \dfrac{\partial\bar{P}}{\partial x_i}\right)$,

Turbulent dissipation: $\bar{\varepsilon}_{ij} = \overline{\left(\tau'_{jk}\dfrac{\partial u''_i}{\partial x_k} + \tau'_{ik}\dfrac{\partial u''_j}{\partial x_k}\right)}$ .

Two formal modifications can be introduced in eq.(11.1), as inferred from the analysis of the constant density situation.

a) In isovolume turbulent fluid motions, the pressure-strain term is traceless $(\Pi^{*(I)}_{ii} \equiv 0)$, which means that it is only responsible for a *redistribution* of energy between the normal stresses, without changing the total amount of turbulence kinetic energy. The same characteristic can be recovered in variable density fluid motions, when considering the *deviatoric* part of the pressure-strain correlation:

$$\overline{\rho}\,\Pi_{ij} = \overline{\rho}\,\Pi^*_{ij} - \frac{2}{3}\overline{p'\vartheta'}\delta_{ij}\,, \qquad (11.2)$$

where $\vartheta' = \partial u'_i/\partial x_i$. It is recalled that $u'_i = u''_i - \overline{u''_i}$ (see Tab.5.2), so that $\overline{p'\vartheta'} \equiv \overline{p'\vartheta''}$.

b) As suggested by Lumley [306], a similar treatment can be applied to the dissipation rate tensor $\bar{\varepsilon}_{ij}$, making use of its *deviatoric* part $\bar{\mathbf{e}}_{ij}$. Neglecting viscosity fluctuations, the expression reads:

$$\bar{\varepsilon}_{ij} = \bar{\rho}\bar{\epsilon}_{ij} \equiv \bar{\rho}(\bar{\mathbf{e}}_{ij} + \frac{2}{3}\bar{\epsilon}\delta_{ij}) ,$$

where $\bar{\epsilon} = \bar{\epsilon}_{ii}/2$ is the dissipation rate of turbulence kinetic energy. Now, according to the distinction between solenoidal and dilatation dissipation introduced in compressible flows (see Chapter 9), the isotropic contribution can be taken as:

$$\bar{\epsilon} = \bar{\epsilon}_s + \bar{\epsilon}_d + \bar{\epsilon}_{nh} . \tag{11.3}$$

The definitions of the dilatation (or compressible), solenoidal and non homogeneous, contributions to the dissipation — respectively denoted by $\bar{\epsilon}_d$, $\bar{\epsilon}_s$ and $\bar{\epsilon}_{nh}$ — are given in eq.(10.7).

### 11.2.2. CLOSURE PECULIARITIES IN VARIABLE DENSITY FLOWS

On substituting the previous equations in eq.(11.1), we can write the open Reynolds stress transport equation as follows:

$$\frac{\partial(\bar{\rho}\widetilde{u''_i u''_j})}{\partial t} + \frac{\partial(\bar{\rho}\widetilde{u''_i u''_j}\tilde{U}_k)}{\partial x_k} = \bar{\rho}P_{ij} - \frac{\partial(T_{ijk})}{\partial x_k} + \bar{\rho}\,\Pi_{ij} - \bar{\rho}\mathbf{e}_{ij} - \frac{2}{3}\bar{\rho}\,\bar{\epsilon}_s\delta_{ij}$$
$$+\Sigma_{ij} + \frac{2}{3}[\overline{p'\vartheta'} - \bar{\rho}(\bar{\epsilon}_d + \bar{\epsilon}_{nh})]\delta_{ij} , \tag{11.4}$$

where the terms that *explicitly* differ from the constant density situation are now grouped in the second line.

Now, as far as the modeling of eq.(11.4) is concerned, it can be observed, since the the production term is exact, that two categories of terms are to be considered:

- the first one consists of transport, pressure-redistribution correlation and dissipation contributions that are present in constant density flows;
- the second one, which has no counterpart in constant density situations, is due to turbulent mass flux coupling, pressure-dilatation correlation and dilatational dissipation.

The modeling issue of variable density fluid flows will be now presented, according to the previous distinction.

## 11.3. RST equation closure schemes

### 11.3.1. DIFFUSION IN INCOMPRESSIBLE FLOWS

The incompressible counterpart $T^I_{ijk}$ of the transport term $T_{ijk}$ in eq.(11.4) includes three additive contributions: (i) diffusion or turbulent transport by fluctuating motions, (ii) pressure-velocity correlation and (iii) molecular or viscous diffusion, viz. formally:

$$T^I_{ijk} = T^{It}_{ijk} + T^{Ip}_{ijk} + T^{I\nu}_{ijk} ,$$

with

$$T^{It}_{ijk} = \rho_0 \overline{u'_i u'_j u'_k} \quad \text{and} \quad T^p_{ijk} = (\overline{p' u'_i}\delta_{jk} + \overline{p' u'_j}\delta_{ik}) .$$

Now, in incompressible fluid motions, it can be demonstrated that the sum of all viscous contributions in the Reynolds stress transport equation — which includes molecular diffusion plus dissipation —, can be written as (see, e.g. Chassaing [84], page 85):

$$\nu \left( \overline{u'_i \frac{\partial^2 u'_j}{\partial x_k \partial x_k}} + \overline{u'_j \frac{\partial^2 u'_i}{\partial x_k \partial x_k}} \right) = \nu \frac{\partial^2 (\overline{u'_i u'_j})}{\partial x_k \partial x_k} - 2\nu \overline{(\frac{\partial u'_i}{\partial x_k})(\frac{\partial u'_j}{\partial x_k})} ,$$

so that the viscous diffusion "flux" is exactly:

$$T^{I\nu}_{ijk} = \nu \frac{\partial}{\partial x_k} (\overline{u'_i u'_j}) , \tag{11.5}$$

which requires no closure at second-order modeling level.

*Diffusion by fluctuating motions*

The modeling of turbulence diffusion is based on simple or generalized gradient diffusion expressions. Some of them are recalled in Table 11.1, where, by convention, the corresponding diffusion term is written:

$$-\frac{\partial T^{It}_{ijk}}{\partial x_k} \equiv -\rho_0 \frac{\partial \overline{u'_i u'_j u'_k}}{\partial x_k} .$$

More elaborate schemes have been proposed by Cormack *et al.* [100] in 1978 or Magnaudet [312] in 1993. They are not detailed here, since their use in variable density fluid motions has not been yet reported.

As observed in Table 11.1, some proposals are based on more tensorially consistent expressions than the simpler gradient diffusion scheme (see e.g. Shir's model) and use either scalar or tensorial diffusivities. However, as demonstrated by Cazalbou & Chassaing [71], some tensorial extensions can exhibit a spurious behavior, even in constant density flows. Moreover, the implementation of the more general tensorial expressions in a stable numerical code is not necessarily an easy task, so that simpler forms are still often preferred. They give rather similar predictions in simple shear flows.

TABLE 11.1. Turbulent diffusion schemes in modeling incompressible fluid motions.

| Author - Ref. | Year | $\overline{u'_i u'_j u'_k}$ | Parameter(s) |
|---|---|---|---|
| Daly & Harlow [117] | 1970 | $-C_S \frac{\overline{k}}{\overline{\epsilon}} \overline{u'_k u'_l} \frac{\partial}{\partial x_l}(\overline{u'_i u'_j})$ | $C_S \approx 0.22$ to 0.25 |
| Donaldson [133] | 1971 | $-L\overline{k}^{1/2}(\frac{\partial \overline{u'_i u'_j}}{\partial x_k} + \frac{\partial \overline{u'_j u'_k}}{\partial x_i} + \frac{\partial \overline{u'_i u'_k}}{\partial x_j})$ | $L$, macro length scale |
| Mellor & Herring [318] | 1972 | $-\frac{2}{3} C_S \frac{\overline{k}^2}{\overline{\epsilon}}(\frac{\partial \overline{u'_i u'_j}}{\partial x_k} + \frac{\partial \overline{u'_j u'_k}}{\partial x_i} + \frac{\partial \overline{u'_i u'_k}}{\partial x_j})$ | $C_S \approx 0.108$ |
| Hanjalić & Launder [205] | 1972 | $-C_S \frac{\overline{k}}{\overline{\epsilon}}(\overline{u'_k u'_l} \frac{\partial \overline{u'_i u'_j}}{\partial x_l} + \overline{u'_i u'_l} \frac{\partial \overline{u'_j u'_k}}{\partial x_l} + \overline{u'_j u'_l} \frac{\partial \overline{u'_i u'_k}}{\partial x_l})$ | $C_S \approx 0.11$ |
| Shir [429] | 1973 | $= -C_S \frac{\overline{k}^2}{\epsilon} \frac{\partial}{\partial x_k} \left(\overline{u'_i u'_j}\right)$ | $C_S \approx 0.04$ |

*Diffusion due to pressure fluctuations*
The pressure contribution is generally neglected or taken as part of the closure scheme of the turbulent transport by fluctuating motions. Some exceptions are given in Table 11.2, that apply to the diagonal contribution of the pressure-diffusion term, i.e.,

$$\frac{\partial T_k^{Ip}}{\partial x_k} = \frac{\partial \overline{p' u'_k}}{\partial x_k}.$$

In Craft & Launder [110] (CL), a non-conventional decomposition of the pressure correlation is adopted:

$$-\left(\overline{u'_i \frac{\partial p'}{\partial x_j} + u'_j \frac{\partial p'}{\partial x_i}}\right) = \phi^*_{ij} + \frac{\overline{u'_i u'_j}}{2\overline{k}} \frac{\partial}{\partial x_l}(\overline{p' u'_l}),$$

where $\phi^*_{ij}$ is a modified pressure-strain tensor, the modeing of which is described in section 11.3.3.
In the expression of the CL incompressible scheme (Table 11.2), $f$ is a function of the second invariant $\mathbf{II}_b = b_{ij} b_{ji}$ of the turbulent stress anisotropy tensor $b_{ij} = \overline{u'_i u'_j}/\overline{k} - 2\delta_{ij}/3$, the turbulence Reynolds number $R_t = \overline{k}^2/(\nu \overline{\epsilon})$ and the two-component limit parameter $\beta = 1 - 9(\mathbf{II}_b - \mathbf{III}_b)/8$.

TABLE 11.2. Pressure-velocity correlation schemes in modeling incompressible pressure fluctuations diffusion.

| Author - Ref. | Year | $\overline{p'u'_i}/\rho$ |
|---|---|---|
| Hirt [216] | 1969 | $\frac{\overline{k}^2}{\overline{\epsilon}}\frac{\partial}{\partial x_j}(\overline{u'_i u'_j})$ |
| Lumley [305] | 1975 | $-\frac{2}{5}\overline{u'_i u'_j u'_j}$ |
| Craft & Launder [110] | 1996 | $-f(\mathbf{II}_b, \beta, R_t) \times (\nu \overline{k}\overline{\epsilon})^{1/2}$ |

11.3.2. DIFFUSION IN VARIABLE DENSITY FLOWS

*Viscous diffusion*

Any change in density is generally associated with variations in diffusive coefficients of the fluid, introducing additional correlations in the exact expression of the mean molecular diffusion terms, due to fluctuations of all molecular transport coefficients (viscosity, thermal conductivity...). Various assumptions can be adopted to simplify the averaged result.

Some models, as in Batten *et al.* [38], are using a formal extension of eq.(11.5), viz.

$$T^{\nu}_{ijk} = \overline{\mu}\frac{\partial \widetilde{u''_i u''_j}}{\partial x_k}.$$

In Speziale & Sarkar [443], the viscous diffusion term is approximated by:

$$T^{\nu}_{ijk} = \overline{\mu}(\frac{\partial \overline{u'_i u'_j}}{\partial x_k} + \frac{\partial \overline{u'_j u'_k}}{\partial x_i} + \frac{\partial \overline{u'_k u'_i}}{\partial x_j}), \tag{11.6}$$

in which fluctuations in viscosity are neglected, as well as some higher-order correlations, particularly those involving dilatational effects.

*Turbulent diffusion*

Concerning turbulent diffusion, most of the closure schemes adopted in variable density fluid motions come from *direct* or *formal* extensions of incompressible expressions. To provide a more substantial basis to this rather heuristic assertion, Shih *et al.* [427] pointed out that the general approach developed by Lumley [303] in modeling constant density flows is also satisfactory in variable density fluid motions. Thus it can be shown that the influence of density variation on the third-order moment equations only appears at second-order level in density fluctuations — $\mathcal{O}(\rho'/\overline{\rho})^2$. Therefore, closure schemes of third-order moments developed in constant density flows can be used in a variable density fluid motions.

*Direct* extension means that compressible "forces" and "fluxes", $\overline{\rho u'_i u'_j u'_k}$ and $\overline{\rho u'_i u'_j}$ for example, are substituted to the incompressible ones, say $\rho_0(\overline{u'_i u'_j u'_k})$ and $\rho_0(\overline{u'_i u'_j})$ respectively. Adopting, for instance, a simple gradient-type relation, this yields:

$$\overline{\rho u''_i u''_j u''_k} = -\alpha_t \frac{\partial}{\partial x_k}(\overline{\rho u''_i u''_j}) \qquad \text{or} \qquad \overline{\rho u'_i u'_j u'_k} = -\alpha_t \frac{\partial}{\partial x_k}(\overline{\rho u'_i u'_j}) ,$$

depending on whether a binary or ternary regrouping is adopted (see Chapter 5). In the previous relations, $\alpha_t$ ($\mathrm{L}^2 \times \mathrm{T}^{-1}$) stands for a scalar diffusion coefficient. With usual Favre notations, the first expression equally reads:

$$\overline{\rho}\, \widetilde{u''_i u''_j u''_k} = -\alpha_t \frac{\partial}{\partial x_k}(\overline{\rho}\, \widetilde{u''_i u''_j}) . \tag{11.7}$$

Eq.(11.7) differs from the following *formal* transposition, also dimensionally correct, of the incompressible gradient-type relation:

$$\overline{\rho} \widetilde{u''_i u''_j u''_k} = -\alpha_t \overline{\rho} \frac{\partial}{\partial x_k}(\widetilde{u''_i u''_j}) \;\text{or simply}\; \widetilde{u''_i u''_j u''_k} = -\alpha_t \frac{\partial}{\partial x_k}(\widetilde{u''_i u''_j}) . \tag{11.8}$$

Since $\quad \dfrac{\partial(\overline{\rho}\widetilde{u''_i u''_j})}{\partial x_k} \equiv \overline{\rho} \dfrac{\partial(\widetilde{u''_i u''_j})}{\partial x_k} + \widetilde{u''_i u''_j} \dfrac{\partial \overline{\rho}}{\partial x_k} ,$

large differences between the consequences of eqs.(11.7) and (11.8) are expected in the presence of strong mean density gradients.

As shown in the following examples, both types of extensions can be found in the literature.

**Varma-Fishburne-Beddini (1977).** In predicting turbulent diffusion flames with a second-order model, Varma *et al.* [475] adopted a simple gradient diffusion scheme for the third-order velocity correlations. It satisfies tensrial symmetry and has a scalar diffusion coefficient:

$$\overline{u'_i u'_j u'_k} = -2 C_S \Lambda \sqrt{\overline{k}} \left( \frac{\partial \overline{u'_i u'_j}}{\partial x_k} + \frac{\partial \overline{u'_j u'_k}}{\partial x_i} + \frac{\partial \overline{u'_k u'_i}}{\partial x_j} \right) ,$$

with $C_S \approx 0.1$ to $0.3$.
$\Lambda$ is a turbulent macroscale determined from, either the mean velocity profile, or the turbulence kinetic energy profile. For axisymmetric flows for instance, $\Lambda = 0.5\, r_{1/2}$, where $r_{1/2}$ is the conventional half-width radius based on mean axial velocity profiles, or $\Lambda = 0.2\, r_*$, where $r_*$ is the distance where $\overline{k} = \overline{k}_{max}/4$.

**Chassaing (1979).** In predicting turbulent mixing of low-speed variable density jets, the following extension of the gradient relation, originally proposed by Daly & Harlow[117], is adopted by Chassaing [80]:

$$\overline{\rho u_i'' u_j'' u_k''} + \overline{p' u_i'}\delta_{jk} + \overline{p' u_j'}\delta_{ik} = -C_S \frac{\overline{k}}{\overline{\rho}\,\overline{\epsilon}} \overline{\rho u_k'' u_l''} \frac{\partial(\overline{\rho u_i'' u_j''})}{\partial x_l},$$

with $C_S$=0.25.

**Shih-Janicka-Lumley (1987).** In predicting low-speed mixing layers of variable composition, the following third-order velocity correlation is adopted by Shih *et al.* [427] :

$$\overline{u_i' u_j' u_k'} = \frac{2}{3\beta_s} \frac{\overline{k}}{\overline{\epsilon}} [-(\overline{u_k' u_p'} \frac{\partial \overline{u_i' u_j'}}{\partial x_p} + \overline{u_i' u_p'} \frac{\partial \overline{u_j' u_k'}}{\partial x_p} + \overline{u_j' u_p'} \frac{\partial \overline{u_k' u_i'}}{\partial x_p})$$

$$+ \frac{\beta_s - 2}{3} \frac{\overline{\epsilon}}{\overline{k}} (\overline{k u_i'}\delta_{jk} + \overline{k u_j'}\delta_{ki} + \overline{k u_k'}\delta_{ij})] \,.$$

Here, $\beta_s$ is a model parameter, taken as a function of the turbulence Reynolds number and the second and third invariant of the Reynolds stress anisotropy tensor.

**Speziale-Sarkar (1991).** The model adopted by Speziale & Sarkar [443], Sarkar & Lakshmanan [413], in predicting various high-speed, compressible flows (isotropic turbulence, homogeneous shear flow, supersonic mixing layer, supersonic flat-plate turbulent boundary layer) is the isotropized version of the gradient transport formulation introduced by Launder, Reece & Rodi [279]. It reads:

$$\widetilde{u_i'' u_j'' u_k''} = -\frac{2}{3} C_S \frac{\widetilde{k}^2}{\widetilde{\epsilon}} (\frac{\partial \overline{u_i' u_j'}}{\partial x_k} + \frac{\partial \overline{u_j' u_k'}}{\partial x_i} + \frac{\partial \overline{u_k' u_i'}}{\partial x_j}) \tag{11.9}$$

with $C_S$=0.11.
When eq.(11.9) is used along with eq.(11.6), each turbulent transport term is coupled with a viscous term of the same form.

**Batten** *et al.* **(1999).** In Batten *et al.* [38], the diffusion terms in the Reynolds stress transport equation used for the prediction of two- and three-dimensional compressible flows are modeled by:

$$\frac{\partial}{\partial x_k}(-\overline{\rho}\, \widetilde{u_i'' u_j'' u_k''} + \widetilde{\mu} \frac{\partial \widetilde{u_i'' u_j''}}{\partial x_k}) - \frac{\widetilde{u_i'' u_j''}}{2\widetilde{k}} \frac{\partial}{\partial x_k}(\overline{p' u_k'}) \,, \tag{11.10}$$

where the following simple gradient-type expression is applied to the turbulent-diffusion contribution:

$$\bar{\rho}\,\widetilde{u_i''u_j''u_k''} = -C_S\frac{\bar{\rho}\tilde{k}}{\tilde{\epsilon}}\widetilde{u_k''u_l''}\frac{\partial\widetilde{u_i''u_j''}}{\partial x_l}\,,$$

with $C_S \approx 0.22$.
The pressure-velocity correlation is modeled according to the original form proposed by Craft & Launder [110] (see Table 11.2). The viscous contribution is obtained from a direct extension of the incompressible formulation, with a locally varying viscosity.

### 11.3.3. PRESSURE-STRAIN CORRELATION IN INCOMPRESSIBLE FLOWS

In incompressible fluid flow, the pressure fluctuation can be obtained by solving a Poisson equation, as detailed in Chapter 3. In this case, the source terms of this equation involve both mean and fluctuating contributions from the velocity field, viz.

$$\Delta p'^I = -2\frac{\partial^2(\rho_0\overline{U}_j u_i')}{\partial x_i\partial x_j} - \frac{\partial^2(\rho_0 u_i'u_j' - \rho_0\overline{u_i'u_j'})}{\partial x_i\partial x_j} + \frac{\partial^2\tau_{ij}'}{\partial x_i\partial x_j}\,, \tag{11.11}$$

where $\rho_0$ is the (constant) density of the fluid.
In high turbulence Reynolds number flows, the presence of linear and quadratic expressions in velocity fluctuations — first two terms in the right-hand-side of eq.(11.11) — classically suggests to consider the pressure fluctuation contribution as a sum of two parts:

$$p' = p'^I_{(1)} + p'^I_{(2)}\,.$$

By construction, each pressure fluctuation satisfies one of the following equations:

$$\Delta p'^I_{(1)} = -2\frac{\partial^2(\rho_0\overline{U}_j u_i')}{\partial x_i\partial x_j}\ (a)\,,\quad \Delta p'^I_{(2)} = -\frac{\partial^2(\rho_0 u_i'u_j' - \rho_0\overline{u_i'u_j'})}{\partial x_i\partial x_j}\ (b)\,. \tag{11.12}$$

In isovolume turbulence, it is well known that the pressure-strain correlations resulting from $p'^I_{(1)}$ and $p'^I_{(2)}$, are associated with two distinct mechanisms. The first one, involving $p'^I_{(1)}$, corresponds to the response of a turbulent field to changes driven by mean strain rates which are explicitly present in eq.(11.12) $(a)$. It introduces the so called "*rapid*" part of the pressure-strain correlation. The second one (associated with $p'^I_{(2)}$) is present even

in purely homogeneous flows. It acts as a relaxation mechanism of the turbulence field toward an isotropic state. It introduces the "return-to-isotropy" or *slow*" part of the pressure-strain correlation.
Hence, the general closure scheme of the pressure-strain correlation is usually taken as:

$$\Pi_{ij}^{I} = \Pi_{ij}^{I\,rapid} + \Pi_{ij}^{I\,slow}\,, \tag{11.13}$$

where

$$\Pi_{ij}^{I\,rapid} = \overline{p'^{I}_{(1)}(\frac{\partial u'_i}{\partial x_j} + \frac{\partial u'_j}{\partial x_i})} \quad \text{and} \quad \Pi_{ij}^{I\,slow} = \overline{p'^{I}_{(2)}(\frac{\partial u'_i}{\partial x_j} + \frac{\partial u'_j}{\partial x_i})}\,.$$

Several closure schemes have been derived for both slow and rapid parts, as reviewed for instance in Schiestel [421], Piquet [366], Pope [370] and Chassaing [84]. Some of them are simply recalled here.
Since the deviatoric dissipation rate tensor has been introduced in the formulation of the Reynolds stress transport equation — eq.(11.4) —, the following expressions apply to all redistributing contributions, viz.

$$L_{ij} = \Pi_{ij} - \overline{\mathbf{e}}_{ij}$$

and not only to the pressure-strain correlation.

*Slow part*
According to Lumley & Newman [308] (see also Schiestel [421], page 111), the general expression in a third order development in anisotropy of $L_{ij}$ is:

$$\frac{L_{ij}}{\overline{\epsilon}} = \alpha_0 \mathbf{II}_a \delta_{ij} + (\alpha_1 + \alpha_2 \mathbf{II}_a)\, a_{ij} - 3\alpha_0 a_{ik} a_{kj}\,, \tag{11.14}$$

where $a_{ij} = \overline{u'_i u'_j}/(2\overline{k}) - \delta_{ij}/3$ is the Reynolds stress anisotropy tensor and $\mathbf{II}_a = a_{ij} a_{ji}$ its second invariant.
According to eq.(11.14), three model parameters are to be prescribed.

***Linear expressions:*** Based on Rotta's ideas in 1951, closure schemes that were first proposed were based on a linear form of eq.(11.14) ($\alpha_0 = 0$), with $\alpha_1 + \alpha_2 \mathbf{II}_a = C^t$, viz.

$$\Pi_{ij}^{I\,slow} = -2A_0 \overline{\epsilon} a_{ij}\,.$$

Some values of the model constant $A_0$ are given in Table 11.3. As suggested from the data in this table, the value of the model "constant" is found to be dependent on the anisotropy level and the turbulence Reynolds number. In 1979, Lumley [306] proposed an interpolated expression for $A_0$, as a function of the turbulence Reynolds number, the second and third invariant of $a_{ij}$.

TABLE 11.3. Model constant of linear return-to-isotropy schemes.

| Author - Ref. | Year | $A_0$ |
|---|---|---|
| Launder, Reece & Rodi [279] | 1975 | 1.5 |
| Lumley & Newman [308] | 1977 | 1.62 |
| Gibson & Launder [182] | 1978 | 1.8 |
| Reynolds [389] | 1984 | 1.25 |

***Non-linear expressions:*** Non-linear schemes were probably first developed by Lumley & Khajeh-Nouri [307] in 1974. Some proposals, which have been derived by the UMIST group, are recalled in Table 11.4, referring to the general expression:

$$\frac{L_{ij}}{\overline{\varepsilon}} = \alpha_0' \mathbf{II}_b \delta_{ij} + \alpha_1' b_{ij} - 3\alpha_0' b_{ik} b_{kj} \,. \tag{11.15}$$

This scheme corresponds to eq.(11.14) with $\alpha_2 = 0$. Some model coefficients are given in Table 11.4, where, according to the notations from the quoted authors, $b_{ij} = 2a_{ij}$, and $\mathbf{II}_b = b_{ij} b_{ji}$ denotes its second invariant. $\beta$ is the two-component limit parameter based on $b_{ij}$ ($\beta = 1 - 9\left(\mathbf{II}_b - \mathbf{III}_b\right)/8$).

TABLE 11.4. Non-linear return-to-isotropy coefficient in eq.(11.15), from the UMIST group.

| Author - Ref. | Year | Model | Coefficient |
|---|---|---|---|
| Fu *et al.* [167] | 1987 | $\alpha_0' = 0.25$ | $\alpha_1' = -1.2 - \mathbf{II}_b$ |
| Launder [275] | 1987 | $\alpha_0' = -0.2\alpha_1'$ | $\alpha_1' = -7.5\mathbf{II}_b\sqrt{\beta}$ |
| Launder [276] | 1991 | $\alpha_0' = -0.4\alpha_1'$ | $\alpha_1' = -3.1\mathbf{II}_b\sqrt{\beta}$ |
| Craft *et al.* [112] | 1996 | $\alpha_0' = -\frac{7}{30}\alpha_1'$ | $\alpha_1' = -(1 + 3.75\sqrt{\mathbf{II}_b})\beta$ |

*Rapid part*

From eq.(11.12a), it can be inferred that the general expression of the rapid part can be taken as:

$$L_{ij}^{I\,rapid} = A_{nj}^{mi} \frac{\partial \overline{U}_m}{\partial x_n} \,. \tag{11.16}$$

A wide variety of models has been developed on this basis. Different proposals result from the choice of the constraints that are applied to the fourth order rank tensor $A^{mi}_{nj}$. Some of them are recalled hereafter.

***Linear expressions:*** These schemes are based on a generic expression of the fourth rank tensor $A^{mi}_{nj}$ which is *linear* in the Reynolds stress and satisfies the following required conditions, due to kinematic and symmetry properties:
$A^{mi}_{ni}=0$ and $A^{mi}_{kk}=2\overline{u'_m u'_i}$.

The resultant model for $L^{I\,rapid}_{ij}$ is expressible as:

$$L^{I\,rapid}_{ij} = (\frac{2}{5}+4C)\overline{k}\times\left(\frac{\partial\overline{U}_i}{\partial x_j}+\frac{\partial\overline{U}_j}{\partial x_i}\right)+\frac{2}{3}(C-1)(P_{ij}-\frac{Q}{3}\delta_{ij}) + \frac{2}{3}(1+8C)(Q_{ij}-\frac{Q}{3}\delta_{ij}), \tag{11.17}$$

or

$$L^{I\,rapid}_{ij} = 2\overline{k}\{\frac{2}{5}\overline{S}_{ij}-3C(b_{ik}\overline{S}_{kj}+b_{jk}\overline{S}_{ki}-\frac{2}{3}b_{mn}\overline{S}_{mn}\delta_{ij}) - \frac{1}{3}(2+7C)(b_{ik}\overline{R}_{kj}+b_{jk}\overline{R}_{ki})\}. \tag{11.18}$$

In eq.(11.17), $Q_{ij}=-\left(\overline{u_i u_k}\frac{\partial\overline{U}_k}{\partial x_j}+\overline{u_j u_k}\frac{\partial\overline{U}_k}{\partial x_i}\right)$ and $Q=Q_{ii}(\equiv P_{ii})$.

In eq.(11.18), $\overline{S}_{ij}$ and $\overline{R}_{ij}$ are the mean strain rate and rotation rate tensors respectively.
Both formulations are equivalent. They were originally derived by Launder *et al.* [279] and Reynolds [388]. As shown in eqs.(11.17) and (11.18), only one model parameter is required for such linear closure schemes. Some values of that parameter are given in Table 11.5.

TABLE 11.5. Rapid part of pressure-strain closure: Model constant $C$ of linear schemes, according to eqs.(11.17) or (11.18).

| Author - Ref. | Year | $C$ |
|---|---|---|
| Launder, Reece & Rodi [279] | 1975 | $-8/55 \approx -0.145$ |
| Lumley [305] | 1975 | $-0.166$ |
| Reynolds [388] | 1976 | $-0.150$ |
| Reynolds [389] | 1984 | $-2/7 \approx -0.286$ |

Two simplified expressions to the previous linear schemes have also been considered, corresponding to the following models:

***Isotropic model:*** $$\Pi_{ij}^{I\,rapid} = \frac{2}{5}\overline{k}\left(\frac{\partial \overline{U}_i}{\partial x_j} + \frac{\partial \overline{U}_j}{\partial x_i}\right) . \tag{11.19}$$

This scheme was first derived by Crow [115] in 1968. It is *exact* in isotropic turbulence.

***Isotropization of production:*** $$\Pi_{ij}^{I\,rapid} = -C_{ip}(P_{ij} - \frac{Q}{3}\delta_{ij}) . \tag{11.20}$$

This scheme was proposed by Naot *et al.* [344] in 1970, with $C_{ip}=3/5$ in order to satisfy the isotropic limit.

***Non-linear expressions:*** A direct counterpart of the attractive simplicity of linear schemes is their potentially serious weakness, even in simple homogeneous shear flows (see ref. [84] page 519 for instance). Moreover, linear schemes are incompatible with two-component turbulent field.
Various attempts have been made to remove these deficiencies, Reynolds [389] in 1984, Shih & Lumley [425] in 1985, UMIST group, see Craft *et al.* [113] in 1989 for example. Speziale *et al.* [444] in 1991, Johansson & Hallback [234] in 1994 also derived non-linear closure schemes. They are reviewed in Chassaing [84], for instance. Here, we shall limit the presentation to those schemes that have been used to predict compressible and/or variable density flows.

*Examples of pressure-strain models*
As a basis for comparison, the following sample of three complete pressure-strain models (including both slow and rapid contributions) will be considered:

- the linear Launder-Reece-Rodi (LRR) model [279];
- the cubic Fu-Launder-Tselepidakis (FLT) model [167];
- the model derived by Speziale-Sarkar-Gatski (SSG) [444] from a dynamical system approach.

The corresponding schemes can be detailed with reference to the following generic expression of the pressure-strain correlation (Johansson & Hallback [234]):

$$\begin{aligned}\frac{L_{ij}}{\overline{k}} &= \alpha_0\frac{\overline{\epsilon}}{\overline{k}}a_{ij} + \alpha_1\frac{\overline{\epsilon}}{\overline{k}}(a_{ik}a_{kj} - \frac{a_{kl}a_{lk}}{3}\delta_{ij}) + \alpha_2\overline{S}_{ij} + \alpha_3\frac{P_{rod}}{\overline{k}}a_{ij} \\ &+\alpha_4(a_{ik}\overline{S}_{kj} + a_{jk}\overline{S}_{ki} - \frac{2}{3}a_{kl}\overline{S}_{lk}\delta_{ij}) + \alpha_5(a_{ik}\overline{R}_{jk} + a_{jk}\overline{R}_{ik}) \\ &+\alpha_6(a_{ik}a_{kl}\overline{S}_{lj} + a_{jk}a_{kl}\overline{S}_{li} - 2a_{kj}a_{li}\overline{S}_{kl}) + \alpha_7(a_{ik}a_{kl}\overline{R}_{jl} + a_{jk}a_{kl}\overline{R}_{il}) \\ &+\alpha_8\left[a_{nl}a_{nl}\left(a_{ik}\overline{R}_{jk} + a_{jk}\overline{R}_{ik}\right) + 3a_{mj}a_{nj}\left(a_{mk}\overline{R}_{nk}\right) + a_{nk}\overline{R}_{mk}\right] ,\end{aligned} \tag{11.21}$$

where $P_{rod} = -\overline{u'_i u'_j}\partial\overline{U}_i/\partial x_j$ stands for the shear production rate of turbulence kinetic energy.

The non-zero model parameters are given in Table 11.6.

TABLE 11.6. Pressure-strain schemes according to eq.(11.21) in LRR, FLT and SSG models.

| Author - Ref. | Year | Model | *Non-zero* parameters |
|---|---|---|---|
| Launder, Reece & Rodi [279] | 1975 | LRR | $\alpha_0 = -3.6$ ; $\alpha_4 = \alpha_5 = 1.2$ |
| Fu, Launder & Tselepidakis [167] | 1987 | FLT | $\alpha_0 = -60\mathbf{II}_a\sqrt{\beta}$ ; $\alpha_1 = -72\mathbf{II}_a\sqrt{\beta}$<br>$\alpha_2 = \alpha_6 = \alpha_7 = 0.8$ ; $\alpha_3 = \alpha_4 = 1.2$<br>$\alpha_5 = 1.73$ ; $\alpha_8 = 11.2$ |
| Speziale, Sarkar & Gatski [444] | 1991 | SSG | $\alpha_0 = -3.4$ ; $\alpha_1 = 4.2$ ; $\alpha_3 = -1.8$<br>$\alpha_2 = 0.8 - 1.3\sqrt{\mathbf{II}_a}$ ; $\alpha_4 = 1.25$ ;<br>$\alpha_5 = 0.4$ |

### 11.3.4. LOW TURBULENCE REYNOLDS NUMBER CORRECTIONS

The closure schemes discussed in the previous sections are mainly concerned with high turbulence Reynolds number flows. Accordingly, molecular transfer can be neglected everywhere in the flow field, as compared with diffusion by continuous turbulent motions. Such an assumption is no longer acceptable near a solid surface where, for example, molecular transfer is the dominant diffusion process and strong modifications of the turbulence field result from the no-slip condition at the wall.

In incompressible fluid motions over a flat solid surface, the modifications of the turbulent field are generally associated with three separate effects: blocking effect, viscous effect and (strong) shear effect. Thanks to DNS, these effects can be analyzed separately or all together, see e.g., Kim *et al.* [249], Spalart [439], Lee *et al.* [282], Perot *et al.* [362], Aronson *et al.* [23].

Different attempts have been made to incorporate such "low-Reynolds-number" effects in second-order modeling. They may be grouped into two families. The first one consists in adding correction terms to the high-Reynolds-number version of the model. It will be referred to as an **explicit** correction method. The second one may be considered as an **implicit** approach, insofar as modifications of the turbulence structure near the wall are recovered from physical parameters included in the model itself.

- **Explicit** corrections typically require knowledge of the local surface topography. In 1973, Shir [429] introduced a correction term based on tensorial products including the unit vector of the normal to the surface. In Launder *et al.* [279] 1975, Gibson & Launder [182] 1978, Launder & Shima [278] 1989, for instance, near-wall corrections to the pressure-strain model are proposed to incorporate the so-called *pressure-echo* or *wall-reflection* effect. The correction consists in additional terms that explicitly involve the distance to the wall;
- In the vicinity of complex geometrical bodies, the normal-to-wall vector has no unique definition, so that corrections based on explicit detection of the surface are not directly tractable. To overcome this difficulty, an **implicit** correction technique should be preferred. It is based on turbulence modifications near the wall. In order to include such modifications, nonlinear schemes are to be developed, including realizability constraints consistent with the approach of turbulence to the two-component limit. An original model, which is claimed to require no pressure-strain wall-related corrections, was developed by Speziale *et al.* [444] in 1991. Another long-time effort was devoted to this issue at UMIST by Launder and co-workers, developing low-Reynolds number extensions of the cubic model by Fu *et al.* [167] to ensure the appropriate wall damping behavior of each individual Reynolds stress. Such extensions are using various inhomogeneity corrections[1] determined without reference to geometry-dependent parameters[2], as proposed for instance by Launder & Li [277], Craft & Launder [110], Eifert *et al.* [144].

As compared to ($k$-$\epsilon$) and *linear* second-order closures, the performance of models incorporating such low-Reynolds corrections, mainly with nonlinear schemes, is significantly better than that of usual linear Reynolds stress models for a wide range of incompressible fluid motions, including separated flow configurations, Eifert *et al.* [144].

### 11.3.5. REDISTRIBUTION IN VARIABLE DENSITY FLOWS

In variable density fluid motions, the Laplacian of pressure fluctuations can be derived from the momentum equation, as detailed in Chapter 3. By simple rearrangement of eq.(3.22), a slightly more compact form of this equation can be obtained when assuming that the external body forces reduce to gravity forces.

[1]An abundant literature exists on this topic which is not reported here. Some non exhaustive proposals are reviewed in Chassaing [84].

[2]such as the turbulence Reynolds number $\overline{k}^2/(\nu\overline{\epsilon})$, the two-component limit parameter...

Using conventional (centered) fluctuations, it reads:

$$\Delta p' = -2\frac{\partial^2(\overline{\rho}u_i'\overline{U}_j)}{\partial x_i \partial x_j} - \frac{\partial^2(\rho u_i' u_j' - \overline{\rho u_i' u_j'})}{\partial x_i \partial x_j} + \frac{\partial^2 \tau_{ij}'}{\partial x_i \partial x_j}$$
$$+g_i\frac{\partial \rho'}{\partial x_i} - 2\frac{\partial^2[(\rho' u_i' - \overline{\rho' u_i'})\,\overline{U}_j]}{\partial x_i \partial x_j} + \frac{\partial^2 \rho'}{\partial t^2} - \frac{\partial^2(\rho'\overline{U}_j\overline{U}_j)}{\partial x_i \partial x_j}. \qquad (11.22)$$

As compared with the constant density equation (11.11), two types of changes can be observed in the right-hand-side of eq.(11.22). Terms in the first line are simply the variable density counterpart of those in eq.(11.11). Specific contributions, which are associated with density *fluctuations*, are grouped in the second line.

Various closure schemes have been derived from eq.(11.22), addressing different types of variable density fluid motions.

*Buoyant flows*

In low speed buoyant flows, under Boussinesq approximations, (see Chapter 3), the second line in eq.(11.22) reduces to the first term, with $\rho' = -\theta'\,\overline{\rho}/\overline{T}$. Hence, gravitational effect on pressure appears as introducing an additional linear source, quite separate from the two contributions present in constant density fluid flows (slow and rapid parts). Accordingly, in high turbulence Reynolds number flows, the pressure-strain correlation can be modeled as a sum of *three* contributions:

$$\Pi_{ij} = \Pi_{ij}^{I\,slow} + \Pi_{ij}^{I\,rapid} + \Pi_{ij}^{B}.$$

Current models (Launder [274]) assume that the rapid and slow parts can be approximated by the same closure schemes as used in non-buoyant, incompressible fluid motions, so that the sole influence of gravity on the pressure-containing correlations is expressed through the buoyant contribution $\Pi_{ij}^B$.

Applying quasi-isotropic assumptions similar to the kinematic constraints adopted when determining the rapid part, the following "isotropization of production" (IP) model is obtained for the buoyant contribution:

$$\Pi_{ij}^B = -C_3(G_{ij} - \frac{1}{3}G_{mm}\delta_{ij}), \qquad (11.23)$$

where the model constant is $C_3 \approx 0.3$ to $0.5$, and $G_{ij}$ denotes the buoyant generation of $\overline{u_i' u_j'}$

$$G_{ij} = -\frac{1}{\overline{T}}(\overline{\theta' u_i'}\,g_j + \overline{\theta' u_j'}\,g_j) \equiv \frac{1}{\overline{\rho}}(\overline{\rho' u_i'}\,g_j + \overline{\rho' u_i'}\,g_j).$$

As pointed out by Craft [109], the IP model corresponds to the leading order term in the cubic development of $\Pi_{ij}^B$.

*Low-speed ideal binary mixing*

This situation is encountered when density variations only result from changes in composition in a quasi-constant pressure and temperature, low-velocity flow (ideal mixing). In this case, Boussinesq's approximations are no longer valid and, when using density weighted averages, body-force contributions disappear from the Reynolds stress transport equation (see Chapter 5-§-5.6).

In 1987, using order of magnitude considerations, Shih *et al.* [427] reached the conclusion that in a low-speed, variable composition (helium-nitrogen) mixing layer, the "incompressible" contributions to the pressure-strain correlation — eq.(11.13) —, are the only significant ones. Consequently, usual closure schemes derived for constant-density fluid motions can also be adopted in this case.

However, in stratified fluid motions, it can be seen that, when Favre's averages are used, the buoyant generation of $\overline{u'_i u'_j}$ can be recovered from mean pressure coupling terms in the transport equation (5.35) governing the Reynolds stresses $\overline{\rho}\,\widetilde{u''_i u''_j}$. Assuming that gravity acts along the $i=3$ component, the mean pressure contribution in this equation yields for instance:

$$-\overline{u''_j}\,\frac{\partial \overline{P}}{\partial x_i}\delta_{i3} \equiv \frac{\overline{\rho' u'_j}}{\overline{\rho}}\,\frac{\partial \overline{P}}{\partial x_i}\delta_{i3} = \overline{\rho' u'_j}\,g_i \delta_{i3}\,.$$

Based on this stratified flow analogy, Chassaing [80] proposed in 1979 the following scheme to account for external body force effects in free variable-density jets, when a mass-weighted formulation is used:

$$\Pi_{ij} = \Pi_{ij}^{I\,slow} + \Pi_{ij}^{I\,rapid} + \Pi_{ij}^{3}\,,$$

with

$$\Pi_{ij}^{3} = b_4 \frac{1}{\overline{\rho}}(\overline{u''_i}\frac{\partial \overline{P}}{\partial x_j} + \overline{u''_j}\frac{\partial \overline{P}}{\partial x_i} - \frac{2}{3}\overline{u''_k}\frac{\partial \overline{P}}{\partial x_k}\delta_{ij})\,,$$

and $b_4 = 0.3$ to $0.6$. This scheme was also adopted by Bailly *et al.* [27] in predicting premixed flames with $b_4=0.3$, and by Vallet [465] with $b_4=0.75$ in jet flows with very high density ratios.

*High-speed compressible flows*

As illustrated in the following examples, most investigators consider that closure schemes derived for pressure-strain correlation in incompressible fluid motions can be adapted to the modeling of the deviatoric part of this correlation in compressible fluid flows.

**Vandromme *et al.* (1983).** Vandromme *et al.* [473] implemented a second-order closure in a *2-D* implicit Navier-Stokes solver to predict the turbulent boundary layer over an adiabatic flat plate with zero axial pressure gradient at $M=3$. When the basic Launder-Reece-Rodi (LRR) [279] and Hanjalic-Launder [204] models are used, deficiencies are noted in predicting the anisotropy of the Reynolds stresses and the kinetic energy level throughout the boundary layer. As shown by the authors, improved predictions of the skin friction and the recovery factor, along with reasonably good velocity profiles in the wake region, can be achieved by simple adjustment of some model parameters, such as the return-to-isotropy constant in Rotta's term and the diffusion coefficient in the dissipation equation.

In 1991, Sarkar *et al.* [413] adopted the LRR scheme for the pressure-strain correlation in predicting the compressible shear layer with a compressible dissipation model.

**Sarkar (1992).** In simulating simple compressible flows, Sarkar [408] adopted the SSG scheme for the deviatoric part of the pressure-strain correlation. The only change consists in substituting the deviatoric part of the strain rate tensor $S^*_{ij} = S_{ij} - S/3\delta_{ij}$, where $S = S_{ii}$, to $S_{ij}$. The model coefficients are the same as those given in Table 11.6.

**Zeman (1993).** The following expression, corresponding to the compressible counterpart of the linear restriction of eq.(11.21), is adopted by Zeman [498] for the pressure gradient-velocity correlation:

$$\Pi^*_{ij} = C_m \widetilde{\epsilon}_s \widetilde{a}_{ij} - 2\widetilde{k}[\frac{2}{5}\widetilde{S}^*_{ij} + \alpha_4(\widetilde{S}^*_{ik}\widetilde{a}_{jk} + \widetilde{S}^*_{jk}\widetilde{a}_{ik} - \frac{2}{3}\widetilde{S}^*_{mn}\widetilde{a}_{mn}\delta_{ij}) + \alpha_5(\widetilde{R}_{ik}\widetilde{a}_{jk} + \widetilde{R}_{jk}\widetilde{a}_{ik})] \, .$$

Here, $\widetilde{\epsilon}_s$ denotes the solenoidal (incompressible) dissipation rate. As in Sarkar [408], $\widetilde{S}^*_{ij}$ is the deviatoric part of the strain rate tensor, $\widetilde{R}_{ij}$ the antisymmetric rotation tensor referring to Favre's averaged velocity field and $\widetilde{a}_{ij}$ is the anisotropy tensor of the turbulent stresses $\widetilde{u''_i u''_j}$. The model parameters are $C_m = 3.25$, $\alpha_4 = 0.3 + f(R_t)$ and $\alpha_5 = 0.3 - f(R_t)$, where $f(R_t)$ is a function of the turbulence Reynolds number $R_t = 4\widetilde{k}/(9\nu\widetilde{\epsilon}_s)$ which is introduced to account for near wall effects.

**Batten *et al.* (1999).** As far as the pressure-strain model is concerned, the "compressible" version adopted by Batten *et al.* [38] is a straightforward, mass-weighted extension of the original Craft-Launder [110] (CL) scheme. It is recalled that this model is not concerned with the "usual" pressure-strain term $\Pi^*_{ij}$, but with a modified pressure-strain tensor, including diffusive transport of the Reynolds stresses by pressure fluctuations, viz. $\Pi^*_{ij} - [(\overline{u'_i u'_j}/(2\overline{\rho}\overline{k}) \times \partial\overline{(p'u'_k)}/\partial x_k]$, where the pressure-velocity correlation has to be modeled separately.

However, as noticed in [38], the favorable performance of the CL model in predicting incompressible fluid motions does not carry over to compressible flows involving shocks. In some circumstances, the CL predictions in this case can be somewhat worse than that given by a ($k$- $\epsilon$) model (Leschziner *et al.* [294] 1997). In shock-induced separation over a channel bump for instance, the inhomogeneity corrections to the "incompressible" pressure-strain model turn out to be inappropriately sensitive to the shock-wave. This is the reason why in 1999, Batten *et al.* [38] presented a variant of the Craft-Launder model, in which *ad hoc* modifications to the inhomogeneity coefficients and their functional dependence on the turbulence Reynolds number, have been designed and tested to give good predictions in compressible flows, involving shock-boundary layer interaction.

### 11.3.6. DISSIPATION

At high turbulence Reynolds number, the dissipation rate tensor is commonly believed to be isotropic:

$$\overline{\epsilon}_{ij} = \overline{\epsilon}\delta_{ij} \qquad \text{or} \qquad \overline{\mathbf{e}}_{ij} = 0\,. \tag{11.24}$$

In incompressible fluid motions, eq.(11.24) is usually adopted, but non-isotropic schemes have been derived by Hallback *et al.* [201], [202], and Speziale *et al.* [442], *inter alia*. Such non-isotropic schemes are not detailed here, since their use in variable-density flows has not been well documented so far.

Many investigators assume that the modeled equation governing the *solenoidal* turbulent dissipation rate is left unchanged from that adopted in incompressible situations.

## 11.4. Turbulent scalar fluxes modeling

### 11.4.1. GRADIENT DIFFUSION SCHEMES

As noticed in the introduction to the present chapter, a simple modeling practice to calculate the turbulent scalar fluxes without using additional transport equations, consists in adopting a generalized gradient diffusion hypothesis. This procedure applies to a wide variety of scalar transport closure, addressing various types of fluid motions: passive contaminant situations, buoyant flows according to Boussinesq's approximations, high speed compressible flows (Ha Minh & Vandromme [332] Sarkar & Balakrishnan [411], Speziale & Sarkar [443] Batten *et al.* [38], for instance). Using Favre's averages and taking the turbulent heat flux $\widetilde{\theta'' u_i''}$ for example, possible expressions of such gradient closures are:

$$\widetilde{\theta'' u_i''} = -D_\theta \frac{\partial \widetilde{T}}{\partial x_i} \quad \text{or} \quad \widetilde{\theta'' u_i''} = -D_{\theta ij} \frac{\partial \widetilde{T}}{\partial x_j}\,,$$

depending on whether a scalar or tensorial diffusivity is introduced.
In [411], [443], a scalar diffusivity is adopted, $D_\theta = C_\theta \widetilde{k}^2/\widetilde{\epsilon}$, with $C_\theta \approx 0.128\,(=0.09/0.7)$.
A tensorial diffusion coefficient is taken in references [332] and [38], where $D_{\theta ij} = C_\theta \widetilde{u_i''u_j''}\widetilde{k}/\widetilde{\epsilon}$, with $C_\theta$=0.3 for the latter.

### 11.4.2. TRANSPORT EQUATION

When *scalar* second-order moments are derived from modeled transport equations, the generic form of the model foundation equations emerges as:

$$\frac{\partial(\overline{\rho}\,\widetilde{\alpha''u_i''})}{\partial t} + \frac{\partial(\overline{\rho}\,\widetilde{\alpha''u_i''}\widetilde{U}_j)}{\partial x_j} = \overline{\rho}P_{\alpha i} - \overline{\rho}\Pi_{\alpha i} - \frac{\partial T_{\alpha ij}}{\partial x_j} + \Sigma_{\alpha i} - \overline{\varepsilon}_{\alpha i}\,, \tag{11.25}$$

$$\frac{\partial(\overline{\rho}\,\widetilde{\alpha''^2})}{\partial t} + \frac{\partial(\overline{\rho}\,\widetilde{\alpha''^2}\widetilde{U}_j)}{\partial x_j} = \overline{\rho}P_{\alpha} - \frac{\partial T_{\alpha j}}{\partial x_j} + \Sigma_{\alpha} - \overline{\varepsilon}_{\alpha}\,, \tag{11.26}$$

where $\alpha''$ stands for any scalar Favrian fluctuation (temperature or mass-fraction). By convention, in eqs.(11.25) and (11.26) and later on, *no summation* is applied to *Greek* indices.
The expressions of *exact* production terms for heat and mass-fraction are:

$$P_{\theta i} = -\widetilde{u_i''u_j''}\frac{\partial \widetilde{T}}{\partial x_j} - \widetilde{\theta''u_j''}\frac{\partial \widetilde{U}_i}{\partial x_j} \quad \text{or} \quad P_{\gamma i} = -\widetilde{u_i''u_j''}\frac{\partial \widetilde{C}}{\partial x_j} - \widetilde{\gamma''u_j''}\frac{\partial \widetilde{U}_i}{\partial x_j}\,,$$

$$P_{\theta} = -2\widetilde{\theta''u_j''}\frac{\partial \widetilde{T}}{\partial x_j} \quad \text{or} \quad P_{\gamma} = -2\widetilde{\gamma''u_j''}\frac{\partial \widetilde{C}}{\partial x_j}\,.$$

All remaining terms in eqs.(11.25) and (11.26) — pressure-correlation ($\Pi_{\alpha i}$), transport ($T_{\alpha ij}$), ($T_{\alpha j}$), d.f.c. coupling ($\Sigma_{\alpha i}$), ($\Sigma_{\alpha}$), and dissipation ($\overline{\varepsilon}_{\alpha i}$), ($\overline{\varepsilon}_{\alpha}$) —, are to be modeled.
The rest of this section aims at outlining some of the most popular closure schemes for such terms in various variable density flows, assuming high turbulence Reynolds number.

*Buoyant flows*

In this case, d.f.c. terms cancel out from eqs.(11.25) and (11.26). Since usual (centered) Reynolds averages are adopted, an explicit buoyant production term similar to $G_{ij}$ is introduced in eq.(11.25), viz.

$$G_{\theta i} = \frac{1}{\overline{\rho}}\overline{\rho'\theta'}\,g_i \quad \text{or} \quad G_{\gamma i} = \frac{1}{\overline{\rho}}\overline{\rho'\gamma'}\,g_i\,.$$

Now, according to Boussinesq's approximations, $\rho' = -\theta'\overline{\rho}/\overline{T}$, so that the production term $G_{\theta i}$, for instance, can be exactly deduced from the temperature variance ($G_{\theta i} \equiv -g_i\overline{\theta'^2}/\overline{T}$).

The derivation of closure schemes for transport equations governing scalar correlations is based on a general procedure which is very similar to that adopted for modeling RST equation in an incompressible fluid flow [112]. Some salient features are recalled here.

a) Considering that major contribution to the dissipation comes from the smallest scale eddies, which are assumed to be almost isotropic[3], the model usually adopted for the dissipation term in the transport equation of the cross scalar-velocity correlation is simply $\overline{\varepsilon}_{\alpha i} = 0$.
The dissipation in the scalar variance equation can be obtained either from a prescribed time-scale ratio (assuming for instance $\overline{\varepsilon}_\alpha \propto \overline{\epsilon}$), or by solving another transport equation for this function.

b) According to the generalized gradient diffusion hypothesis, and adopting for instance Daly-Harlow proposal [117], the triple correlation $\overline{u'_i u'_j \alpha'}$ can be modeled as:

$$\overline{u'_i u'_j \alpha'} = -C_\alpha \frac{\overline{k}}{\overline{\epsilon}} \overline{u'_j u'_l} \frac{\partial \overline{\alpha' u'_i}}{\partial x_l},$$

which yields the following expression of the diffusion term:

$$-\frac{\partial}{\partial x_j} T_{\alpha ij} = C_\alpha \frac{\partial}{\partial x_j}\left(\frac{\overline{k}}{\overline{\epsilon}} \overline{u'_j u'_l} \frac{\partial \overline{\alpha' u'_i}}{\partial x_l}\right). \tag{11.27}$$

Similarly, the diffusion term in eq.(11.26) can be modeled as:

$$-\frac{\partial}{\partial x_j} T_{\alpha j} = C'_\alpha \frac{\partial}{\partial x_j}\left(\frac{\overline{k}}{\overline{\epsilon}} \overline{u'_j u'_l} \frac{\partial \overline{\alpha'^2}}{\partial x_l}\right). \tag{11.28}$$

As noted by Launder [274] and Craft [109], more complex and seemingly more rigorous models for the triple moments have been proposed, but eqs.(11.27) and (11.28) are found to be quite convenient for most practical predictions of thermally buoyant flows, with $C_\alpha = C'_\alpha (\equiv C_\theta) \approx 0.18$ (Craft [109]).

c) We turn now to the pressure-strain correlation. The detailed expression of the source terms in the Poisson equation governing the fluctuating pressure in this case, suggests a closure scheme made of three additive parts (Launder [274]):

$$\Pi_{\theta i} = \Pi_{\theta i}^{slow} + \Pi_{\theta i}^{rapid} + \Pi_{\theta i}^{B}.$$

[3]As shown in Chapter 7, this assumption may correspond to a crude approximation of the physics.

Here, $\Pi_{\theta i}^{slow}$ is the "turbulence" contribution or "slow" part, corresponding to the "return to isotropy" process. $\Pi_{\theta i}^{rapid}$, which involves mean-strain contributions, is the counterpart of the so-called "linear" or "rapid" contribution. The last term ($\Pi_{\theta i}^{B}$) arises from gravitational or buoyancy effects due to fluctuating body forces.

The simplest linear expression, equivalent to Rotta's scheme for the *return-to-isotropy* contribution is:

$$\Pi_{\theta i}^{slow} = -C_{1\theta}\frac{\overline{\epsilon}}{\overline{k}}\overline{\theta' u_i'}\,,$$

with $C_{1\theta} = 3$. It corresponds to the simplification ($C_{1\theta}'=0$) in the following commonly proposed linear form:

$$\Pi_{\theta i}^{slow} = -C_{1\theta}\frac{\overline{\epsilon}}{\overline{k}}(\overline{\theta' u_i'} + C_{1\theta}'\, b_{ij}\,\overline{\theta' u_j'})\,.$$

A quadratic extension in $b_{ij}$ has been proposed by Craft *et al.* [113] in 1989:

$$\Pi_{\theta i}^{slow} = -C_{1\theta}\frac{\overline{\epsilon}}{\overline{k}}(\overline{\theta' u_i'} + C_{1\theta}'\, b_{ij}\,\overline{\theta' u_j'} + C_{1\theta}''\, b_{ik}b_{kj}\overline{\theta' u_j'}) + d_{1\theta}\overline{k}\, b_{ij}\frac{\partial \overline{T}}{\partial x_j}\,. \quad (11.29)$$

Although largely empirical, the presence of the last term in the previous scheme, including mean temperature gradients, emerged as an essential feature in the model [113]. It is introduced to handle flows where the normalized generation rates of $\overline{k}$ and $\overline{\theta'^2}$ are greatly different (Jones & Musonge [240]). Equation (11.29) has been adopted in 1996 by Craft *et al.* [112], taking the model parameters as functions of the Reynolds stress anisotropy invariants and the time-scale ratio between scalar and mechanical dissipation rates: $R=2\overline{\epsilon}_\theta\overline{k}/\overline{\theta'^2}\overline{\epsilon}$.

A simple expression of the *rapid* contribution, corresponding to the isotropization of production is:

$$\Pi_{\theta i}^{rapid} = C_{2\theta}\overline{\theta' u_k'}\frac{\partial \overline{U}_i}{\partial x_k}\,,$$

where the model coefficient is empirically set to $C_{2\theta} \approx 0.5$.

A quasi-isotropic expression, similar to the one considered in LRR scheme (11.17) reads:

$$\Pi_{\theta i}^{rapid} = 0.8\overline{\theta' u_k'}\frac{\partial \overline{U}_i}{\partial x_k} - 0.2\overline{\theta' u_k'}\frac{\partial \overline{U}_k}{\partial x_i}\,.$$

Nonlinear developments have been derived by Shih & Lumley [425] and Craft *et al.* [113]. They differ in some details related to the way realizability and two-component limit constraints are handled.

According to Craft [109], the *buoyant* contribution including all possible cubic terms and satisfying the two-component limit takes the form:

$$\Pi^B_{\theta i} = \frac{1}{\overline{\overline{T}}}(\frac{1}{3}g_i - b_{ik}g_k)\overline{\theta'^2}\,. \qquad (11.30)$$

The simplification, corresponding to the isotropization of production, is:

$$\Pi^B_{\theta i} = C_{B\theta}\frac{1}{\overline{\overline{T}}}\overline{\theta'^2}g_i\,,$$

with an empirically adjusted coefficient $C_{B\theta} \approx 0.5$.
Eq.(11.30) is adopted in Craft *et al.* [112] to predict several types of buoyant flows, including the downward-directed warm jet, the stratified mixing layer and the buoyancy-affected grid decaying turbulence.

*Low-speed ideal binary mixing*
Let us consider now ideal mixing situations where the temperature of the mixture remains constant, so that scalar correlations only involve mass-fraction fluctuations. In this case, it can be seen from the exact transport equations that $\Sigma_\alpha \equiv 0$ and $\Sigma_{\alpha i} \equiv -\overline{\gamma''}\partial\overline{P}/\partial x_i$. Hence, when using the analytical relationship derived in Chapter 10, the latter d.f.c. contribution is obtained exactly, within the frame of second-order closure.

In 1979, Chassaing [80] proposed the following closure schemes to the modeling of eqs. (11.25) and (11.26):

$$\text{Pressure strain:}\quad \overline{\rho}\,\Pi_{\gamma i} = -C_0\frac{\widetilde{\epsilon}}{\widetilde{k}}\overline{\rho\gamma''u''_i} + C_2\overline{\rho\gamma''u''_k}\frac{\partial\widetilde{U}_i}{x_k}$$

$$+C_3\overline{\rho\gamma''u''_k}\frac{\partial\widetilde{U}_k}{x_i} + C_4\overline{\gamma''^2}\frac{\partial\overline{P}}{\partial x_i}\,,$$

$$\text{Diffusion:}\; T_{\gamma ij} = -c_5\frac{\widetilde{k}}{\widetilde{\epsilon}}\overline{\rho u''_j u''_l}\frac{\partial\overline{\rho u''_i\gamma''}}{\partial x_l} \quad\text{and}\quad T_{\gamma j} = -d_5\frac{\widetilde{k}}{\widetilde{\epsilon}}\overline{\rho u''_j u''_l}\frac{\partial\overline{\rho\gamma''^2}}{\partial x_l}\,,$$

$$\text{Dissipation:}\; \overline{\varepsilon}_{\alpha i} = 0 \quad\text{and}\quad \overline{\varepsilon}_\alpha = C_6\frac{\widetilde{\epsilon}}{\widetilde{k}}\overline{\rho\gamma''^2}\,.$$

With the following values of the model coefficients, $C_0 = 2.1$, $C_2 = 0.8$, $C_3 = -0.2$, $C_4 = 0.5$, $c_5 = 0.3$, $d_5 = 0.15$ $C_6 = 0.7$, the model yields reasonably good predictions of a free 80% CO2-20% air jet discharging into a quiescent atmosphere at an exit Reynolds number of 54,100 (Chassaing & Ha Minh [85]). Similar closure schemes have also been adopted by Chibat [93] and Ruffin [401] dealing with the same type of flows.

*High-speed compressible flows*
In 1993, Zeman [498] derived a new model for super/hypersonic turbulent boundary layers in quasi-equilibrium, based on a simplified expression of eq.(11.25):

$$\frac{\partial(\widetilde{\theta'' u_i''})}{\partial t} + \frac{\partial(\widetilde{\theta'' u_i''}\widetilde{U}_j)}{\partial x_j} = P_{\theta i} - \Pi_{\theta i} - \frac{1}{\overline{\rho}}\frac{\partial T_{\theta y}}{\partial y},$$

where $y$ is the coordinate normal to the plane wall.
No explicit contribution due to d.f.c. coupling $\Sigma_{\theta i}$ is introduced in the model.
The closure scheme for the pressure term ($\Pi_{\theta i}$) is:

$$\Pi_{\theta i} = \frac{C_\theta}{2}\frac{\overline{\epsilon}}{\widetilde{k}}\widetilde{\theta'' u_i''} - [P_{\theta i} + \frac{1.11 C_\theta}{2}\frac{\overline{\epsilon}}{\widetilde{k}}\frac{P^*_{rod}}{|\mathbf{S}^*|^2}\frac{\partial \widetilde{T}}{\partial x_i}],$$

where the model constant is $C_\theta = 2.25$. The terms in brackets correspond to the rapid contribution, where $P^*_{rod} = -\widetilde{u_i'' u_j''}\widetilde{S}^*_{ij}$ is the turbulence kinetic energy production rate based on the trace-free deformation tensor, $\widetilde{S}^*_{ij} = \frac{1}{2}(\partial\widetilde{U}_i/\partial x_j + \partial\widetilde{U}_j/\partial x_i - \frac{2}{3}\partial\widetilde{U}_l/\partial x_l\delta_{ij})$.
The transport term is modeled as:

$$T_{\theta y} = -C_d\,\overline{\rho}\frac{\widetilde{k}}{\overline{\epsilon}}\widetilde{v''^2}\frac{\partial(\widetilde{\theta'' v''})}{\partial y},$$

with $C_d$=0.15.

## 11.5. Examples of second-order models predictions of variable density flows

### 11.5.1. BUOYANT FLOWS

The following two examples concern buoyancy affected turbulent flows. The predictions are taken from Craft *et al.* [112]. Strictly speaking, and as shown in Chapters 3 and 4, such flows are not addressing variable density turbulence. They are used here as providing clear illustration about the superiority of second-order modeling over eddy-viscosity closures.
In that regard, the first example illustrates the poor performance of the standard ($k$-$\epsilon$) to predicting a saline-mixing layer. The profiles of mass-fraction of salt water in a spatially evolving horizontal mixing layer are shown in figure 11.1. The flow is stably stratified.
From the comparison with experimental data, it can be concluded that the ($k$-$\epsilon$) model is insufficiently sensitive to buoyant damping, which results in

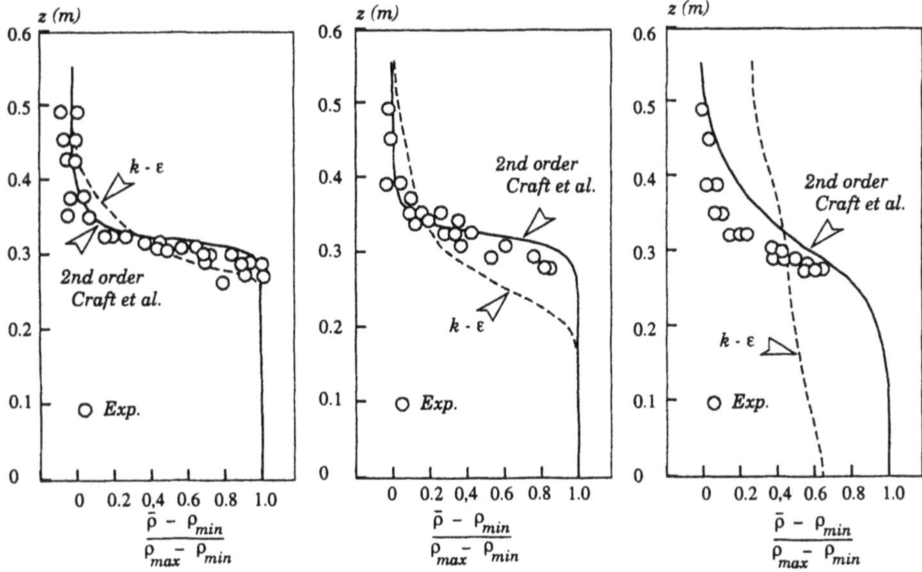


*Figure 11.1.* Mass-fraction of salt water profiles in a spatially evolving horizontal stably stratified saline-mixing layer, adapted from [112].

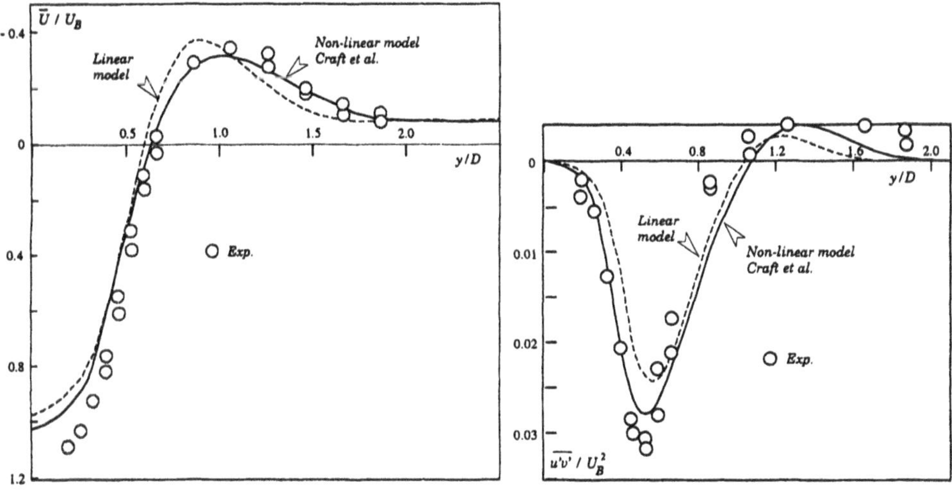


*Figure 11.2.* Comparison of predictions from linear and non-linear second-order models in an axisymmetric, downward directed buoyant jet: Mean velocity (left) and shear stress profiles (right) at x/D=1 (adapted from [112]).

a high mixing level and a nearly uniform salinity profile at the final location shown in Fig. 11.1. The non-linear second-order closure developed by Craft *et al.* [112] yields far better agreement with measurements.

The second example deals with free plume/jet flows. An exact self-preserving condition in the downstream development of the flow is achieved for constant density fluid motions (round and plane free jets). Self-preservation is approximately observed in round and plane plumes at a given distance from the exit, as discussed in Chen & Rodi [92] and Joly [235] for instance. Some recommended experimental values of the corresponding growth rate, in the self-preserving region of such flows, based on the half-width velocity thickness, are reported in Table 11.7.

TABLE 11.7. Growth rate of some free shear flows (self-preserving region).

| Type of flow | Exp. | Linear model | Non-linear model [112] |
|---|---|---|---|
| Round jet | 0.093 | 0.105 | 0.101 |
| Plane jet | 0.110 | 0.100 | 0.110 |
| Round plume | 0.112 | 0.088 | 0.122 |
| Plane plume | 0.120 | 0.078 | 0.118 |

From Table 11.7, it is evident that a far better overall agreement is achieved with non-linear closure [112] than with basic linear second-order models. The discrepancies between both types of models are illustrated in figure 11.2. It concerns the case of a hot jet discharging vertically downwards into a cold water flow moving upwards at less than 2% of the jet velocity.

From mean velocity and shear stress profiles given in Fig.11.2, it can be concluded that the anisotropy-dependent non-linear model does better in capturing the effects of shear and buoyancy, both in the core region of the jet and in the recirculating zone of this complex flow.

### 11.5.2. VARIABLE COMPOSITION MIXING

It has long been observed that density variations strongly affect all global properties of low-speed, free jets discharging into a quiescent atmosphere of different temperature or composition. Concerning the entrainment rate in axisymmetric turbulent jets for instance, the measurements of Ricou & Spalding [393] show that the mean mass flux $Q_m(x)$ at a downstream location $x$ is given by:

$$\frac{Q_m(x)}{Q_{m0}} = C_Q \frac{x}{s_\rho \sqrt{D_0}} . \qquad (11.31)$$

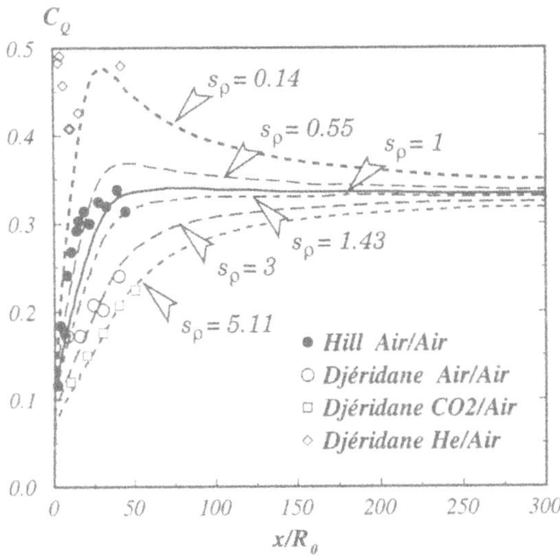


*Figure 11.3.* Influence of the inlet density ratio on the entrainment rate of free round jets (adapted from [210]).

Here, $D_0$ is the exit diameter of the jet and $s_\rho = \rho_{jet}/\rho_{air}$ the density ratio at the jet exit. The value of the empirical coefficient $C_Q$ is about 0.32, for a wide variety of gases. The quantity $s_\rho\sqrt{D_0}$ can be considered as an equivalent diameter $D_e$, as introduced by Thring & Newby [461].
The previous expression (11.31) only applies in the downstream region of the jet where quasi self-preserving conditions can be achieved. In the near field region, a strong dependence of $C_Q$ on the density ratio can be observed, as shown in Fig.11.3.

In addition to the experimental data of Hill [214] for an air/air jet and Djéridane [132] for air/air, CO2/air and He/air jets, the results of numerical predictions of Harran *et al.* [210] are also plotted in the figure. To analyze d.f.c. effects, gravity effects have been discarded from the numerical simulation and the predictions are obtained from a ternary formulation of the mean motion equations (see Chapter 5), with a linear second-order closure, as detailed in Harran *et al.* [210] and Chassaing *et al.* [86].
As clearly shown in Fig.11.3, different behaviors can be observed in "light" ($s_\rho < 1$) and "heavy" ($s_\rho > 1$) jets. Such departures are closely associated with the downstream evolution of second order characteristics, such as the turbulence kinetic energy or the mass fraction variance. When normalized by the inlet velocity of the jet, the streamwise variation of the turbulence kinetic energy along the axis exhibits a maximum, the level and location of which are both a function of the density ratio, Pitts [368], Sautet [417]. When normalized by local mean centerline references, the trend towards a

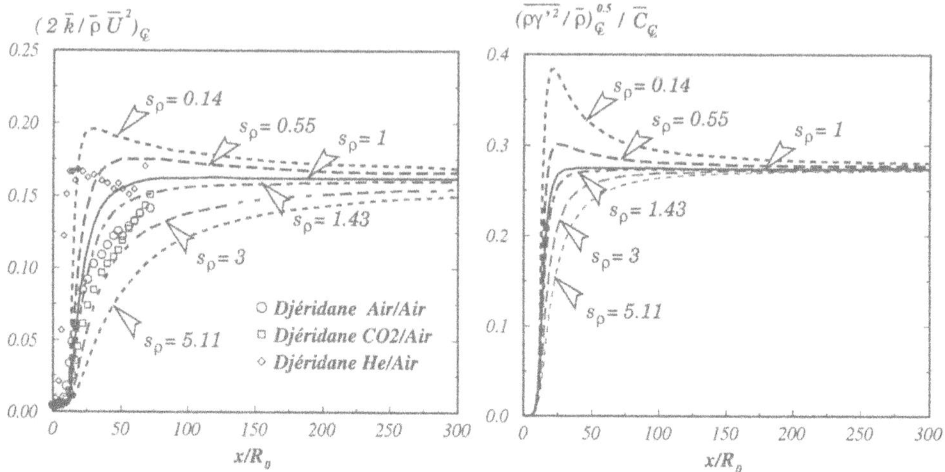


*Figure 11.4.* Turbulence kinetic energy and mass fraction variance along the axis of a variable density free round jet.

quasi self-preserving state can be observed in figure 11.4, for both turbulence kinetic energy and mass fraction variance. However, here again, a different kind of departure from the constant density situation can be observed in "light" and "heavy" jets. In heavy jets, a monotonously increasing curve is observed from the exit, while the light jet evolution overshoots the asymptotic level.

As demonstrated in [86], such different behaviors between "light" and "heavy" jets originate from d.f.c. and are closely related to distinct roles of such terms in these situations. It should be recalled that, in an ideal mixing, d.f.c. can be explicitly obtained from density weighted second-order moments, viz. (see Chapters 4 & 6):

$$\frac{\overline{\rho\gamma'}}{\overline{\rho\gamma'^2}} = \frac{\overline{\rho u_i'}}{\overline{\rho\gamma' u_i'}} = \frac{s_\rho - 1}{s_\rho - (s_\rho - 1)\overline{C}},$$

where $\overline{C}$ is the mean mass fraction. Hence d.f.c. effects can be directly reflected by second-order closure.

Even though the absolute amount of such d.f.c. terms is weak in the present flow configuration (any d.f.c. is zero at the exit of the jet and tends to zero far downstream), it has been observed from numerical simulations that:

- quantitatively different predictions can be obtained, depending on whether d.f.c. terms are disregarded or not from the mean balance equations;

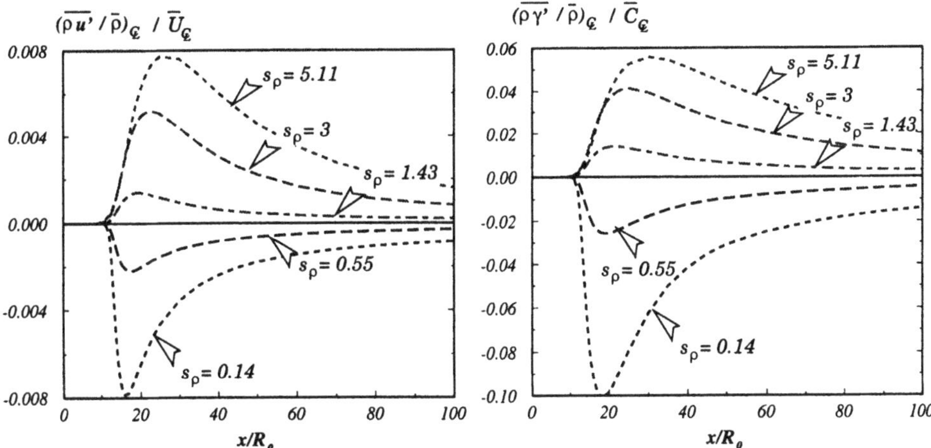


*Figure 11.5.* Axial velocity and mass-fraction d.f.c. along the axis of a variable density free round jet.

- d.f.c. effects, which occur within the first fifteen diameters from the exit, are not removed further downstream.

According to the analytical conclusions drawn in Chapter 6, the turbulent mass fluxes $\overline{\rho u'}$, $\overline{\rho v'}$ and the density mass-fraction fluctuation correlation $\overline{\rho \gamma'}$ are positive when $s_\rho > 1$ and negative when $0 < s_\rho < 1$. Now, as shown in Fig.11.5, any d.f.c., say $\overline{\rho' f'}/\bar{\rho}$ exhibits a distinct evolution from the corresponding mean value, say $\bar{F}$. Thus it can be concluded that d.f.c. behaviors are not "symmetrical" around $s_\rho = 1$. In particular, they bring different contributions to the turbulence kinetic energy budget. Hence a heavy jet ($s_\rho = 1 + \alpha$, $\alpha > 0$) discharging downwards is not equivalent to a light one with $s_\rho = 1 - \alpha$ discharging upwards, as far as both mean and second-order turbulence characteristics are considered.

### 11.5.3. COMPRESSIBLE FLOWS

The last example is the 2-dimensional transonic flow over a bump, as investigated by Délery [127], [130], [128], [129] and intensively used as a test case, since its submission to the 1981 Stanford conference data base (Délery case C).

A bump is mounted on the lower wall of a square section wind tunnel, in the presence of a transonic flow. The incoming flow (at an inlet Mach number $M = 0.63$) is accelerating over the bump until an asymptotic Mach number value, close to 1.4 is reached. The downstream pressure, which is controlled by a second adjustable throat, is adjusted so as to produce a normal shock wave above the bump trailing edge. The pressure gradient is strong enough

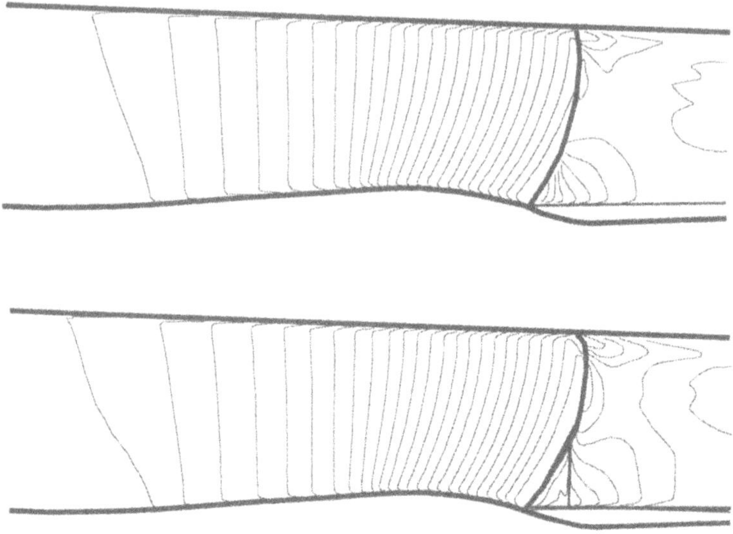

*Figure 11.6.* Shock-wave structure of the transonic flow through a plane channel with a bump - Délery, Case C. Top: ($k$- $\epsilon$) model. Bottom: modified second-order Launder-Shima model (Adapted from [294]).

to induce boundary layer separation. Due to the large extension of the recirculating zone, an oblique shock wave is generated at the separation point, downstream of which the flow remains supersonic. The two shock waves merge above the interaction zone into a single quasi-normal shock wave, yielding a typical "$\lambda$" shape shock.

As first shown by Ha Minh *et al.* [199] and Vandromme & Ha Minh [474], eddy-viscosity models, based on either mixing-length or two-equation ($k$- $\epsilon$) closures, are unable to correctly predict this shock-wave structure. Algebraic mixing-length model prediction results in a unique curved shock, which is not located at the right position. When a ($k$- $\epsilon$) model is adopted, the outer part of the shock and its location are in closer agreement with experimental observations, but the "$\lambda$" shape is not captured and no recirculating zone is predicted. As shown by the same authors [199], [474], the previous flaws are not present in predictions obtained from second-order closure.

These conclusions, which were reached by Ha Minh and Vandromme from simple linear models, have been confirmed later by several authors. They are still valid when considering more elaborate nonlinear schemes, as reviewed for instance in [294], and revisited more recently by Batten *et al.* [38], in 1999. The overall behavior of various types of turbulence models in predicting this compressible flow is illustrated in Fig.11.6 from the iso-Mach contours, as adapted from Leschziner *et al.* [294].

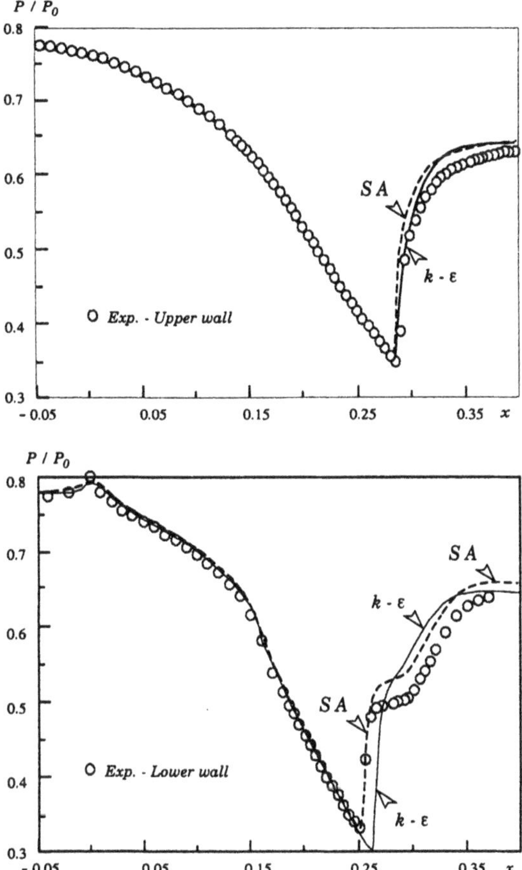


*Figure 11.7.* Fully turbulent transonic flow through a plane channel with a bump (Délery, Case C). Comparison of standard ($k$-$\epsilon$) and Spalart-Allmaras (SA) model predictions with measurements.

A more quantitative assessment of different models performances can be drawn from the pressure distributions plotted in figures 11.7 and 11.8. The first figure confirms that the recirculating region, associated with the plateau-like pattern on the curve, is not well predicted by the standard ($k$-$\epsilon$) model. A better agreement can be observed with the Spalart-Allmaras model [440], with an overpredicted level of the plateau.

Rather surprisingly, the original model of Craft & Launder [110] does worse than the standard ($k$-$\epsilon$) model to predicting the pressure evolution at the lower wall (Fig.11.8). As pointed out by Batten *et al.* [38] the inhomogeneity corrections to the pressure-strain term in the CL model respond by damping the normal fluctuations in the vicinity of the shock, as if it were a wall or free surface. Thus, such corrections are inappropriately

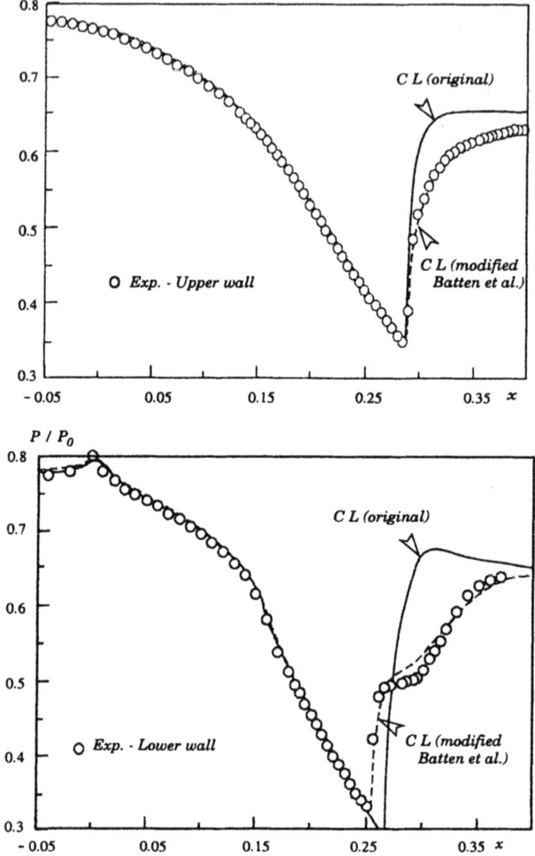


*Figure 11.8.* Fully turbulent transonic flow through a plane channel with a bump (Délery, Case C). Comparison of measurements with predictions from original Craft-Launder model [110] and Batten *et al.* [38] modified version.

sensitive to the shock wave, so that the CL model is unable to predict the "$\lambda$-structure" for instance. This is no longer the case with the modified variant of cubic Reynolds-stress model developed by Batten *et al.* [38], as shown in Fig.11.8. In particular, it is seen that this model returns a considerably much pronounced pressure-plateau region, in agreement with the experimental data.

## 11.6. Concluding remarks

As compared with the incompressible situation, second-order closure in variable density fluid motions is concerned with two types of terms. They reflect the way compressibility or variable density affect any transport

equations for second-order moments, and in particular, the Reynolds stress transport equation.
– The first type is concerned with terms that have a formal counterpart in the incompressible version of the corresponding transport equations. Thus, and assuming that closure schemes derived for constant density flows are not drastically changed when density varies, the modeling issue is basically addressing new *modifications* and/or *additional* contributions to the incompressible schemes that are needed to account for variable density and/or compressibility effects.
– On the other hand, the second type of terms introduces *specific* compressibility contributions that have no equivalent expressions in the constant density situation. They mainly refer to the so-called dilatational contributions.

Assessing the influence of each type of contributions and appreciating the pertinence of incorporating all of them in a complete model are still open questions. To enlighten this discussion, some salient points can be added to the arguments allready given in the present chapter:

- Since the publication by Zeman [495] in 1990 of the first specific closure scheme for the dilatational dissipation, a lot of work has been devoted to deriving new models and assessing the effects of dilatational contributions;
- However, almost concomitantly, a controversy emerged over the need for such specific modeling and the actual effects of the associated compressibility corrections (see for instance the report by Leschziner *et al.* [294] on the UMIST workshop dedicated in 1997 to the shock/boundary-layer interaction);
- As quoted by the previous authors [293] in 2000, the commonly adopted statement on the incidence of specific modeling of pressure-dilatation and dilatational dissipation in *wall-bounded* flows was in 1996, as summarized by Marvin & Huang "*indeed experience has shown that for the prediction of subsonic and supersonic flows, these two modifications degrade the results and are not recommended*";
- As shown by DNS of homogeneous isotropic turbulence, [359], [149], [416], [284], compressible uniform shear flow, [412], [47], [409], [431] and compressible turbulent channel flow, [221], profound differences exist in the level of such dilatational terms in free and wall-bounded flows without shock;
- As pointed out by Sarkar in Chapter 9 "*the contribution of dilatational terms to the turbulence kinetic energy balance remains small at Mach numbers at which substantial reduction in turbulence levels and thickness growth rate is observed.*"

Hence, in modeling compressible, wall-bounded turbulent flows, even in presence of shock-boundary layer interaction and separation, recent second-order models, Gerolymos & Vallet [179] do not incorporate specific compressibility correction terms.

Finally, and from a more general point of view, it can be concluded from the comparison of the capabilities of first- and second-order modeling in predicting variable density fluid turbulence that:

- At least in low speed flows, second-order modeling is theoretically well fitted to dealing with d.f.c.,

- As it is the case in incompressible fluid motions, second-order modeling can be expected to better reflect near-wall flows with anisotropy-resolving turbulence closure,

- The previous statement can be generalized to any situation where the anisotropy of the turbulence field plays a dominant role in variable density and/or compressible effects,

- In addition to the general limitations of single point modeling, some other limitations already mentioned in conclusion to the review of first-order models (Chapter 10) also apply to second-order, (dissipation equation Batchelor's assumption) and should, at least receive more attention in the future.

## References

1. N.A. Adams. Direct simulation of the turbulent boundary layer along a compressible ramp at M=3 and $Re_\theta$=1685. *J. Fluid Mech.*, 420:pages 47–83, 2000.
2. O. Andreassen, P.O. Hvidsten, D.C. Fritts, and S. Arendt. Vorticity dynamics in a breaking internal gravity wave. part 1. initial instability evolution. *J. Fluid Mech.*, 367:pages 27–46, 1998.
3. Y Andreopoulos J.H. Agui and G. Briassulis. Shock wave-turbulence interactions. *Ann. Rev. Fluid Mech.*, 32:pages 309–345, 2000.
4. F. Anselmet Y. Gagne E.J. Hopfinger and R.A. Antonia. High-order velocity structure functions in turbulent shear flows. *J. Fluid Mech.*, 140:pages 63–89, 1984.
5. R.A. Antonia. Reynolds number dependence of velocity and temperature increments in turbulent flows. In Y. Nagano K. Hanjalic and T. Tsuji, editors, *Turbulence, Heat and Mass Transfer 3*, pages 3–14, Aichi Shuppan, Nagoya, 2000.
6. R.A. Antonia and J. Kim. Similarity between turbulent kinetic energy and temperature spectra in the near-wall region. *Phys. Fluids*, 3:pages 989–991, 1991.
7. R.A. Antonia and J. Kim. Turbulent Prandtl number in the near-wall region of a turbulent channel flow. *Int. Jl. Heat Mass Trans.*, 34:pages 1905–1908, 1991.
8. R.A. Antonia and B.R. Pearson. Effect of initial conditions on the mean energy dissipation rate and the scaling exponent. *Phys. Rev. E*, (to appear), 2000.
9. R.A. Antonia and R.J. Smalley. Scaling range exponents from X-wire measurements in the atmospheric surface layer. *Boundary-Layer Meteorol.*, (submitted), 2000.
10. R.A. Antonia and R.J. Smalley. Velocity and temperature scaling in a rough wall boundary layer. *Phys. Rev. E*, 61:pages 640–646, 2000.
11. R.A. Antonia A.J. Chambers C.A. Friehe and C.W. Van Atta. Temperature ramps in the atmospheric surface layer. *J. Atmos. Sci.*, 36:pages 99–108, 1979.
12. R.A. Antonia A.J. Chambers C.W. Van Atta C.A. Friehe and K.N. Helland. Skewness of temperature derivative in a heated grid flow. *Phys. Fluids*, 21:pages 509–510, 1978.
13. R.A. Antonia B.R. Satyaprakash and A.J. Chambers. Reynolds number dependence of high order moments of velocity structure functions in turbulent shear flows. *Phys. Fluids*, 25:pages 29–37, 1982.
14. R.A. Antonia L.W.B. Browne and L. Fulachier. Spectra of velocity and temperature fluctuations in the intermittent region of a turbulent wake. *PhysicoChemical Hydrodyn.*, 8:pages 125–135, 1987.
15. R.A. Antonia M. Ould-Rouis F. Anselmet and Y. Zhu. Analogy between predictions of Kolmogorov and Yaglom. *J. Fluid Mech.*, 332:pages 395–409, 1997.
16. R.A. Antonia and C.W. Van Atta. On the correlation between temperature and velocity dissipation fields in a heated turbulent jet. *J. Fluid Mech.*, 67:pages 273–288, 1975.
17. R.A. Antonia and C.W. Van Atta. Structure functions of temperature fluctuations in turbulent shear flows. *J. Fluid Mech.*, 84:pages 561–580, 1978.
18. R.A. Antonia T. Zhou and G. Xu. Correlation between energy and temperature dissipation rates in turbulent flows. In T. Kambe, editor, *Geometry and Statistics of Turbulence*. Kluwer, 2000.
19. R.A. Antonia T. Zhou and G. Xu. Second-order temperature and velocity structure functions : Reynolds number dependence. *Phys. Fluids*, 12:pages 1509 1517, 2000.
20. R.A. Antonia T. Zhou L. Danaila and F. Anselmet. Streamwise inhomogeneity of decaying grid turbulence. *Phys. Fluids*, 12:pages 3086–3089, 2000.
21. R.A. Antonia Y. Zhu F. Anselmet and M. Ould-Rouis. Comparison between the sum of second-order velocity structure functions and the second-order temperature structure function. *Phys. Fluids*, 8:pages 3105–3111, 1996.
22. A. Arneodo C. Baudet F. Belin R. Benzi et al. Structure functions in turbulence

in various flow configurations, at Reynolds number between 30 and 5000, using extended self-similarity. *Europhys. Lett.*, 34:pages 411–416, 1996.

23. D. Aronson A.V. Johansson and L. Lofdahl. Shear-free turbulence near a wall. *J. Fluid Mech.*, 338:pages 363–385, 1997.
24. B. Aupoix G.A. Blaisdell W.C. Reynolds and O. Zeman. Modeling the turbulent kinetic energy equation for compressible, homogeneous turbulence. In P. Moin W. Reynolds J. Kim, editor, ***Studying turbulence using numerical simulation databases***, Proceedings Summer Program, pages 63–74, Center for Turbulence Research, 1990. NASA-Ames/Stanford Univ.
25. C. Auriault. *Intermittence en turbulence 3D: statistiques de la température et de son transfert.* Thèse de doctorat, Université Joseph Fourier, Grenoble, 1999.
26. T. Aurier and C. Rey. Modelisation of variable volume turbulent flows, a comparison between a cold and hot plume. In Anselmet F. Chassaing P. and Pietri L., editors, *Proceedings International Conférence on Variable Density Turbulent Flows*, pages 175–186, Banyuls, June 2000. Press. Univ. Perpignan.
27. P. Bailly M. Champion and D. Garréton. Counter-gradient diffusion in a confined turbulent premixed flame. *Phys. Fluids*, 9:pages 766–775, 1996.
28. B.S. Baldwin and T.J. Barth. A one-equation turbulence transport model for high Reynolds number wall-bounded flows. *NASA*, TM-102847, 1990.
29. G. Barakos and D. Drikakis. Investigation of nonlinear eddy-viscosity turbulence models in shock/boundary-layer interaction. *AIAA Jl.*, 38(3):pages 461–469, 2000.
30. S. Barre C. Quine and J.-P. Dussauge. Compressibility effects on the structure of supersonic mixing layers : experimental results. *J. Fluid Mech.*, 259:pages 47–78, 1994.
31. S. Barre D. Alem and J.-P. Bonnet. Experimental study of a normal shock/ homogeneous turbulence interaction. *AIAA J.*, 34(5):pages 968–974, 1996.
32. S. Barre D. Alem and J.-P. Bonnet. Reply to H.S. Ribner. *AIAA J.*, 36(3):page 495, 1998.
33. G.K. Batchelor. The condition for dynamical similarity of motion of a frictionless perfect-gas atmosphere. *Quart. J. R. Meteor. Soc.*, 79:pages 224–235, 1953.
34. G.K. Batchelor. Small-scale variations of convected quantities like temperature in turbulent fluid. Part 1. General discussion and the case of small conductivity. *J. Fluid Mech.*, 5:pages 113–133, 1959.
35. G.K. Batchelor. *An introduction to Fluid Dynamics.* Cambridge Univ. Press, reprinted edition, 1970.
36. G.K. Batchelor I.K. Howells and A.A. Townsend. Small-scale variations of convected quantities like temperature in turbulent fluid. Part 2. The case of large conductivity. *J. Fluid Mech.*, 5:pages 134–139, 1959.
37. G.K. Batchelor V.M. Canuto and J.R. Chasnov. Homogeneous buoyancy-driven trubulence. *J. Fluid Mech.*, 235:pages 349–378, 1992.
38. P. Batten T.J. Craft M.A. Leschziner and H. Loyau. Reynolds-stress-transport modeling for compressible aerodynamics applications. *AIAA J.*, 37(7):pages 785–797, 1999.
39. P.T. Bauer G.W. Zumwalt and L.J. Fila. A numerical method and an extension of the Korst jet mixing theory for multispecies jet mixing. In *6th Aerospace Science Meeting*, AIAA Paper No. 68-112, New-York, 1968.
40. B.J. Bayly C.D. Levermore and T. Passot. Density variations in weakly compressible flows. *Phys. Fluids A*, 4(5):pages 945–954, 1992.
41. C. Béguier I. Dekeyser and B.E. Launder. Ratio of scalar and velocity dissipation time scales in shear flow turbulence. *Phys. Fluids*, 21:pages 307–310, 1978.
42. D.M. Bell and J.H. Ferziger. Turbulent boundary layer DNS with passive scalars. In R. M. C. So C. G. Speziale and B. E. Launder, editors, *Near-Wall Turbulent Flows*, pages 327–336, Amsterdam, 1993. Elsevier Science Publishers.
43. R. Benzi S. Ciliberto C. Baudet G. Ruiz-Chavarria and R. Tripiccione. Extended self-similarity in the dissipation rate of fully developed turbulence. *Europhys. Lett.*,

24:pages 275–279, 1993.
44. L.P. Bernal and A. Roshko. Streamwise vortex structure in plane mixing layers. *J. Fluid Mech.*, 170:pages 499–525, 1986.
45. S. Birch and J. Eggers. Free turbulent shear flows. *NASA SP-321*, 1, 1972.
46. R.B. Bird W.E. Stewart and E.N. Lightfoot. *Transport phenomena.* John Wiley and Sons, 1960.
47. G.A. Blaisdell N.N. Mansour and W.C. Reynolds. Compressibility effects on the growth and structure of homogeneous turbulent shear flow. *J. Fluid Mech.*, 256:pages 443–485, 1993.
48. G.A. Blaisdell W.C. Reynolds and N.N. Mansour. Compressibility effects on the growth and structure of homogeneous turbulent shear flow. In *Proceedings 8th Symposium on Turbulent Shear Flows*, pages 1.1.1–1.1.6, Munich, 1991.
49. D.W. Bogdanoff. Compressibility effects in turbulent shear layers. *AIAAJl.*, 21(6):pages 926–927, 1983.
50. J. Borée. Behaviour of highly accelerated variable-density turbulent flows. application to mixing preparation in internal combustion engines. *International Conférence on Variable Density Turbulent Flows, Banyuls, France*, pages pages 141–152, 2000.
51. J. Boussinesq. Théorie analytique de la chaleur. *Gauthiers-Villars*, 2, 1903.
52. P. Bradshaw. Compressible turbulent shear layers. *Ann. Rev. Fluid Mech.*, 9:pages 33–54, 1977.
53. P. Bradshaw and D.H. Ferriss. Applications of a general method of calculating turbulent shear layers. *J. Basic Eng.Trans. ASME*, pages pages 345–352, 1972.
54. P. Brancher. *Étude numérique des instabilités secondaires de jets.* Thèse, École Polytechnique. Laboratoire d'Hydrodynamique, 1996.
55. P. Brancher J.M. Chomaz and P. Huerre. Direct numerical simulations of round jets: Vortex induction and side jets. *Phys. Fluids*, 6(5):pages 1768–1774, 1994.
56. R. Breidenthal. The sonic eddy - a model for compressible turbulence. *AIAA Paper 90-0495*, 1990.
57. G. Brethouwer B.J. Boersma M.B.J.M. Pourquié and F.T.M. Nieuwstadt. Direct numerical simulation of turbulent mixing of a passive scalar in pipe flow. *Euro. J. Mech. B/Fluids*, 18:pages 739–756, 1999.
58. G. Briassulis and J. Andreopoulos. Compressibility effects in grid generated turbulence. *AIAA Paper*, 96-2055, 1996.
59. M. Brouillette and B. Sturtevant. Experiments on the Richtmyer-Meshkov instability: single-scale perturbations on a continuous interface. *J. Fluid Mech.*, 263:pages 271–292, 1994.
60. G. Brown. The Entrainment and Large Structure in Turbulent Mixing Layers. *5th Australasian Conference on Hydraulics and Fluid Mechanics*, pages 352–359, 1974.
61. G.L. Brown and A. Roshko. On density effects and large scale structure in turbulent mixing layers. *J. Fluid Mech.*, 64:pages 775–816, 1974.
62. L.W.B. Browne R.A. Antonia and D.A. Shah. Turbulent energy dissipation in a wake. *J. Fluid Mech.*, 179:pages 307–326, 1987.
63. G. Bruhat. *Thermodynamique.* Masson, 5ème Ed. 1962.
64. Y. Bury. *Structure de jets légers ou lourds en écoulement externe fortement pulsé.* Thèse, INPT $n^o$ 1694, 2000.
65. D.M. Bushnell. Turbulence modeling in aerodynamic shear flow - Status and problems. In *Proceedings 29th Aero. Sci. Meeting*, AIAA 91-0214, Reno, Nevada, jan. 1991.
66. X.D. Cai E.E. O'Brien and F. Ladeinde. Advection of mass fraction in forced, homogeneous, compressible turbulence. *Phys. Fluids*, 10(9):pages 2249–2259, 1998.
67. X.D. Cai E.E.O'Brien and F. Ladeinde. Thermodynamic behavior in decaying, compressible turbulence with initially dominant temperature fluctuations. *Phys. Fluids*, 9(6):pages 1754–1763, 1997.

68. C. Cambon and A. Simone. A structural approach to the effects of compressibility in turbulent flows subjected to external compression or high shear. In L. Fulachier J.L. Lumley and F. Anselmet, editors, *Variable density Low-Speed Turbulent Flows*, IUTAM Symposium, pages 333–342, Marseille, 1997. Kluwer Acad. Pub.
69. C. Cambon G. Coleman and N.N. Mansour. Rapid distortion theory analysis and direct simulation of compressible homogeneous turbulence at finite mach number. *J. Fluid Mech.*, 257:pages 641–665, 1993.
70. S. Candel. *Mécanique des fluides*. Dunod Univ., Bordas, Paris, 1990.
71. J.-B Cazalbou and P. Chassaing. The structure of the solution obtained with reynolds-stress-transport models at the free-stream edges of turbulent flows. *Phys. Fluids*, (to appear), 2002.
72. T. Cebeci and A.M.O Smith. *Analysis of turbulent boundary layers*. Academic Press, 1974.
73. T. Cebeci and A.M.O. Smith. Analysis of turbulent boundary layers. *Academic Press*, 1974.
74. T.O. Cebeci A.M. Smith andS.G. Mosinskis. Calculations of compressible adiabatic turbulent boundary layers. *AIAAJl.*, 8(11):pages 1974–1982, 1970.
75. A.J. Chambers and R.A. Antonia. Atmospheric estimates of power-law exponents $\mu$ and $\mu_\theta$. *Boundary-Layer Meteorol.*, 18:pages 399–410, 1984.
76. O. Chambres, S. Barre, and J. Bonnet. Detailed turbulence characteristics of a highly compressible supersonic turbulent plane mixing layer. *J. Fluid Mech.*, 1998. submitted.
77. O. Chambres S. Barre and J.P. Bonnet. Balance of kinetic energy in supersonic mixing layer compared to subsonic mixing layer and subsonic jets with variable density. In L. Fulachier J.L. Lumley and F. Anselmet, editors, *Variable Density Low Speed Turbulent Flows*, pages 303–308, Marseille, 1996. IUTAM Symposium, Kluwer Acad. Pub.
78. V.G. Chapin, M. Malandra, and P. Chassaing. A new control strategy of the mixing in two-dimensional jets. *AIAA Paper, 99-3407*, 1999.
79. D. Chapman. Computational aerodynamics and outlook. *AIAA J.*, 17:pages 1293–1313, 1979.
80. P. Chassaing. *Mélange turbulent de gaz inertes dans un jet de tube libre*. Thèse doctorat ès Sciences no. 42, Inst. Nat. Poly. de Toulouse, 1979.
81. P. Chassaing. Une alternative à la formulation des équations du mouvement turbulent d'un fluide à masse volumique variable. *Jl. Méc. Théo. Appl.*, 4(3):pages 375–389, 1985.
82. P. Chassaing. Some problems on single point modeling of turbulent, low-speed, variable density fluid motions. In L. Fulachier J.L. Lumley and F. Anselmet, editors, *Variable Density Low Speed Turbulent Flows*, pages 65–84, Marseille, 1996. IUTAM Symposium, Kluwer Acad. Pub.
83. P. Chassaing. *Mécanique des fluides. Éléments d'un premier parcours*. CEPADUES Éditions, 1997.
84. P. Chassaing. *Turbulence en Mécanique des fluides. Analyse du phénomène en vue de sa modélisation à l'usage de l'ingénieur*. CEPADUES Éditions, 2000.
85. P. Chassaing and H. Ha Minh. Some problems on second order modelling of mass transfer in a turbulent gas mixture. In Z.P. Zarič, editor, *Structure of turbulence in heat and mass transfer*, pages 509–528. Hemis. Pub. Corp., 1983.
86. P. Chassaing, G. Harran and L. Joly. Density fluctuation correlations in turbulent binary mixing. *J. Fluid Mech.*, 279:pages 239–278, 1994.
87. P. Chassaing H. Ha Minh and H. Boisson. Mass transport in turbulent round jet. In *Proceedings 2nd Symposium on Turbulent Shear Flow*, pages 1.20–1.25, London, 1979.
88. P. Chassaing S. Castaldi G. Harran and L. Joly. Variable density mixing in kinematically homogeneous turbulence. In F. Durst B.E. Launder F.W.

Schmidt J.H. Whitelaw, editor, *Eleventh Symposium on Turbulent Shear Flows*, volume 2, pages 12–24;12–28, Grenoble, France, 1997.

89. C.-J. Chen and S.-Y. Jaw. *Fundamentals of turbulence modeling*. Combustion: An International Series. Taylor & Francis, 1998.
90. C.H.P. Chen and R.F. Balckwelder. The large scale motion in a turbulent boundary layer : a study using temperature contamination. *J. Fluid Mech.*, 89:pages 1–31, 1978.
91. C.J. Chen and C.H. Chen. On the prediction and unified correlation for decay of vertical buoyant jets. *Jl. Heat Transfer Trans. ASME*, 101:pages 532–537, 1979.
92. C.J. Chen and W. Rodi. Vertical turbulent buoyant jets - A review of experimental data. *HMT-4*, pergamon, 1980.
93. M. Chibat. *Modélisation au second ordre des équations en moyenne pondérée du mélange turbulent. Application au jet isotherme de gaz non réactifs*. Thèse, INPT, IMFT, 1995.
94. K.-Y. Chien. Predictions of channel and boundary-layer flows with a low-Reynolds-number turbulence model. *AIAA J.*, 20(1):pages 33–38, 1982.
95. N. Chinzei, G. Masuya, T. Komuro, T. Murakami, and K. Kudou. Spreading of two-stream supersonic mixing layer. *Phys. Fluids*, 29:1345–1347, 1986.
96. N. T. Clemens and M. G. Mungal. Large-scale structure and entrainment in the supersonic mixing layer. *J. Fluid Mech.*, 284:171–216, 1995.
97. G.M. Corcos and S.J. Lin. The mixing layer : deterministic models of a turbulent flow. Part 2. the origin of the three-dimensional motion. *J. Fluid Mech.*, 139:pages 67–95, 1984.
98. G.M. Corcos and F.S. Sherman. Vorticity concentration and the dynamics of unstable free shear layers. *J. Fluid Mech.*, 73:pages 241–264, 1976.
99. G.M. Corcos and F.S. Sherman. The mixing layer : deterministic models of a turbulent flow. part 1. introduction and the two-dimensional flow. *J. Fluid Mech.*, 139:pages 29–65, 1984.
100. D.E. Cormack L.G. Leal and J.H. Seinfeld. An evaluation of mean Reynolds stress turbulence models: The triple velocity correlation. *Jl. of Fluid Engineering Trans. ASME*, 100:pages 47–54, 1978.
101. S. Corrsin. On the spectrum of isotropic temperature fluctuations in an isotropic turbulence. *J. Appl. Phys.*, 22:pages 469–473, 1951.
102. S. Corrsin. Turbulence: experimental methods. In S. Flügge and C. Truesdell, editors, *Handbuch der Physik*, VIII/2, pages 524–590. Springer, 1963.
103. S. Corrsin. Limitations of gradient transport models in random walks and in turbulence. *Adv. in Geophysics, A*, 18:pages 25–60, 1974.
104. S. Corrsin and M.S. Uberoi. Further experiments on the flow and heat transfer in a heated turbulent air jet. *NACA TN 1865.*, 1949.
105. A.B. Cortesi, G. Yadigaroglu, and S. Banerjee. Numerical investigation of the formation of three-dimensional structures in stably-stratified mixing layers. *Phys. FLuids*, 10(6):pages 1449–1473, 1998.
106. J. Cousteix. *Turbulence et couches limites*. Aérodynamique en fluide visqueux. ENSAE, 1988.
107. J. Cousteix and B. Aupoix. Turbulence models for compressible flows. In *Three-dimensional supersonic and hypersonic flows including separation*, AGARD-FDP-VKI Special course, 1989.
108. J. Cousteix V. Saint-Martin R. Messing H. Bézard and B. Aupoix. Development of the k-$\phi$ turbulence model. In F. Durst B.E. Launder F.W. Schmidt J.H. Whitelaw, editor, *Eleventh Symposium on Turbulent Shear Flows*, volume 2, pages 13–24;13–29, Grenoble, France, 1997.
109. T.J. Craft. *Second-moment modelling of turbulent scalar transport*. PhD dissertation, University of Manchester, Department of Mechanical Engineering, april 1991. TFD/91/3.
110. T.J. Craft and B.E. Launder. A Reynolds stress closure designed for complex

geometries. *Int. Jl. Heat and Fluid Flow*, 17:pages 245–254, 1996.

111. T.J. Craft B.E. Launder and K. Suga. Development and application of a cubic eddy-viscosity model of turbulence. *Int. Jl. Heat and Fluid Flow*, 17:pages 108–115, 1996.
112. T. Craft N.Z. Ince and B.E. Launder. Recent developments in second moment closure for buoyancy-affected flows. *Dyn. of Atmos. and Oceans*, 23:pages 99–114, 1996.
113. T. Craft S. Fu B.E. Launder and D.P. Tselepidakis. Developments in modelling the turbulent second-moment pressure correlations. TFD/89/1 7, Manchester University, UMIST-Dep. Mech. Eng., oct. 1989.
114. A. Craya. *Contribution à l'analyse de la turbulence associée à des vitesses moyennes*. Thèse doct. ès Sciences, Faculté des Sciences de l'Université de Grenoble, 1957.
115. S.C. Crow. Viscoelastic properties of fine-grained incompressible turbulence. *J. Fluid Mech.*, 33(part 1):pages 1–20, 1968.
116. S.C. Crow and F.H. Champagne. Orderly structure in jet turbulence. *J. Fluid Mech.*, 48:pages 547–591, 1971.
117. B.J. Daly and F.H. Harlow. Transport equations in turbulence. *Phys. Fluids*, 13(11):pages 2634–2649, 1970.
118. L. Danaila F. Anselmet T. Zhou and R.A. Antonia. Turbulent energy scale-budget equations in a fully developed channel flow. *J. Fluid Mech.*, (to appear), 2000.
119. L. Danaila P. Le Gal F. Anselmet F. Plaza and J.F. Pinton. Some new features of the passive scalar mixing in a turbulent flow. *Phys. Fluids*, 11:pages 636–646, 1999.
120. L. Danaila R.A. Antonia F. Anselmet and T. Zhou. A generalization of Yaglom's equation which accounts for the large-scale forcing in heated grid turbulence. *J. Fluid Mech.*, 391:pages 359–372, 1999.
121. K. Dang and Y.F. Morchoisne. Numerical simulation of homogeneous compressible turbulence. In *2nd International Symposium on Transport Phenomena in Turbulent Flows*, pages 25–29, Tokyo, 1987.
122. R.F. Davey and A. Roshko. The two-dimensional mixing region. *J. Fluid Mech.*, 53:pages 523–543, 1972.
123. B.I. Davidov. On the statistical dynamics of an incompressible fluid. *Doklady AN. SSSR*, 136, 1961.
124. G. De Moor. *Les théories de la turbulence dans la couche-limite atmosphérique*, volume 3 of *Cours et Manuels*. Ministère des Transports - Direction de la Météorologie, Boulogne Billancourt, Oct 1983.
125. J.F. Debiève P. Dupont A.J. Smits and J.-P. Dussauge. Compressibility vs density variations and the structure of turbulence. In L. Fulachier J.L. Lumley and F. Anselmet, editors, *Variable density Low-Speed Turbulent Flows*, IUTAM Symposium, pages 309–316, Marseille, 1997. Kluwer Acad. Pub.
126. J.R. Debisschop and J.P. Bonnet. Mean and fluctuating velocity measurements in supersonic mixing layers. In W. Rodi and F. Martelli, editors, *Engineering Turbulence Modeling and Experiments 2*. Elsevier Science Publishers, 1993.
127. J. Délery. Analyse du décollement résultant d'une interaction choc-couche limite turbulente en transsonique. *La Reche. Aéro.*, 6:pages 305–320, 1978.
128. J. Délery. Investigation of strong shock boundary layer interaction in 2-D transonic flow with emphasis in turbulence phenomena. *AIAA paper*, 81-1245, 1981.
129. J. Délery. Experimental investigation of turbulence properties in transonic shock-wave/boundary-layer interactions. *AIAA Jl.*, 21:pages 180–185, 1983.
130. J.M. Délery and P. Le Diuzet. Décollement résultant d'une interaction onde de choc-couche limite turbulente. Technical Paper 1979-146, ONERA, 29, Av. de la Division Leclerc, 92320 Chatillon, FRANCE, 1979.
131. P. Dimotakis. Two-dimensional shear layer entrainment. *AIAA J.*, 24(11):pages 1791–1796, 1984.

132. T. Djeridane. Contribution à l'étude expérimentale de jets turbulents axisymétriques à densité variable. *Thèse Aix-Marseille II (IRPHE)*, 1994.
133. C. du P. Donaldson. A progress report on an attempt to construct an invariant model of turbulent shear flows. In *AGARD Conference Proceedings on Turbulent shear flows*, C.P. 93, London, 1971.
134. J.F. Driscoll R.W. Schefer and R.W. Dibble. Mass fluxes $\overline{\rho' u'}$ and $\overline{\rho' u'}$ measured in a turbulent nonpremixed flame. In *Proceedings Nineteenth Symposium (International) on Combustion*, pages 477–485. The Comb. Inst., 1982.
135. D.G. Dritschel, P.H. Haynes, M.N. Juckes, and T.G. Sheperd. The stability of a two-dimensional vorticity filament under uniform strain. *J. Fluid Mech.*, 230:pages 647–665, 1991.
136. P. Dupont P. Muscat and J.-P. Dussauge. Time and space-time statistics in a supersonic mixing layer. In *Symposium on Transitional and turbulent compressible flows*, ASME Fluid Eng. Conference FED-Vol.224, pages 117–122, 1995.
137. S. Duranti and F. Pittaluga. Navier-Stokes prediction of internal flows with a three-equation turbulence model. *AIAAJl.*, 38(6):pages 1098–1100, 2000.
138. P.A. Durbin and O. Zeman. Rapid distortion theory for homogeneous compressed turbulence with application to modelling. *J. Fluid Mech.*, 242:pages 349–370, 1992.
139. P.A. Durbin N.N. Mansour and Z. Yang. Eddy viscosity transport model for turbulent flow. *Phys. Fluids*, Vol.6(No.2):pages 1007–1015, 1994.
140. D.R. Durran. Improving the anelastic approximation. *Jl. of Atm. Sci.*, 46(11):pages 1453–1461, 1989.
141. J.-P. Dussauge and C. Quine. A second ordre closure for supersonic turbulent flows - application to the supersonic mixing. In *The Physics of Compressible Turbulent Mixing*, pages 407–416, Princeton, 1988.
142. J.-P. Dussauge H. Fernholz R.W. Smith P.J. Finley A.J. Smits and E.F. Spina. Turbulent boundary layers in subsonic and supersonic flow. Edited by William S. Saric, *AGARD-AG*-335, 1996.
143. J.A. Dutton and G.H. Fichtl. Approximate equations of motion for gases and liquids. *Jl. of Atm. Sci.*, 26:pages 241–254, 1969.
144. C. Eifert D.G. Pfuderer and J. Janicka. Simulation of flow through circular to rectangular transition duct using non-linear second-moment closures. In F. Durst B.E. Launder F.W. Schmidt J.H. Whitelaw, editor, *Eleventh Symposium on Turbulent Shear Flows*, volume 1, pages 3–1;3–6, Grenoble, France, 1997.
145. M. Elena and J. Gaviglio. La couche limite turbulente compressible : méthodes d'étude et résultats, synthèse. *La Rech. Aérospatiale*, 2:pages 1–21, 1993.
146. G.S. Elliott and M. Samimy. Compressibility effects in free shear layers. *Phys. Fluids A*, 2(7):pages 1231–1240, 1990.
147. G.S. Elliott M. Samimy and S.A. Arnette. The characteristics and evolution of large-scale structures in compressible mixing layers. *Phys. Fluids A*, 7(4):pages 864–876, 1995.
148. H.W. Emmons. Shear flow turbulence. In *Proceedings 2nd U.S. Congress of Applied Mechanics*, pages 1–12, Ann Arbor, June 1954. ASME.
149. G. Erlebacher, M.Y. Hussaini, H.O. Kreiss, and S. Sarkar. The Analysis and Simulation of Compressible Turbulence. *Theor. Comput. Fluid Dynamics*, 2:73–95, 1990.
150. G. Erlebacher and S. Sarkar. Statistical analysis of the rate of strain tensor in compressible homogeneous turbulence. *Phys. Fluids*, 5(12):pages 3240–3254, 1993.
151. A. Favre. Équations statistiques des gaz turbulents. *C. R. Acad. Sci. Paris*, 246: Masse, quantité de mouvement: pages 2576–2579; Énergie totale, énergie interne: pages 2723–2725; Énergie cinétique, énergie cinétique du mouvement macroscopique, énergie cinétique de la turbulence: pages 2839–2842; Enthalpie, entropie, température: pages 3216–3219, 1958.
152. A. Favre. Équations des gaz turbulents compressibles. I. Formes générales.

*J. de Méc.*, 4(3):pages 361–390, 1965.

153. A. Favre. Équations des gaz turbulents compressibles. II. Méthode des vitesses moyennes; méthode des vitesses macroscopiques pondérées par la masse volumique. *J. de Méc.*, 4(4):pages 391–421, 1965.
154. A. Favre. Équations statistiques aux fluctuations d'entropie, de concentration, de rotationnel dans les écoulements compressibles. *C. R. Acad. Sci. Paris*, 273:pages 1289–1294, 1971.
155. A. Favre. Équations statistiques des fluides turbulents compressibles. In *5ème Congrès Canadien de Mécanique Appliquée*, pages G3–G34, Fredericton, 1975. New Brunswick Univ.
156. A. Favre. Équations statistiques des fluides à masse volumique variable en écoulements turbulents. In F. Anselmet L. Fulachier I. Gökalp, editor, *2èmes Journées d'Étude des Écoulements Turbulents à Masse Volumique Variable*, Orléans, 1992.
157. A. Favre L.S.G. Kovasznay R. Dumas J. Gaviglio and M. Coantic. *La turbulence en mécanique des fluides. Bases théoriques et expérimentales méthodes statistiques.* Gauthier-villars, 1976.
158. M. Favre-Marinet and E. Camano. Mixing in coaxial jets with large density differences. In L. Fulachier J.L. Lumley and F. Anselmet, editors, *Variable Density Low Speed Turbulent Flows*, pages 127–135, Marseille, 1996. IUTAM Symposium, Kluwer Acad. Pub.
159. A.T. Fedorchenko. A model of unsteady subsonic flow with acoustics excluded. *J. Fluid Mech.*, 334:pages 135–155, 1997.
160. W.J. Feireisen E. Shirani J.H. Ferziger and W.C. Reynolds. Direct simulation of homogeneous turbulent shear flows on the Illiac IV computer. In F.Durst R.Friedrich B.E.Launder F.W. Schmidt U. Schumann J.H. Whitelaw, editor, *Selected Papers from 3th International Symposium on Turbulent Shear Flows*, pages 309–319. Springer-Verlag, Berlin, 1982.
161. Y. Fouillet. *Contribution à l'étude par expérimentation numérique des écoulements cisaillés libres: Effets de compressibilité.* Thèse, Institut de Mécanique de Grenoble, 1991.
162. J.B. Freund, P. Moin, and S.K. Lele. *Compressibility Effects in a Turbulent Annular Mixing Layer.* Stanford Report No. TF-72, 1997.
163. J.B. Freund S.K. Lele and P. Moin. Compressibility effects in a turbulent annular mixing layer. Part 1. Turbulence and growth rate. *J. Fluid Mech.*, 421:pages 229–267, 2000.
164. R. Friedrich and J. Peinke. Description of a turbulent cascade by a Fokker-Planck equation. *Phys. Rev. Lett.*, 78:pages 863–866, 1997.
165. U. Frish. *Turbulence : The Legacy of A. N. Kolmogorov.* Cambridge, CUP, 1995.
166. U. Frish A. Mazzino A. Noullez and M. Vergassola. Lagrangian method for the multiple correlations in passive scalar advection. *Phys. Fluids*, 11:pages 2178–2186, 1999.
167. S. Fu B.E. Launder and D.P. Tselepidakis. Accomodating the effects of high strain rates in modelling the pressure-strain correlation. Technical Report Rep. TFD/87/5, Manchester University, UMIST - Mech. Eng. Dep., 1987.
168. L. Fulachier. *Contribution à l'étude des analogies des champs dynamique et thermique dans une couche limite turbulente — effet de l'aspiration.* Thèse docteur ès sciences physiques, Université de Provence, Marseille, 1972.
169. L. Fulachier and R.A. Antonia. Turbulent Reynolds and Péclet numbers re-defined. *Int. Comm. Heat Mass Transfer*, 10:pages 435–439, 1983.
170. L. Fulachier and R.A. Antonia. Spectral analogy between temperature and velocity fluctuations in several turbulent flows. *Int. Jl. Heat Mass Trans.*, 27:pages 987–997, 1984.
171. L. Fulachier and R. Dumas. Spectral analogy between temperature and velocity

fluctuations in a turbulent boundary layer. *J. Fluid Mech.*, 77:pages 257–277, 1976.

172. L. Fulachier J.L. Lumley and F. Anselmet, editor. *IUTAM Symposium on Variable density low-speed turbulent flows.* Number 41 in Fluid mechanics and its applications. Kluwer Acad.Pub., Dordrecht / Boston / London, 1997.

173. L. Fulachier R. Borghi F. Anselmet and P. Parenthoen. Influence of density variations on the structure of low-speed turbulent flows : A report on Euromech 237. *J. Fluid Mech.*, 203:pages 577–593, 1989.

174. J.M. Galmes J.-P. Dussauge and I. Dekeyser. Couches limites turbulentes supersoniques soumises à un gradient de pression: calcul à l'aide d'un modèle k-$\epsilon$. *Jl. Méc Théo et Appl.*, 2(4), 1983.

175. T.B. Gatski and C.G. Speziale. On explicit algebraic models for complex turbulent flows. *J. Fluid Mech.*, 254:pages 59–78, 1993.

176. J. Gaviglio. La turbulence dans les écoulements compressibles des gaz. In A. Favre L.S.G. Kovasznay R. Dumas J. Gaviglio and M. Coantic, editor, *La turbulence en mécanique des fluides. Bases théoriques et expérimentales méthodes statistiques*, chapter V. Gauthier-villars, Paris, 1976.

177. J. Gaviglio. Reynolds analogies and experimental study of heat transfer in the supersonic boundary layer. *Int. Jl. Heat Mass Trans.*, 30:pages 911–926, 1987.

178. W.K. George and H.J. Hussein. Locally axisymmetric turbulence. *J. Fluid Mech.*, 233:pages 1–23, 1991.

179. G.A. Gerolymos and I. Vallet. Wall-normal-free near-wall Reynolds-stress closure for three-dimensional compressible separated flows. *AIAA J.*, 39(10):pages 1833–1842, 2001.

180. C.H. Gibson. Kolmogorov similarity hypotheses for scalar fields : sampling intermittent turbulent mixing in the ocean and galaxy. *Proceedings Roy. Soc. Lond. A*, 434:pages 149–164, 1991.

181. M.M. Gibson and B.E. Launder. On the calculation of horizontal, turbulent, free shear flows under gravitational influence. *Jl. Heat Transfer Trans. ASME*, Feb.:pages 81–87, 1976.

182. M.M. Gibson and B.E. Launder. Ground effects on pressure fluctuations in the atmospheric boundary layer. *J. Fluid Mech.*, 83(part 3):pages 491–511, 1978.

183. C.H. Gibson C.A. Friehe and S.O. McConnell. Structure of sheared turbulent fields. *Phys. Fluids*, 20:pages S156–S167, 1977.

184. C.H. Gibson G.R. Stegen and R.B. Williams. Statistics of the fine structure of turbulent velocity and temperature fields measured at high Reynolds number. *J. Fluid Mech.*, 41:pages 153–167, 1970.

185. C.H. Gibson W.T. Ashurst and A.R. Kerstein. Mixing of strongly diffusive passive scalars like temperature by turbulence. *J. Fluid Mech.*, 194:pages 261–293, 1988.

186. S.G. Goebel and J.C. Dutton. Experimental study of compressible turbulent mixing layers. *AIAA Jl.*, 29(4):pages 538–546, 1991.

187. F. Grasso and C.G. Speziale. Supersonic flow computations by two-equation turbulence modeling. AIAA, 89-1951-CP, 1989.

188. D.D. Gray and A. Giorgini. The validity of the Boussinesq approximation for liquids and gases. *Int. Jl. Heat Mass Trans.*, 19:pages 545–551, 1976.

189. F.F. Grinstein, E.S. Oran, and J.P. Boris. Pressure field, feedback and global instabilitie of subsonic spatially developing mixing layers. *Phys. FLuids A*, 3(10):pages 2401–2409, 1991.

190. S. Grossmann. Asymptotic dissipation rate in turbulence. *Phys. Rev. E*, 51:pages 6275–6277, 1995.

191. D. Guézengar H. Guillard and J.-P. Dussauge. Modeling dissipation equation in supersonic turbulent mixing layers with high-density gradients. *AIAAJl.*, 38(9):pages 1650–1655, 2000.

192. A.N. Gulyaev V. Ye. Kozlov and A.N. Secundov. Universal turbulence model $\nu_t - 92$. *ECOLEN Sci. Res. Center (Moscou)*, Preprint No.3, 1993.

193. H. Ha Minh B.E. Launder and J. MacInnes. A new approach to the analysis of

turbulent mixing in variable density flows. In L.S.J. Bradbury F. Durst B.E. Launder F.W. Schmidt J.H. Whitelaw, editor, *Proceedings 3rd Symposium on Turbulent Shear Flows*, pages 19.19 - 19.25, Davis, 1981. Univ. of California.

194. H. Ha Minh B.E. Launder and J. MacInnes. The turbulence modelling of variable density flows - A mixed-weighted decomposition. In L.S.J. Bradbury F. Durst B.E. Launder F.W. Schmidt J.H. Whitelaw, editor, *Selected papers from 3rd Symposium on Turbulent Shear Flows*, pages 291–308, Davis, 1981. Univ. of California, Springer-Verlag, Berlin.
195. H. Ha Minh D. Vandromme and R.W. MacCormack. Computation of complex compressible turbulent flows. In *Preprints The 1980-81 AFOSR-HTTM Stanford Conference*, volume 1 of *Complex Turbulent Flows - Comparison of computation and experiment. Cases 8101, 8201 and 8612*, Stanford Univ., 1981.
196. H. Ha Minh and D. Vandromme. Etude numérique du comportement des contraintes de Reynolds dans une compression-détente supersonique. In *XXI ème Colloque d'Aérodynamique Appliquée*, Lille, FRANCE, 1985. Association Aéronautique et Astronautique de France.
197. H. Ha Minh and D. Vandromme. Modelling of compressible turbulent flows: Present possibilities and perspectives. In *IUTAM Symposium*, Paris, sept. 1985.
198. H. Ha Minh and D. Vandromme. Modélisation de la turbulence en situations compressibles: les aspects physiques et leurs incidences sur la prédétermination numérique des écoulements. In *Congrès Ecoulements Compressibles*, Poitiers, 1986.
199. H. Ha Minh M.W. Rubesin D. Vandromme and J.R. Viegas. On the use of second order closure modelling for the prediction of turbulent boundary layers/shock wave interactions: Physical and numerical aspects. In *International Compressible Fluid Dynamic Symposium*, Tokyo, 1985.
200. J.L. Hall, P.E. Dimotakis, and H. Rosemann. Experiments in Nonreacting Compressible Shear Layers. *AIAA J.*, 31(12):2247–2254, 1993.
201. M. Hallback J. Groth and A.V. Johansson. A Reynolds stress closure for the dissipation in anisotropic turbulent flow. In *Seventh Symposium on Turbulent Shear Flows*, pages 17.2.1–17.2.6. Stanford Univ., 1989.
202. M. Hallback J. Groth and A.V. Johansson. An algebraic model for nonisotropic turbulent dissipation rate in Reynolds stress closures. *Phys. Fluids A*, 2(10):pages 1859–1866, 1990.
203. F. Hamba. Effects of pressure fluctuations on turbulence growth in compressible homogeneous shear flow. *Phys. Fluids*, 11(6):pages 1623–1635, 1999.
204. Hanjalić and K. B.E. Launder. Contribution towards a Reynolds stress closure for low Reynolds number turbulence. *J. Fluid Mech.*, 74(part 4):pages 593–610, 1976.
205. K. Hanjalić and B.E. Launder. Fully developed asymetric flow in a plane channel. *J. Fluid Mech.*, 51(part 2):pages 301–335, 1972.
206. R. Hannapel and R. Friedrich. Interaction of isotropic turbulence with a normal shock wave. *Appl. Scien. Res.*, 51:507, 1993.
207. J.-L. Harion. *Influence de variations de densité importantes sur les propriétés de transfert d'une couche limite turbulente.* Thèse, INP Grenoble, 1994.
208. J.L. Harion R. Riva G. Binder M. and Favre-Marinet. Absolute instability in variable density plane jets. In F. Durst R. Friedrich B.E. Launder F.W. Schmidt U. Schumann J.H. Whitelaw, editor, *Eighth Symposium on Turbulent Shear Flows*, pages 1–6, Munich, 1991.
209. F.H. Harlow and P.I. Nakayama. Turbulence transport equations. *Phys. Fluids*, 10(11), 1967.
210. G. Harran P. Chassaing L.Joly and M. Chibat. Étude numérique des effets de densité dans un jet de mélange turbulent en microgravité. *Rev. Gén Therm.*, 35:pages 151–176, 1996.
211. J. Hecht U. Alon and D. Shvarts. Potential flow models Rayleigh-Taylor and Richtmyer-Meshkov bubble fronts. *Phys. Fluids*, 6(12):pages 4019–4030, 1994.

212. J.C. Hermanson and B.M. Cetegen. Shock-induced mixing of nonhomogeneous density turbulent jets. *Phys. Fluids*, 12(5):pages 1210–1225, 2000.

213. R. Hermouche. *Étude expérimentale des champs statistiques et des spectres de la zone proche d'un jet Helium-Air*. Thèse, INPT, 1996.

214. B.J. Hill. Measurements of local entrainment rate in the initial region of axisymmetric turbulent jets. *J. Fluid Mech.*, 51(part 4):pages 1234–1246, 1972.

215. R.J. Hill. Applicability of Kolmogorov's and Monin's equations of turbulence. *J. Fluid Mech.*, 353:pages 67–81, 1997.

216. C.W. Hirt. Generalized turbulence transport equations. In *International Seminar of the International Center of Heat and Mass Transfer*, Herceg Novi, Yugoslavia, 1969.

217. A. Honkan and J. Andreopoulos. Experiments in a shock wave/homogeneous turbulence interaction. *AIAA Paper*, 90-1647, 1996.

218. C.C. Horstman. Prediction of hypersonic shock wave-turbulent boundary layer interaction. *AIAA Paper*, 87-1367, 1987.

219. T.Y. Hou X.-H. Wu S. Chen and Y. Zhou. Effect of finite computational domain on turbulence scaling law in both physical and spectral spaces. *Phys. Rev. E*, 58:pages 5841–5844,, 1998.

220. P.G. Huang and G.N. Coleman. Van Driest transformation and compressible wall-bounded flows. *AIAA J.*, 32(10):pages 2110–2113, 1994.

221. P.G. Huang G.N. Coleman and P. Bradshaw. Compressible turbulent channel flows: DNS results and modelling. *J. Fluid Mech.*, 305:pages 185–218, 1995.

222. P. Huerre. The nonlinear stability of a free shear layer in the viscous critical layer regime. *Philos. Trans. Roy. Soc. London*, 293(1408):pages 643–675, 1980.

223. P. Huerre. On the Landau constant in mixing layers. *Proceedings Roy. Soc. London A*, 409:pages 369–381, 1987.

224. P. Huerre and P.A. Monkewitz. Absolute and convective instabilities in free shear layers. *J. Fluid Mech.*, 159:pages 151–168, 1985.

225. P. Huerre and P.A. Monkewitz. Local and global instabilities in spatially developing flows. *Annu. Rev. Fluid Mech.*, 22:pages 473–537, 1990.

226. P. Huerre and J.F. Scott. Effects of critical structure on the nonlinear evolution of waves in free shear layers. *Proceedings Roy. Soc. London A*, 371:pages 509–524, 1980.

227. P. Huerre K. Amram and J.M. Chomaz. Instabilities and bifurcations in variable density flows. In L. Fulachier J.L. Lumley and F. Anselmet, editors, *Variable Density Low Speed Turbulent Flows*, pages 3–8, Marseille, 1996. IUTAM Symposium, Kluwer Acad. Pub.

228. M.Y. Hussaini and G. Erlebacher. Interaction of an entropy spot with a shock. *AIAA J.*, 37(3):pages 346–356, 1999.

229. Y. Iritani N. Kasagi and M. Hirata. Heat transfer mechanism and associated turbulence structure in the near-wall region of a turbulent boundary layer. In L. J. S. Bradbury F. Durst B.E. Launder F. W. Schmidt and J. H. Whitelaw, editor, *Proceedings Fourth Turbulent Shear Flows*, page 223, Berlin, 1985. Springer.

230. L. Jacquin E. Blin and P. Geffroy. Experiments on free turbulence/shock wave interaction. In *Proceedings 8th Symposium on Turbulent Shear Flows*, pages 1.2.1–1.2.6, Munich, 1991.

231. S. Jamme. *Étude de l'interaction entre une turbulence homogène isotrope et une onde de choc*. Thèse No. 1510, I.N.P.T., Dépt. Méc. des Fluides de l'ENSICA, Déc. 1998.

232. J. Janicka and J.L. Lumley. Second-order modeling in non-constant density flows. Report FDA-81-01, Cornell Univ., 1980.

233. T. Janicka and W. Kollmann. A two-variables formalism for the treatment of chemical reactions in turbulent H2-Air diffusion flames. In *Proceedings Seventeenth Symposium (International) on Combustion*, Colloquium on Turbulent-Combustion

Interactions, pages 421–430. The Combustion Institute, 1979.

234. A.V. Johansson and M. Hallback. Modelling of rapid pressure-strain in Reynolds-stress closures. *J. Fluid Mech.*, 269:pages 143–168, 1994.

235. L. Joly. *Écoulements cisaillés libres à masse volumique variable. Analyse physique et modélisation.* Thèse Méc. des fluides, I.N.P.T., ENSICA, 1994.

236. L. Joly and A. Purwanto. Les équations de la statistique en un point d'un écoulement à masse volumique variable. Rapport interne DEF-01/93, ENSICA, 1993.

237. W.P. Jones. Models for turbulent flows with variable density and combustion. In W. Kollmann, editor, *Prediction methods for turbulent flows*, Lecture series 1979-2. Von Kármán Institute, 1979.

238. W.P. Jones and B.E. Launder. The prediction of laminarization with a two-equation model of turbulence. *Int. Jl. Heat Mass Trans.*, 15:pages 301–313, 1972.

239. W.P. Jones and B.E. Launder. The calculation of low-Reynolds-number phenomena with a two-equation model of turbulence. *Int. Jl. Heat Mass Trans.*, 16:pages 1119–1130, 1973.

240. W.P. Jones and P. Musonge. Modelling of scalar transport in homogeneous turbulent flows. In *Proceedings Fourth Symposium on Turbulent Shear Flows*, pages 17.18–17.24. University of Karlsruhe, 1983.

241. N. Kasagi and N. Shikazono. Contribution of direct numerical simulation to understanding and modelling turbulent transport. *Proceedings R. Soc. Lond. A*, 451:pages 257–292, 1995.

242. H. Kawamura H. Abe and Y. Matsui. DNS of turbulent heat transfer in channel flow with respect to Reynolds and Prandtl number effects. *Int. J. Heat Fluid Flow*, 20:pages 196–207, 1999.

243. H. Kawamura K. Ohsaka H. Abe and K. Yamamoto. DNS of turbulent heat transfer in channel flow with low to medium-high Prandtl number fluid. *Int. J. Heat Fluid Flow*, 19:pages 482–491, 1998.

244. J. Keller and W. Merzkirch. Interaction of a normal shock wave with a compressible turblent flow. *Exp. in Fluids*, 8:241, 1990.

245. J.H. Kent and R.W. Bilger. The prediction of turbulent diffusion flame fields and nitric oxide formation. In *Proceedings Sixteenth Symposium (International) on Combustion.* The Combustion Institute, 1977.

246. S. Kida and S.A. Orszag. Enstrophy budget in decaying compressible turbulence. *Jl. Sci. Comp.*, 5(1):pages 1–34, 1990.

247. J. Kim. Investigation of heat and momentum transport in turbulent flows via numerical simulations. In M. Hirata and N. Kasagi, editors, *Transport Phenomena in Turbulent Flows*, pages 715–730, New York, 1988. Hemisphere.

248. J. Kim and P. Moin. Transport of passive scalars in a turbulent channel flow. In F. W. Schmidt F. Durst, B. E. Launder and J. H. Whitelaw, editors, *Turbulent Shear Flows 6*, pages 35–96, Berlin, 1989. Springer.

249. J. Kim P. Moin and R. Moser. Turbulence statistics in fully developed channel flow at low Reynolds number. *J. Fluid Mech.*, 177:pages 133–166, 1987.

250. G.P. Klaassen and W.R. Peltier. The influence of stratification on secondary instability in free shear layers. *J. Fluid Mech.*, 227:pages 71–106, 1991.

251. S. Klainerman and A. Majda. Compressible and incompressible fluids. *Comm. Pure and Appl. Math.*, 35:pages 629–651, 1982.

252. O. Knio and A. Ghoniem. The three-dimensionnal structure layers under non-symmetric conditions. *J. Fluid Mech.*, 243:pages 353–392, 1992.

253. O.M. Knio and A. Ghoniem. Three-dimensional vortex simulation of rollup and entrainment in a shear layer. *J. Comp. Phys.*, 97:172, 1991.

254. A.N. Kolmogorov. Energy dissipation in locally isotropic turbulence. *Dokl. Akad. Nauk. SSSR*, 32:pages 19–21, 1941.

255. A.N. Kolmogorov. The local structure of turbulence in an incompressible fluid for

very large Reynolds numbers. *Dokl. Akad. Nauk. SSSR*, 30:pages 299–303, 1941.

256. A.N. Kolmogorov. Equations of turbulent motion of an incompressible fluid. *Izvest. Acad. Sci., SSSR Physics*, 6(1 et 2), 1942.
257. A.N. Kolmogorov. A refinement of previous hypotheses concerning the local structure of turbulence in a viscous incompressible fluid at high Reynolds number. *J. Fluid Mech.*, 13:pages 82–85, 1962.
258. N.N. Korchashkin. The effect of fluctuations of energy dissipation and temperature dissipation on locally isotropic turbulent fields. *Izv. Atmos. Ocean. Phys.*, 6:pages 947–949, 1970.
259. L.S.G. Kovasznay. Turbulence in supersonic flow. *Jl. of the Aeronaut. Sci.*, 20(10):pages 657–682, 1953.
260. S.L.G. Kovasznay. Turbulence in compressible and electrically conductive media. In A. Favre, editor, *Mécanique de la Turbulence*, pages 357–365, Marseille, 1961. C.R.N.S.
261. R.H. Kraichnan. Anomalous scaling of a randomly advected passive scalar. *Phys. Rev. Lett.*, 72:pages 1016–1019, 1994.
262. L.V. Krishnamoorthy and R.A. Antonia. Temperature dissipation measurements in a turbulent boundary layer. *J. Fluid Mech.*, 176:pages 265–281, 1987.
263. A. Kumar, D.M. Bushnell and M.Y. Hussaini. Mixing augmentation technique for hypervelocity scramjets. *Jl. of Propulsion and Power*, 5(5):pages 514–522, 1989.
264. S. Kurien and K.R. Sreenivasan. Anisotropic scaling contributions to high-order structure functions in high-Reynolds-number turbulence. *Phys. Rev. E*, 62:pages 2206–2212, 2000.
265. V.R. Kuznetsov A.A Praskovsky and V.A. Sabelnikov. Fine-scale turbulence structure of intermittent shear flows. *J. Fluid Mech.*, 243:pages 595–622, 1992.
266. D.M. Kyle and K.R. Sreenivasan. The instability and breakdown of a round variable-density jet. *J. Fluid Mech.*, 249:pages 619–664, 1993.
267. L. Landau and E. Lifchitz. *Physique théorique*. MIR, Moscou, 2nd edition, 1989.
268. L.D. Landau and E.M. Lifshitz. *Fluid Mechanics*. Pergamon Press, second edition, 1987.
269. J.C. Larue and P.A. Libby. Measurements in the turbulent boundary layer with slot injection of helium. *Phys. Fluids*, 20, 1977.
270. J.C. Larue and P.A. Libby. Further results related to the turbulent boundary layer with slot injection of helium. *Phys. Fluids*, 23, 1980.
271. J. Laufer. The structure of turbulence in fully developed pipe flow. Tech. Report 1174, NACA, 1954.
272. J. Laufer J.E. Ffowcs Williams and S. Childress. Mechanism of noise generation in the turbulent boundary layer. Adv. Gr. Aero. Res. and Dev. AGARDograph 90, NATO, 64 rue de Varenne, Paris VII, 1964.
273. B.E. Launder. Heat and mass transport. In P. Bradshaw, editor, *Topics in Applied Physics*, Vol.12: Turbulence, pages 231–287, Berlin, 1976. Springer.
274. B.E. Launder. Second-moment closure: Methodology and practice. In J. Mathieu D. Jeandel B.E. Launder W.C. Reynolds W. Rodi, editor, *Turbulence models and their aplications*, number 2 in Collection Direction Etudes et Recherches, pages 1–147. Ecole d'été d'analyse numérique CEA-EDF-INRIA, Editions Eyrolles, 1984.
275. B.E. Launder. An introduction to single-point closure methodology. In *Lecture notes for Fluid Dynamic*, TFD/87/7, Rhode-St-Genese, 1987. Von Kármán Institute.
276. B.E. Launder. Introduction to the modelling of turbulence: A new form of second-moment closure. In *Lecture series for Fluid Dynamic*, 1991-02, Rhode-St-Genese, 1991. Von Kármán Institute.
277. B.E. Launder and S.-P. Li. On the elimination of wall-topography parameters from second-moment closure. *Phys. Fluids*, 6(2):pages 999–1006, 1994.
278. B.E. Launder and N. Shima. Second-moment closure for the near-wall sublayer: Development and application. *AIAA J.*, 27(10):pages 1319–1325, 1989.

279. B.E. Launder G.J. Reece and W. Rodi. Progress in the development of a Reynolds-stress turbulence closure. *J. Fluid Mech.*, 68(part 3):pages 537–566, 1975.

280. Y. Lebret D. Vandromme and Ha Minh. Structure and modelling in strongly sheared turbulent compressible flow. In F. Durst R. Friedrich B.E. Launder F.W. Schmidt U. Schumann J.H. Whitelaw, editor, *Eighth Symposium on Turbulent Shear Flows*, pages 29–4–1; 29–4–5, Tech. Univ. Munich, 1991.

281. S. Lee, S.K. Lele and P. Moin. Interaction of isotropic turbulence with shock waves: effect of shock strength. *J. Fluid Mech.*, 340:pages 225–247, 1997.

282. M.J. Lee J. Kim and P. Moin. Structure of turbulence at high shear rate. *J. Fluid Mech.*, 216:pages 561–583, 1990.

283. S. Lee S.K. Lele and P. Moin. Direct numerical simulation and analysis of shock/turbulence interaction. *AIAA Paper*, 91-0523, 1991.

284. S Lee S.K. Lele and P. Moin. Eddy shocklets in decaying compressible turbulence. *Phys. FluidsA*, 3:pages 657–664, 1991.

285. S. Lee S.K. Lele and P. Moin. Direct numerical simulation of isotropic turbulence interacting with a weak shock wave. *J. Fluid Mech.*, 251:pages 533–562, 1993.

286. C. Lejeune and A. Kourta. Modeling extra-compressibility high speed turbulent flows. In F. Durst B.E. Launder F.W. Schmidt J.H. Whitelaw, editor, *Eleventh Symposium on Turbulent Shear Flows*, pages P3–71;P3–76, Grenoble, France, 1997.

287. C. Lejeune A. Kourta and P. Chassaing. Modelling of high speed turbulent flows. In *27th AIAA Fluid Dyn. Conference*, AIAA-96-2041, New-Orleans - USA, 1996.

288. S.K. Lele. Direct numerical simulation of compressible free shear flows. Ann. Res. Briefs N89-22827, Center for Turbulence Research, 1988.

289. S.K. Lele. Direct numerical simulation of compressible free shear flows. *AIAA paper*, 89-0374, 1989.

290. S.K. Lele. Notes on the effects of compressibility on turbulence. CTR Mamuscript 145, NASA, Aug. 1993.

291. S.K. Lele. Compressibility effects on turbulence. *Ann. Rev. Fluid Mech.*, 26:pages 211–154, 1994.

292. S.K. Lele. Flows with density variations and compressibility : Similarities and differences. In L. Fulachier J.L. Lumley and F. Anselmet, editors, *Variable Density Low Speed Turbulent Flows*, Marseille, 1997. IUTAM Symposium, Kluwer Acad. Pub.

293. M.A. Leschziner, P. Batten and H. Loyau. Modelling shock-affected near-wall flows with anisotropy-resolving turbulence closures. *Int. Jl.Heat Fluid Flows*, 21:pages 239–251, 2000.

294. M.A. Leschziner P. Batten and H. Loyau. SIG on shock/boundary-layer interaction workshop - UMIST, march 25-26, 1997. *ERCOFTAC Bull.*, 34:pages 21–24, 1997.

295. M.J. Lighthill. On sound generated aerodynamically. I General theory. *Proceedings Roy. Soc. A*, Vol.211:pages 564–587, 1952.

296. M.J. Lighthill. On sound generated aerodynamically. II Turbulence as a source of sound. *Proceedings Roy. Soc. A*, Vol.222:pages 1–32, 1954.

297. S.J. Lin and G.M. Corcos. The mixing layer : deterministic models of a turbulent flow. Part 3. The effect of plane strain on the dynamics of streamwise vortices. *J. Fluid Mech.*, 141:pages 139–178, 1984.

298. E. Lindborg. A note on Kolmogorov's third order structure-function law, the local isotropy hypothesis and pressure-velocity correlation. *J. Fluid Mech.*, 326:pages 343–356, 1996.

299. W.W. Liou and T.-H. Shih. On the basic equations for the second-order modeling of compressible turbulence. Tech. Mem. 105277 ICOMP-91-19, NASA, October 1991.

300. W.W. Liou T.-H Shih and B.S. Duncan. A multiple-scale model for compressible turbulent flows. *Phys. Fluids*, 7(3):pages 658–666, 1995.

301. E.J. List. Tubulent jets and plumes. *Annual Review of Fluid Mech.*, 14:pages 189–212, 1982.

302. F.C. Lockwood and A.S. Naguib. The prediction of the fluctuations in the properties of free, round-jet, turbulent diffusion flames. *Comb. and Flame*, 24:pages 109–124, 1975.

303. J. Lumley. Computational modeling of turbulent flows. *Adv. Appl. Mech.*, 18:pages 123–176, 1978.

304. J.L. Lumley. Toward a turbulent constitutive relation. *J. Fluid Mech.*, 41(part 2):pages 413–434, 1970.

305. J.L. Lumley. Pressure-strain correlation. *Phys. Fluids*, 18(6):pages 750–7515, 1975.

306. J.L. Lumley. Second order modeling of turbulent flow. In *Predictions methods for turbulent flows*, Lecture series 1979-2, pages 1–30, Rhode-St-Genese, Jan. 1979. Von Kármán Institute.

307. J.L. Lumley and B. Khajeh-Nouri. Computational modeling of turbulent transport. *Advances in Geophys.*, 18(A):pages 169–192, 1974.

308. J.L. Lumley and G.R. Newman. The return to isotropy of homogeneous turbulence. *J. Fluid Mech.*, 82(part 1):pages 161–178, 1977.

309. J.L. Lumley and H. Panofsky. *The Structure of Atmospheric Turbulence.* Interscience Publishers, 1964.

310. T. Lumpp. *Simulations numériques de couches de mélange compressibles au moyen de schémas d'ordres élevés du type ENO.* Thèse Sofia Antipolis, Univ. de Nice, 1994.

311. S.L. Lyons T.J. Hanratty and J.B. McLaughlin. Direct numerical simulations of passive heat transfer in a turbulent channel flow. *Int. Jl. Heat Mass Trans.*, 34:pages 1149–1161, 1991.

312. J. Magnaudet. Modelling of inhomogeneous turbulence in the absence of mean velocity gradients. In *Advances in Turbulence IV*, 51, pages 525–531. Kluwer Academic Publishers, 1993.

313. K. Mahesh S.K. Lele and P. Moin. The influence of entropy fluctuations on the interaction of turbulence with a shock wave. *J. Fluid Mech.*, 334:pages 353–379, 1997.

314. N.N. Mansour and T.H. Lundgren. Three-dimensional instability of rotating flows with oscillating axial strain. *Phys. Fluids A*, 2(12):pages 2089–2091, 1990.

315. P. Marcq and A. Naert. A Langevin equation for the energy cascade in fully developed turbulence. *Physica D*, 124:pages 368–381, 1998.

316. J.E. Martin and E. Meiburg. Numerical investigation of three-dimensionally evolving jets subject to axisymmetric and azimuthal perturbations. *J. Fluid Mech.*, 230:pages 271–318, 1991.

317. S.A. Maslowe and R.E. Kelly. Inviscid instability of an unbounded heterogeneous shear layer. *J. Fluid Mech.*, 48:pages 405–415, 1971.

318. Mellor and G.L. H.J. Herring. A survey of the mean turbulent field closure models. *AIAA J.*, 11(5):pages 590–599, 1972.

319. C. Meneveau. Transition between viscous and inertial-range scaling of turbulence structure functions. *Phys. Rev. E*, 54:pages 3657–3663, 1996.

320. C. Meneveau K.R. Sreenivasan P. Kailasnath and M.S. Fan. Joint multifractal measures : theory and applicaitons to turbulence. *Phys. Rev. A*, 41:pages 894–913, 1990.

321. P. Mestayer. *De la structure fine des champs turbulents dynamique et thermique pleinement développés en couches limites.* Thèse doct. ès sciences physiques, Université d'Aix-Marseille II, 1980.

322. P. Mestayer. Local isotropy and anisotropy in a high-Reynolds-number turbulent boundary layer. *J. Fluid Mech.*, 125:pages 475–503, 1982.

323. P. Mestayer C.H. Gibson M. Coantic and A.S. Patel. Local anisotropy in heated and cooled turbulent boundary layers. *Phys. Fluids*, 19:pages 1279–1283, 1976.

324. J. Mi and R.A. Antonia. Effect of large-scale intermittency on scaling range exponents in a turbulent jet. *Phys. Rev. E*, (submitted), 2000.

325. A. Michalke. Survey on jet instability theory. *Prog. Aerospace Sci.*, 21:pages 159–

199, 1984.

326. R. Michel C. Quémard and J. Cousteix. Méthode pratique de prévision des couches limites turbulentes bi- et tri-dimensionnelles. *La Rech. Aéro.*, 1:pages 1–14, 1972.

327. R. Michel C. Quemard and R. Durant. Application d'un schéma de longueur de mélange à l'étude des couches limites d'équilibre. *ONERA Note Tech.*, 154, 1969.

328. K.O. Mikaelian. Density gradient stabilization of the Richtmyer-Meshkov instability. *Phys. Fluids A*, 3(11):pages 2638–2643, 1991.

329. K.O. Mikaelian. Freeze-out and the effect of compressibility in the Richtmyer-Meshkov instability. *Phys. Fluids*, 6(1):pages 356–368, 1994.

330. K.O. Mikaelian. Numerical simulation of Richtmyer-Meshkov instabilities in finite-thickness fluid layers. *Phys. Fluids*, 8(5):pages 1269–1292, 1996.

331. F. Milinazzo and P.G. Turbulence model predictions for the inhomogeneous mixing layer. *Studies in Appl. Math.*, 55:pages 45–63, 1976.

332. Ha Minh and H. D. Vandromme. Modelling of compressible turbulent flows: Present possibilities and perspectives. In J. Délery, editor, *IUTAM Symposium on Turbulent shear layer/shock wave interaction*, pages 13–26, Palaiseau, France, 1985. Springer-Verlag, Berlin.

333. A.S. Monin and A.M. Yaglom. *Statistical fluid mechanics.* Cambridge MIT Press, 1971.

334. P.A. Monkewitz, D.W. Bechert, B. Barsikow, and B. Lehmann. Self-excited oscillations and mixing in a heated round jet. *J. Fluid Mech.*, 213:pages 611–639, 1990.

335. P.A. Monkewitz and E. Pfizenmaier. Mixing by side jets in strongly forced and self-excited round jets. *Phys. FLuids A*, 3(5):pages 1356–1361, 1991.

336. P.A. Monkewitz and K.D. Sohn. Absolute instability in hot jets and their control. In *10th Aeroacoustics Conference*, AIAA Paper No.86-1882, Seattle WA, july 1983. AIAA.

337. P.A. Monkewitz and K.D. Sohn. Absolute instabilities in hot jets. *AIAA J.*, 26:pages 911–916, 1988.

338. M. V. Morkovin. Transition at hypersonic speeds. *ICASE Interim Report 1*, 1987.

339. M.V. Morkovin. Compressible effects on turbulence. In A. Favre, editor, *Mécanique de la Turbulence*, pages 367–380, Marseille, 1961. C.R.N.S.

340. P. J. Morris, M. G. Giridharan, and G. M. Lilley. On the turbulent mixing of compressible free shear layers. *Proceedings R. Soc. Lond. A*, 431:219–243, 1990.

341. I. Müller. Thermodynamics of mixtures of fluids. *Jl. de Méc.*, 14(2):pages 267–303, 1975.

342. L. Mydlarski and Z. Warhaft. On the onset of high-Reynolds-number grid-generated wind tunnel turbulence. *J. Fluid Mech.*, 320:pages 331–368, 1996.

343. L. Mydlarski and Z. Warhaft. Passive scalar statistics in high-Péclet-number grid turbulence. *J. Fluid Mech.*, 358:pages 135–175, 1998.

344. D. Naot S. Shavit and M. Wolfshtein. Interactions between components of the turbulent velocity correlation tensor due to pressure fluctuations. *Israel Jl. Tech.*, 8(3):pages 259–269, 1970.

345. V.W. Nee and L.S.G. Kovasznay. Simple phenomenological theory of turbulent shear flows. *Phys. Fluids*, Vol.12(3):pages 473–484, 1969.

346. R.H. Nichols. A two-equation model for compressible flows. In *28th Aero. Sci. Meeting*, AIAA-90-0494, Reno, Nevada, January 1990. AIAA J.

347. J.L. Novotny and T.F. Irvine. Thermal conductivity and Prandtl number of carbon dioxide and carbon dioxide-air mixtures at one atmosphere. *Jl. Heat Transf. Trans. ASME*, 83-C(2):pages 125–132, 1961.

348. S. Obata and J.C. Hermanson. Numerical simulation of shock-enhanced mixing in nonuniform density jets. *AIAAJl.*, 38(11):pages 2113–2119, 2000.

349. A.M. Obukhov. Structure of the temperature field in a turbulent flow. *Izv. Akad. Nauk. SSSR, Geogr. i Geofiz.*, 13:pages 58–69, 1949.

350. A.M. Obukhov. Some specific features of atmospheric turbulence. *J. Fluid Mech.*,

13:pages 77–81, 1962.
351. Y. Ogura and N.A. Phillips. Scale analysis of deep and shallow convection in the atmosphere. *Jl. of Atm. Sci.*, 19:pages 173–179, 1962.
352. N.R. Panchapakesan and J.L. Lumley. Turbulence measuremants in axisymmetric jets of air and helium. Part 2. helium jet. *J. Fluid Mech.*, 246:pages 225–247, 1993.
353. N.R. Panchapakesan and J.L. Lumley. Turbulence measurements in axisymmetric jets of air and helium. Part 1 Air jet. and Part 2 Helium jet. *J. Fluid Mech.* **246**, *197*, 246:pages 197–223 and 225–247, 1993.
354. C. Pantano and S. Sarkar. A study of compressibility effects in the high-speed turbulent shear layer using direct numerical simulation. *J. Fluid Mech.*, 1999. submitted.
355. D. Papamoschou. Evidence of shocklets in a counterflow supersonic shear layer. *Phys. Fluids*, 7(2), 1995.
356. D. Papamoschou and S.K. Lele. Vortex-induced disturbance field in a compressible shear layer. *Phys. Fluids A*, 5:1412–1419, 1993.
357. D. Papamoschou and A. Roshko. Observations of supersonic free shear layers. *24th Aerospace Sciences Meeting*, AIAA-86-0162, 1986.
358. D. Papamoschou and A. Roshko. The compressible turbulent shear layer : an experimental study. *J. Fluid Mech.*, 197:pages 453–477, 1988.
359. T. Passot and A. Pouquet. Numerical simulation of compressible homogeneous flows in the turbulent regime. *J. Fluid Mech.*, 181:pages 441–466, 1987.
360. B.R. Pearson. *Experiments on small scale turbulence.* PhD thesis, University of Newcastle, Australia, 1999.
361. B.R. Pearson and R.A. Antonia. Reynolds number dependence of turbulent velocity and pressure increments. *J. Fluid Mech.*, (submitted), 2000.
362. B. Perot and P. Moin. Shear-free turbulent boundary layers. Part 1. Physical insights into near-wall turbulence. *J. Fluid Mech.*, 295:pages 199–227, 1995.
363. O.M. Phillips. On the aerodynamic surface sound from a plane turbulent boundary layer. *Proceedings Roy. Soc. A*, Vol.234:pages 327–335, 1956.
364. D. Phong-Anant R.A. Antonia A.J. Chambers and S. Rajagopalan. Features of the organized motion in the atmospheric surface layer. *J. Geophys. Res.*, 85:pages 424-432, 1980.
365. R.T. Pierrehumbert and S.E. Widnall. The two- and three-dimensional instabilities of a spatially periodic mixing layer. *J. Fluid Mech.*, 114:pages 59–82, 1982.
366. J. Piquet. *Turbulent Flows. Models and Physics.* Springer-Verlag, Berlin, 1999.
367. W.M. Pitts. Effects of global density and Reynolds number variations on mixing in turbulent, axisymmetric jets. NBSIR-86/3340, Nat. Bur. Stand., Washington D.C., march 1986.
368. W.M. Pitts. Effects of global density ratio on the centerline behavior of axisymmetric turbulent jets. *Exp. in Fluids*, 11:pages 125–134, 1991.
369. S.B. Pope. A more general effective-viscosity hypothesis. *J. Fluid Mech.*, 72(2):pages 331–340, 1975.
370. S.B. Pope. *Turbulent Flows.* Cambridge Univ. Press, 2000.
371. A. Pouquet. Étude de la turbulence compressible. Contrat DRET - Rapport de synthèse final 92/1202A, Observatoire de la Côte d'Azur, 1994.
372. L. Prandtl. Über ein neues formelsystem für die ausgebildete Turbulenz. *Nachrichten Akad. Wissenschaft Göttingen*, Math-Phys. Kl:pages 6–19, 1945.
373. C.H.B. Priestley. *Turbulent Transfer in the Lower Atmosphere.* The University of Chicago Press, 1959.
374. A. Pumir. A numerical study of the mixing of a passive scalar in three dimensions in the presence of a mean gradient. *Phys. Fluids*, 6:pages 2118–2132, 1994.
375. A. Pumir. Small-scale properties of scalar and velocity differences in three-dimensional turbulence. *Phys. Fluids*, 6:pages 3974–3984, 1994.
376. A. Purwanto. Modèles de turbulence en écoulement basse vitesse et forte

inhomogénéité thermique. Contrat SNECMA - Rapport d'avancement IV, ENSICA, 1994.

377. A. Purwanto. *Modélisation d'écoulements turbulent-basse vitesse à forte variation de masse volumique. Application aux schémas de fermeture k-ε.* Thèse Méc. des fluides, I.N.P.T., ENSICA, 1994.
378. L. Raynal J.-L. Harion M. Favre-Marinet and G. Binder. The oscillatory instability of plane variable-density jets. *Phys. Fluids*, 8(4):pages 993–1006, 1996.
379. R.G. Rehm and H.R. Baum. The equations of motions for thermally driven, buoyant flows. *J. Res. Natl. Bur. Stand.*, 83(3):pages 297–308, 1978.
380. J. Reinaud. *Analyse physique par simulations numériques lagrangiennes de couches de mélange à densité variable.* Thèse, INPT *n°* 1689, 2000.
381. J. Reinaud, L. Joly, and P. Chassaing. The baroclinic instability of the two-dimensional shear layer. *TSFP-1, Santa-Barbara*, page 727, 1999.
382. J. Reinaud, L. Joly, and P. Chassaing. The baroclinic secondary instability of the two-dimensional shear layer. *Phys. FLuids*, 12(10):pages 2489–2505, 2000.
383. C. Rey. *Effets du nombre de Prandtl, de la gravité et de la rugosité sur les spectres de turbulence cinématique et scalaires.* Thèse doct. ès Sciences, Université Claude Bernard, Lyon, 1977.
384. C. Rey. Écoulements turbulents compressibles et variables aléatoires centrées. In *10ème Congrès Francais de Mécanique*, Paris, 1991.
385. C. Rey and J.M. Rosant. Influence of density variations on small turbulent structures of temperature in strongly heated flows. In *Proceedings 9ème Conférence Internationale sur le Transfert de Chaleur*, 4-MC-10, pages 405–409, Jerusalem, 1990.
386. O. Reynolds. On the dynamical theory of incompressible viscous fluids and the determination of the criterion. *Philos. Trans. Roy. Soc. London*, A 186:pages 123–164, 1895.
387. O. Reynolds. On the dynamical theory of incompressible viscous fluids and the determination of the criterion. *Papers on mechanical and physical subjects*, II:pages 535–577, 1901.
388. W.C. Reynolds. Computation of turbulent flows. *Ann. Rev. Fluid Mech.*, 8:pages 183–208, 1976.
389. W.C. Reynolds. Physical and analytical foundations, concepts, and new directions in turbulence modeling and simulation. In J. Mathieu D. Jeandel B.E. Launder W.C. Reynolds W. Rodi, editor, *Turbulence models and their aplications*, number 2 in Collection Direction Etudes et Recherches., pages 149–294. Ecole d'été d'analyse numérique CEA-EDF-INRIA, Editions Eyrolles, 1984.
390. H.S. Ribner. Convection of a pattern of vorticity through a shock wave. Lewis Flight Propulsion Laboratory, *NACA TN 2864* 1953.
391. H.S. Ribner. Shock-turbulence interaction and the generation of noise. Lewis Flight Propulsion Laboratory, *NACA TN 3255-* Report 1233, 1954.
392. C.D. Richards and W.M. Pitts. Global density effects on the self-preservation behaviour of turbulent free jets. *J. Fluid Mech.*, 254:pages 417–435, 1993.
393. F.P. Ricou and D.B. Spalding. Measurement of entrainment in axisymmetric turbulent jets. *J. Fluid Mech.*, 11:pages 21–32, 1961.
394. J.R. Ristorcelli. A pseudo-sound constitutive relationship for the dilatational covariances in compressible turbulence. *J. Fluid Mech.*, 347:pages 37–70, 1997.
395. R. Riva G. Binder M. and Favre-Marinet. Jets of air-helium mixtures in air. In P.H. Alfredsson A.V.Johansson, editor, *Advances in Turbulence 3*, pages 227–234. Springer Verlag, 1991.
396. M.M. Rogers and R.D. Moser. The three-dimensional evolution of a plane mixing layer : the kelvin-helmholtz rollup. *J. Fluid Mech.*, 243:pages 183–226, 1992.
397. M.W. Rubesin. A one-equation model of turbulence for use with the compressible Navier-Stokes equations. *NASA*, TM X-73-128, 1976.

398. M.W. Rubesin. Extra compressibility terms for Favre-averaged two-equation models of inhomogeneous turbulent flows. NASA Contractor Report 177556, MCAT Institute - NASA Ames, Moffett Field, 1990.
399. M.W. Rubesin and W.C. Rose. The turbulent mean-flow, Reynolds-stress, and heat-flux equations in mass-averaged dependent variables. *NASA T.M.*, 62-248, 1973.
400. R. Rubinstein and J.M. Barton. Nonlinear Reynolds stress models and the renormalization group. *Phys. Fluids A,*, 2(8):pages 1472–1476, 1990.
401. E. Ruffin. *Étude de jets turbulents à densitè variable à l'aide de modèles de transport au second ordre.* Thèse réf. 207-94-59, Univ. Aix-Marseille II, IMST, Sept. 1994.
402. S.G. Saddoughi and S.V. Veeravalli. Local isotropy in turbulent boundary layers at high Reynolds number. *G-Animal's Journal*, 268:pages 333–372, 1994.
403. P.G. Saffman. A model for inhomogeneous turbulent flow. *Proceedings Roy. Soc. London, Series A*, 317(1530), 1970.
404. P.G. Saffman. Development of a complete model for the calculation of turbulent shear flows. In *Symposium on Turbulence and Dynamical Systems*, Durham, NC, 1976. Duke Univ.
405. M Samimy M.F. Reeder and G.S. Elliott. Compressibility effects on large structures in free shear flows. *Phys. Fluids A*, 4(6):pages 1251–1258, 1992.
406. N. D. Sandham and W. C. Reynolds. Three-dimensional simulation of large eddies in the compressible mixing layer. *J. Fluid Mech.*, 224:133–158, 1991.
407. D.L. Sandoval. *The dynamics of variable density turbulence.* La-13037-thesis, Los Alamos Nat. Lab., 1995.
408. S. Sarkar. The pressure-dilatation correlation in compressible flows. *Phys. Fluids A*, 4(12):pages 2674–2682, 1992.
409. S. Sarkar. The stabilizing effect of compressibility in turbulent shear flows. *J. Fluid Mech.*, 282:pages 163–186, 1995.
410. S. Sarkar. On density and pressure fluctuations in uniformly sheared compressible flow. In L. Fulachier J.L. Lumley and F. Anselmet, editors, *Variable Density Low Speed Turbulent Flows*, pages 325–332, Marseille, 1996. IUTAM Symposium, Kluwer Acad. Pub.
411. S. Sarkar and L. Balakrishnan. Application of a Reynolds stress turbulence model to compressible shear layer. In *21st Fluid Dyn., Plasma Dyn., and Lasers Conference*, AIAA-90-1465, Seattle, WA, June 1990.
412. S. Sarkar, G. Erlebacher, and M. Y. Hussaini. Direct simulation of compressible turbulence in a shear flow. *Theor. Comput. Fluid Dyn.*, 2:291–305, 1991.
413. S. Sarkar and B. Lakshmanan. Application of a Reynolds stress turbulence model to the compressible shear layer. *AIAA Jl.*, 29(5):pages 743–749, 1991.
414. S. Sarkar and C. Pantano. Effects of density variation in compressible turbulent shear flows. In P. Chassaing F. Anselmet and L. Pietri, editors, *Proceedings International Conférence on Variable Density Turbulent Flows*, Banyuls, June 2000. Press. Univ. Perpignan.
415. S. Sarkar G. Erlbacher and M.Y. Hussaini. Compressible homogeneous shear : Simulation and modeling. In F. Durst R. Friedrich B.E. Launder F.W. Schmidt U. Schumann J.H. Whitelaw, editor, *Selected Papers from 8th International Symposium on Turbulent Shear Flows*, pages 249–267. Springer-Verlag, Sept. 1991.
416. S. Sarkar G. Erlebacher M.Y. Hussaini and H.O. Kreiss. The analysis and modelling of dilatational terms in compressible turbulence. *J. Fluid Mech.*, 227:pages 473–493, 1991.
417. J.-C. Sautet. Effets des différences de densité sur le développement scalaire et dynamique des jets turbulents. *Thése Univ. de Rouen (Fac. des Sciences et des Techniques)*, 1992.
418. J.C. Sautet and D. Stepowski. Single-shot laser Mie scattering measurements

of the scalar profiles in the near field of turbulent jets with variable densities. *Exp. in Fluids*, 16:pages 353–367, 1994.

419. R. Sauvage. *Étude des effets de compressibilité à grands nombres de Mach par simulation numérique directe d'une couche cisaillée*. Thèse Méc. des fluides, I.N.P.T., IMFT, 1994.
420. R.W. Schefer S.C. Johnston R.W. Dibble F.C. Gouldin and W. Kollmann. Nonreacting turbulent mixing flows: A literature survey and data base. Sandia Report SAND86-8217, Sandia Nat. Lab., 1986.
421. R. Schiestel. *Modélisation et simulation des écoulements turbulents*. Traité des Nouvelles Technologies. Hermes, 1993.
422. H. Schlichting. *Boundary-layer theory*. McGraw-Hill, third edition, 1979.
423. D.G. Schowalter, C.W. Van Atta, and J.C. Lasheras. A study of streamwise vortex structure in a stratified shear layer. *J. Fluid Mech.*, 281:pages 247–292, 1994.
424. A.N. Secundov. Application of a differential equation for turbulent viscosity to the analysis of plane non self-similar flows. *Soviet Research*, 5:pages 828–840, 1971.
425. T.-H. Shih and J.L. Lumley. Modeling of pressure correlation terms in Reynlods stresses and scalar flux equations. Fda-85-3, Cornell University, Sibley School of Mech. and Aero. Eng., 1985.
426. T.-H. Shih J. Zhu and J.L. Lumley. A realizable Reynolds stress algebraic equation model. Tech. Mem. 105993 ICOMP-92-27, NASA, 1993.
427. T.-H. Shih J.L. Lumley and J. Janicka. Second-order modelling of a variable-density mixing layer. *J. Fluid Mech.*, 180:pages 93–116, 1987.
428. Y. Shimomura. Turbulent transport modeling in low Mach number flows. *Phys. Fluids*, 11(10):pages 3136–3149, 1999.
429. C.C. Shir. A preliminary numerical study of atmospheric turbulent flows in the idealized planetary boundary layer. *J. of Atmos. Sci.*, 30:pages 1327–1339, 1973.
430. B.I. Shraiman and E.D. Siggia. Scalar turbulence. *Nature*, 405:pages 639–646, 2000.
431. A. Simone G.N. Coleman and C. Cambon. The effect of compressibility on turbulent shear flow : a rapid-distortion-theory and direct-numerical-simulation study. *J. Fluid Mech.*, 330:pages 307–338, 1997.
432. M. Siriex and J.L. Solignac. Contribution à l'étude expérimentale de la couche de mélange turbulente isobare d'un écoulement supersonique. In *Separated Flows*, number 23 in AGARD Conference Proceedings, pages 241–270, 1968.
433. L. Sirovich L. Smith and V. Yakhot. Energy spectrum of homogeneous and isotropic turbulence in far dissipation range. *Phys. Rev. Lett.*, 72:pages 344–347, 1994.
434. A.J. Smits and J.-P. Dussauge. *Turbulent shear layers in supersonic flow*. American Institute of Physics Press, Woodbury, New-York, 1996.
435. R.M.C. So H.S. Zhang T.B. Gatski and C.G. Speziale. Logarithmic laws for compressible turbulent boundary layers. *AIAA Jl.*, 32(11):pages 2162–2168, 1994.
436. R.M.C. So J.Y. Zhu V Ötügen and B.C. Hwang. Some measurements in a binary gas jet. *Exp. in Fluids*, 9:pages 273–284, 1990.
437. R.M.C. So T.B. Gatski and T.P. Sommer. Morkovin hypothesis and the modeling of wall-bounded compressible turbulent flows. *AIAA J.*, 36(9):pages 1583–1592, 1998.
438. M.C. Soteriou and A.F. Ghoniem. Effects of the free-stream density ratio on free and forced spatially developing shear layers. *Phys. Fluids A*, 7(8):2036, 1995.
439. P.R. Spalart. Direct simulation of a turbulent boundary layer up to $R_\theta$=1410. *J. Fluid Mech.*, 187:pages 61–98, 1988.
440. P.R. Spalart and S.R. Allmaras. A one-equation turbulence transport model for aerodynamic flows. *AIAA paper*, 92–0439, 1992.
441. D.B. Spalding. A two-equation model of turbulence. *VDI Forsch-Heft*, 549, 1972.
442. C.G. Speziale and T.B. Gatski. Analysis and modelling of anisotropies in the

dissipation rate of turbulence. *J. Fluid Mech.*, 344:pages 155–180, 1997.

443. C.G. Speziale and S. Sarkar. Second-order closure models for supersonic turbulent flows. In *Proceedings 29th Aero. Sci. Meeting*, AIAA 91-0217, Reno, Nevada, Jan. 1991.
444. C.G. Speziale S. Sarkar and T.B. Gatski. Modelling the pressure-strain correlation of turbulence: An invariant dynamical system approach. *J. Fluid Mech.*, 227:pages 245–272, 1991.
445. E.F. Spina A.J. Smits and S.K. Robinson. The physics of supersonic turbulent boundary layers. *Ann. Rev. Fluid Mech.*, 26:pages 287–319, 1994.
446. K.R. Sreenivasan. The passive scalar spectrum and the Obukhov-Corrsin constant. *Phys. Fluids*, 8:pages 189–196, 1996.
447. K.R. Sreenivasan and R.A. Antonia. The phenomenology of small-scale turbulence. *Ann. Rev. Fluid Mech.*, 29:pages 435–472, 1997.
448. K.R. Sreenivasan and B. Dhruva. Is there scaling in high-Reynolds-number turbulence ? *Prog. Theoret. Phys. Supp.*, 130:pages 103–120, 1998.
449. K.R. Sreenivasan and S. Tavoularis. On the skewness of the temperature derivative in turbulent flows. *J. Fluid Mech.*, 101:pages 783–795, 1980.
450. K.R. Sreenivasan R.A. Antonia and D. Britz. Local isotropy and large structures in a heated turbulent jet. *J. Fluid Mech.*, 94:pages 745–776, 1979.
451. K.R. Sreenivasan R.A. Antonia and H.Q. Danh. Temperature dissipation in fluctuations in a turbulent boundary layer. *Phys. Fluids*, 20:pages 1238–1249, 1977.
452. C. Staquet. Two-dimensional secondary instabilities in a strongly stratified shear layer. *J. Fluid Mech.*, 296:pages 73–126, 1995.
453. G. Stolovitzky P. Kailasnath and K.R. Sreenivasan. Refined similarity hypotheses for passive scalars mixed by turbulence. *J. Fluid Mech.*, 297:pages 275–291, 1995.
454. D. Taulbee and J. VanOsdol. Modeling turbulent compressible flows: The mass fluctuating velocity and squared density. In *19th Aerospace Sciences Meeting*, AIAA-91-0524, pages 1–9, Reno - Nevada USA, January 1991.
455. S. Tavoularis and S. Corrsin. Experiments in nearly homogeneous turbulent shear flow with a uniform mean temperature gradient. Part 1. *J. Fluid Mech.*, 104:pages 311–347, 1981.
456. G.I. Taylor. The instability of liquids surfaces when accelerated in a direction perpendicualr to their planes. *Proceedings Roy. Soc.London*, Ser.A 201:pages 192–196, 1950.
457. R.J. Taylor. Thermal structures in the lowest layers of the atmosphere. *Aust. J. Phys.*, 11:pages 168–176, 1958.
458. C.M. Tchen. On the spectrum of energy in turbulent shear flow. *J. Res. Nat. Bur. Stds.*, 50:pages 51–62, 1953.
459. H. Tennekes and J.L. Lumley. *A First course in Turbulence.* The MIT Press, Cambridge, Massachusetts and London, England, third edition, 1974.
460. S.A. Thorpe. A method of producing a shear flow in a stratified fluid. *J. Fluid Mech.*, 32:pages 693–704, 1968.
461. M. W. Thring and M. P. Newby. Combustion length of enclosed turbulent jet flames. In *Fourth. Symposium (Intl.) on Combustion*, pages 789–796, Baltimore, 1953. The Williams and Wilkins Co.
462. D.J. Tritton. *Physical fluids dynamics.* Van Nostrand Reinhold Comp., New-York - Cincinnati - Toronto - London - Melbourne, 1977.
463. J.S. Turner. *Buoyancy Effects in Fluids.* Cambridge University Press, Cambridge, 2nd edition, 1973.
464. S. Vaienti M. Ould-Rouis F. Anselmet and P. Le Gal. Statistics of temperature increments in fully developed turbulence. Part 1. *Physica D*, 73:pages 99–112, 1994.
465. A. Vallet. *Contribution à la modélisation de l'atomisation d'un jet liquide haute pression.* Thèse, Faculté des Sciences - Univ. de Rouen, Déc. 1997.

466. C.W. Van Atta. Influence of fluctuations in local dissipation rates on turbulent scalar characteristics in the inertial subrange. *Phys. Fluids*, 14:pages 1803–1804, 1971 [and Erratum: 1973 Vol.16, page 574].
467. C.W. Van Atta. Influence of fluctutions in dissipation rates on some statistical properties of turbulent scalar fields. *Izv. Atmos. Ocean. Phys.*, 10:pages 712–719, 1974.
468. E.R. Van Driest. Turbulent boundary layer in compressible fluids. *J. Aeronaut. Sci.*, Vol.18:pages 145–160, 1951.
469. D. Vandromme and H. Ha Minh. Solution of the compressible Navier-Stokes equations: Applications to complex turbulent flows. In *Lecture Series on CFD*. Von Kármán Institute, 1983.
470. D. Vandromme and H. Ha Minh. Physical analysis of turbulent boundary-layer/shock-wave interactions using second order closure predictions. In J. Délery, editor, *IUTAM Symposium on Turbulent shear layer/shock wave interaction*, Palaiseau, 1985.
471. D. Vandromme and H. Ha Minh. Coupling of turbulence models with Reynolds averaged compressible Navier-Stokes equations. Application to shock interactions. In *Proceedings IMA/SMAI Conference*, Reading, 1987.
472. D. Vandromme and H. Ha Minh. Turbulence modeling for compressible flows. In *Introduction to the modeling of turbulence*, Lecture Series 1987-06. Von Kármán Institute, 1987.
473. D. Vandromme H. Ha Minh J.R. Viegas M.W. Rubesin and W. Kollmann. Second order closure for the calculation of compressible wall-bounded flows with an implicit Navier-Stokes solver. In *Proceedings Fourth Symposium on Turbulent Shear Flows*, pages 1–1;1–6c. University of Karlsruhe, 1983.
474. D. Vandromme and H. Ha Minh. About the coupling of turbulence closure models with averaged Navier-Stokes equations. *Jl. Comp. Phys.*, 65(2):pages 386–409, 1986.
475. A.K. Varma E.S. Fishburne and R.A. Beddini. A second-order closure analysis of turbulent diffusion flames. Cr-145226, NASA, Langley Research Center, June 1977.
476. J.R. Viegas and C.C. Horstman. Comparison of multi-equation turbulence models for several shock separated boundary layer interaction. *AIAA Paper*, 78-1165, 1978.
477. A.W. Vreman N.D. Sandham and K.H. Luo. Compressible mixing layer growth rate and turbulence characteristics. *J. Fluid Mech.*, 320:pages 235–258, 1996.
478. L. Wang. Frame-indifferent and positive-definite Reynolds stress-strain relation. *J. Fluid Mech.*, 352:pages 341–358, 1997.
479. L.-P. Wang S. Chen J.G. Brasseur and J.C. Wyngaard. Examination of hypotheses in the Kolmogorov refined turbulence theory through high-resolution simulations. Part 1. Velocity field. *J. Fluid Mech.*, 309:pages 113–156, 1996.
480. L.-P. Wang S. Chen and J.G. Brasseur. Examination of hypotheses in the Kolmogorov refined turbulence theory through high-resolution simulations. Part 2. Passive scalar field. *J. Fluid Mech.*, 400:pages 163–197, 1999.
481. Z. Warhaft. Passive scalars in turbulent flows. *Ann. Rev. Fluid Mech.*, 32:pages 203–240, 2000.
482. Z. Warhaft and J.L. Lumley. An experimental study of the decay of temperature fluctuations in grid-generated turbulence. *J. Fluid Mech.*, 88:pages 659–684, 1978.
483. D.C. Wilcox. Reassessment of the scale-determining equation for advanced turbulence models. *AIAA J.*, 26(11):pages 1299–1310, 1988.
484. D.C. Wilcox. *Turbulence modeling for CFD*. DCW Industries, Inc., second edition, 1994.
485. D.C. Wilcox and M.W. Rubesin. Progress in turbulence modeling for complex flow fields including effects of compressibility. *NASATP-1517*, 1980.
486. D.C. Wilcox and R.M. Traci. A complete model of turbulence. *AIAA Paper*,

76-351, 1976.

487. H.Y. Wong. One-equation turbulence model of Spalart and Allmaras in supersonic separated flows. *AIAA J.*, 37(3):pages 391–393, 1999.

488. C.-T. Wu J.H. Ferziger and D.R. Chapman. Simulation and modeling of homogeneous compressed turbulence. Report TF-21, Stanford University, Thermosciences div. Dept. Mech. Eng., May 1985.

489. G. Xu. *Small-scale measurements in turbulent shear flows.* PhD thesis, University of Newcastle, Australia, 1999.

490. G. Xu R.A. Antonia and S. Rajagopalan. Scaling of mean temperature dissipation rate. *Phys. Fluids*, 12:pages 3090–3093, 2000.

491. A.M. Yaglom. On the local structure of a temperature field in a turbulent flow. *Dokl. Akad. Nauk. SSSR*, 69:pages 743–746, 1949.

492. A. Yoshizawa W.W. Liou N. Yokoi and T.-H. Shih. Modeling of compressible effects on the Reynolds stress using a Markovianized two-scale method. *Phys. Fluids A*, 9(10):pages 3024-3036, 1997.

493. M.-H Yu and P.A. Monkewitz. The effect of nonuniform density on the absolute instability of two dimensional inertial jets and wakes. *Phys. Fluids A*, 2(7):pages 1175–1181, 1990.

494. G.P. Zank and W.H. Matthaeus. The equations of nearly incompressible fluids. I. Hydrodynamics, turbulence and waves. *Phys. Fluids A*, 3(1):pages 69–82, 1991.

495. O. Zeman. Dilatation dissipation : The concept and application in modeling compressible mixing layers. *Phys. Fluids A*, 2(2):pages 178–188, 1990.

496. O. Zeman. Compressible turbulence subjected to shear and rapid compression. In F. Durst R. Friedrich B.E. Launder F.W. Schmidt U. Schumann J.H. Whitelaw, editor, *Eighth Symposium on Turbulent Shear Flows*, pages 21.4.1–21.4.6, Munich, Sept. 1991.

497. O. Zeman. On the decay of compressible isotropic turbulence. *Phys. Fluids A*, 3(5):pages 951–955, 1991.

498. O. Zeman. A new model for super/hypersonic turbulent boundary layers. In *Proceedings 31th Aero. Sci. Meeting & Exhibit*, AIAA 93-0897, Reno, Nevada, jan. 1993.

499. O. Zeman and G.N. Coleman. Compressible turbulence subjected to shear and rapid compression. In F. Durst R. Friedrich B.E. Launder F.W. Schmidt U. Schumann J.H. Whitelaw, editor, *Selected papers from the eighth International Symposium on Turbulent Shear Flows*, T.S.F. 8, pages 283–296, Munich, Germany, September 1991. Springer-Verlag.

500. T. Zhou R.A. Antonia L. Danaila and F. Anselmet. Approach to the four-fifths 'law' for grid turbulence. *J. of Turbulence*, 1:pages 005, 2000.

501. T. Zhou R.A. Antonia L. Danaila and F. Anselmet. Transport equations for the mean energy and temperature dissipation rates in grid turbulence. *Expts. in Fluids*, 28:pages 143–151, 2000.

502. Y. Zhu R.A. Antonia and I. Hosokawa. Refined similarity hypothesis for turbulent velocity and temperature fields. *Phys. Fluids*, 7:pages 1637–1648, 1995.

503. J.Y. Zhu R.M.C. So and M.V. Ötügen. Mass transfer in a binary gas jet. *AIAA Jl.*, 27(8):pages 1132–1135, 1989.

# INDEX

# Mechanics

## *FLUID* MECHANICS AND ITS APPLICATIONS

*Series Editor*: R. Moreau

46. U. Frisch (ed.): *Advances in Turbulence VII.* Proceedings of the Seventh European Turbulence Conference, held in Saint-Jean Cap Ferrat, 30 June–3 July 1998. 1998 ISBN 0-7923-5115-0
47. E.F. Toro and J.F. Clarke: *Numerical Methods for Wave Propagation.* Selected Contributions from the Workshop held in Manchester, UK. 1998 ISBN 0-7923-5125-8
48. A. Yoshizawa: *Hydrodynamic and Magnetohydrodynamic Turbulent Flows.* Modelling and Statistical Theory. 1998 ISBN 0-7923-5225-4
49. T.L. Geers (ed.): *IUTAM Symposium on Computational Methods for Unbounded Domains.* 1998 ISBN 0-7923-5266-1
50. Z. Zapryanov and S. Tabakova: *Dynamics of Bubbles, Drops and Rigid Particles.* 1999 ISBN 0-7923-5347-1
51. A. Alemany, Ph. Marty and J.P. Thibault (eds.): *Transfer Phenomena in Magnetohydrodynamic and Electroconducting Flows.* 1999 ISBN 0-7923-5532-6
52. J.N. Sørensen, E.J. Hopfinger and N. Aubry (eds.): *IUTAM Symposium on Simulation and Identification of Organized Structures in Flows.* 1999 ISBN 0-7923-5603-9
53. G.E.A. Meier and P.R. Viswanath (eds.): *IUTAM Symposium on Mechanics of Passive and Active Flow Control.* 1999 ISBN 0-7923-5928-3
54. D. Knight and L. Sakell (eds.): *Recent Advances in DNS and LES.* 1999 ISBN 0-7923-6004-4
55. P. Orlandi: *Fluid Flow Phenomena.* A Numerical Toolkit. 2000 ISBN 0-7923-6095-8
56. M. Stanislas, J. Kompenhans and J. Westerveel (eds.): *Particle Image Velocimetry.* Progress towards Industrial Application. 2000 ISBN 0-7923-6160-1
57. H.-C. Chang (ed.): *IUTAM Symposium on Nonlinear Waves in Multi-Phase Flow.* 2000 ISBN 0-7923-6454-6
58. R.M. Kerr and Y. Kimura (eds.): *IUTAM Symposium on Developments in Geophysical Turbulence* held at the National Center for Atmospheric Research, (Boulder, CO, June 16–19, 1998) 2000 ISBN 0-7923-6673-5
59. T. Kambe, T. Nakano and T. Miyauchi (eds.): *IUTAM Symposium on Geometry and Statistics of Turbulence.* Proceedings of the IUTAM Symposium held at the Shonan International Village Center, Hayama (Kanagawa-ken, Japan November 2–5, 1999). 2001 ISBN 0-7923-6711-1
60. V.V. Aristov: *Direct Methods for Solving the Boltzmann Equation and Study of Nonequilibrium Flows.* 2001 ISBN 0-7923-6831-2
61. P.F. Hodnett (ed.): *IUTAM Symposium on Advances in Mathematical Modelling of Atmosphere and Ocean Dynamics.* Proceedings of the IUTAM Symposium held in Limerick, Ireland, 2–7 July 2000. 2001 ISBN 0-7923-7075-9
62. A.C. King and Y.D. Shikhmurzaev (eds.): *IUTAM Symposium on Free Surface Flows.* Proceedings of the IUTAM Symposium held in Birmingham, United Kingdom, 10–14 July 2000. 2001 ISBN 0-7923-7085-6
63. A. Tsinober: *An Informal Introduction to Turbulence.* 2001 ISBN 1-4020-0110-X; Pb: 1-4020-0166-5
64. R.Kh. Zeytounian: *Asymptotic Modelling of Fluid Flow Phenomena.* 2002 ISBN 1-4020-0432-X
65. R. Friedrich and W. Rodi (eds.): *Advances in LES of Complex Flows.* Prodeedings of the EUROMECH Colloquium 412, held in Munich, Germany, 4-6 October 2000. 2002 ISBN 1-4020-0486-9
66. D. Drikakis and B.J. Geurts (eds.): *Turbulent Flow Computation.* 2002 ISBN 1-4020-0523-7
67. B.O. Enflo and C.M. Hedberg: *Theory of Nonlinear Acoustics in Fluids.* 2002 ISBN 1-4020-0572-5

# Mechanics

## *FLUID* MECHANICS AND ITS APPLICATIONS

*Series Editor*: R. Moreau

68. I.D. Abrahams, P.A. Martin and M.J. Simon (eds.): *IUTAM Symposium on Diffraction and Scattering in Fluid Mechanics and Elasticity.* Proceedings of the IUTAM Symposium held in Manchester, (UK, 16-20 July 2000). 2002 ISBN 1-4020-0590-3
69. P. Chassaing, R.A. Antonia, F. Anselmet, L. Joly and S. Sarkar: *Variable Density Fluid Turbulence.* 2002 ISBN 1-4020-0671-3

Kluwer Academic Publishers – Dordrecht / Boston / London